Frontiers in Clinical Drug Research
Hematology
(Volume 2)

Editor

Atta-ur-Rahman, *FRS*

Kings College
University of Cambridge
Cambridge
UK

CONTENTS

PREFACE

Volume 2 of Frontiers in Clinical Drug Research-Hematology comprises of five comprehensive chapters written by eminent experts covering recent advances in the treatment of lymphoma, bone marrow transplantation, chronic myeloid leukemia and various hematological malignancies.

In the first chapter Marfe and colleagues reviewed the literature related to molecular biology of chronic myeloid leukemia with particular emphasis on agents that target at the molecular level and their resistance mechanisms in order to combat disease.

The treatment of cancer by targeted biologics such as monoclonal antibodies has been shown to be a promising approach. According to recent studies, bi- and multi-specific antibodies have been found to be more effective in fighting against tumor cells. In the second chapter of this eBook, Stanglmaier and Hess present an overview of the current development of bi- and multi-specific antibodies for treatment of B-cell malignancies.

In the third chapter, Martín-Antonio *et al.,* highlight the strengths and weaknesses of Natural Killer cells in fighting Multiple Myeloma and hematological malignancies. They also propose certain guidelines for the development of future trials.

Janousek and colleagues, in chapter 4, shed light on a potentially revolutionary topic i.e., nanoparticles and nanomaterials. They review the role of nanoparticles in the management of hematological and malignant disorders. Phipps and Teo in the last chapter have drawn attention to the importance of new advancements in delivery methods of Monoclonal Antibodies to target the antigens that are present on lymphoma cells.

I would like to acknowledge the efforts of all the contributors for their outstanding contributions. I am also thankful to the team of Bentham Science Publishers, especially Dr. Faryal Sami and Mr. Shehzad Naqvi led by Mr. Mahmood Alam, Director Bentham Science Publishers for their efforts.

Atta-ur-Rahman, FRS
Honorary Life Fellow
Kings College
University of Cambridge
Cambridge
UK

CONTRIBUTORS

Alvaro Urbano-Ispizua   Department of Hematology, Hospital Clinic, IDIBAPS, Barcelona, Spain; Josep Carreras Leukaemia Research Institute, Barcelona, Spain

Arvind Shukla   School of Biotechnology and Bioinformatics, D.Y. Patil University, Plot No.50, Sector- 15, C.B.D. Belapur, Navi Mumbai, 400614, Maharastra, India

Beatriz Martín-Antonio   Department of Hematology, Hospital Clinic, IDIBAPS, Barcelona, Spain; Josep Carreras Leukaemia Research Institute, Barcelona, Spain

Carla Di Stefano   Department of Hematology, "Tor Vergata" University, Viale Oxford 81, 00133 Rome, Italy

Ciril Rozman   University of Barcelona, Barcelona, Spain and Josep Carreras Leukaemia Research Institute, Barcelona, Spain

Colin Phipps   Department of Hematology, Singapore General Hospital, Singapore

Dagmar Jirova   Center of Toxicology and Health Safety, National Institute of Public Health, Srobarova 48, 100 42, Prague, Czech Republic

Esmeralda Chi-yuan Teo   Department of Hematology, Singapore General Hospital, Singapore

Gabriella Marfe   Department of Biochemistry and Biophysics, Second University of Naples, via De Crecchio 7, Naples 80138, Italy

Giovanna Mirone   Department of Medical Oncology B, Regina Elena National Cancer Institute, via Elio Chianesi 53, 00144, Rome, Italy

Juergen Hess   Lerchenstraße 3, D-85368 Moosburg, Germany

Kristina Kejlova   Center of Toxicology and Health Safety, National Institute of Public Health, Srobarova 48, 100 42, Prague, Czech Republic

Marketa Dvorakova   Center of Toxicology and Health Safety, National Institute of Public Health, Srobarova 48, 100 42, Prague, Czech Republic

Michael Stanglmaier   Lerchenstraße 3, D-85368 Moosburg, Germany

Nuria Martínez-Cibrian   Department of Hematology, Hospital Clinic, IDIBAPS, Barcelona, Spain; Josep Carreras Leukaemia Research Institute, Barcelona, Spain

Stanislav Janousek   Center of Toxicology and Health Safety, National Institute of Public Health, Srobarova 48, 100 42, Prague, Czech Republic

Frontiers in Clinical Drug Research
Hematology

2

**Frontiers in Clinical Drug Research -
Hematology**

First published in 2016

Volume: 2

Editor: Prof. Atta-ur-Rahman

ISSN (Online): 2352-3239

ISSN: Print: 2467-9585

ISBN (eBook): 978-1-68108-181-6

ISBN (Print): 978-1-68108-182-3

**BENTHAM
SCIENCE** Bentham 🔵 **Books**

CHAPTER 1

Molecular and Therapeutic Clues in Chronic Myeloid Leukemia

Gabriella Marfe[1,*], Giovanna Mirone[2], Arvind Shukla[3] and Carla Di Stefano[4]

[1]*Department of Biochemistry and Biophysics, Second University of Naples, via De Crecchio 7, Naples 80138, Italy;* [2]*Department of Medical Oncology B, Regina Elena National Cancer Institute, via Elio Chianesi 53, 00144 Rome, Italy;* [3]*School of Biotechnology and Bioinformatics, D.Y. Patil University, Plot No.50, Sector- 15, C.B.D. Belapur, Navi Mumbai, 400614, Maharastra, India and* [4]*Department of Hematology, "Tor Vergata" University, Viale Oxford 81, 00133 Rome, Italy*

Abstract: Chronic myeloid leukemia (CML) originates from pluripotent hematopoietic stem cells that acquire translocation between the BCR gene on chromosome 22 and the ABL proto-oncogene on chromosome 9. Such rearrangement leads to the the Philadelphia (Ph) chromosome and the oncogenic fusion protein formation. This oncoprotein induces cell proliferation, causes abnormal migration, and reduces apoptosis. The suppression of BCR-ABL protein with imatinib and other specific tyrosine kinase inhibitors (TKIs) has revolutionized CML treatment and is currently regarded as the gold standard of targeted cancer therapy. However, drug resistance to anti-cancer chemotherapy is a significant barrier to the treatment of leukemia patients. Many times, resistance results from molecular adaptation to drug exposure, such as genetic mutation of key enzymes, upregulation of pro-survival compensatory signaling pathways, and altered drug transport. In this chapter, we reviewed the literature on the CML molecular biology, molecular targeted agents and their mechanisms of resistance.

Keywords: Leukemia, resistance, TKIs, BCR-ABL, Philadelphia (Ph) chromosome, leukemia stem cells (LSCs), imatinib.

1. HISTORICAL OVERVIEW

Leukemia constitutes a heterogeneous group of malignant neoplasms of the blood-forming tissue. The word leukemia is Greek for "white blood", and can be considered such as an accumulation of abnormal leukocytes. There are several types of leukemias: acute leukemia is typified by a production of immature hematopoietic cells (blasts) in the bone marrow (BM) and peripheral blood (PB), whereas chronic leukemia typically displays a slow build up of relatively mature

*****Correspondence author Gabriella Marfe:** Department of Biochemistry and Biophysics, Second University of Naples, *via* De Crecchio 7, Naples 80138, Italy; Tel: +390815667522; Fax: +390815667608; E-mail: gabmarfe@alice.it

blood cells. The acute and chronic leukemias may be further classified into lymphoid or myeloid leukemias according to the origin of the leukemic cells, where the main types are referred to as acute lymphoid leukemia (ALL), acute myeloid leukemia (AML), chronic lymphoid leukemia (CLL), and chronic myeloid leukemia (CML). In 1845, the pathologist John Hughes Bennett reported the first patients with CML in the Edinburgh Medical Journal. At the same time, Rudolf Virchow observed a patient with similar features. He tried to distinguish between splenic and lymphatic form of leukemia but only in 1887, Paul Ehrlich classified leukemia in myeloid and lymphatic type [1, 2]. In 1951, William Dameshek proposed that myeloproliferative disease was characterized by a high proliferation rate of BM cells, due to an unknown stimulation [3]. The discovery of the Philadelphia Chromosome by Peter Nowell and David Hungerford took place in 1959 analyzing peripheral blood samples derived from different patients affected by leukemia. The first CML patients were two men thus the Ph chromosome was believed to be a deleted or aberrant Y chromosome [4]. Later, it was considered as the smallest autosome (namely chromosome 21) following the larger number of studied patients [5, 6]. After the development of chromosome banding techniques in the late 1960's, it was possible to describe the Ph chromosome such as an abnormally small chromosome [7, 8]. Thirteen years later, a reciprocal translocation between chromosome 9 and 22, (named t(9;22)(q34;q11)) was discovered in such chromosome [9-11]. This leads to formation of fusion gene, BCR-ABL: the Abl1 gene (the human cellular homologue of the transforming sequence of Abelson murine leukemia virus A-MuLV) from chromosome 9 (region q34) links to the BCR ("breakpoint cluster region") gene on chromosome 22 (region q11) [12-16]. This rearrangement encodes a new chimeric phosphoprotein p210 that plays a crucial role in the onset of CML [17-20]. The BM cells, transfected with a retroviral vector expressing the BCR-ABL oncogene, caused hematologic diseases such as CML in mice that have received these cells after radiation treatment [21-23].

2. DISEASE INCIDENCE, COURSE AND PROGRESSION

CML has an incidence of about 1 in 50,000 per year, accounts for 15% of all adult leukaemias and does not present with significant ethnic or geographical predisposition [24]. The onset of this disease is about 45 to 55 years, with the majority of patients being asymptomatic during diagnosis and discovered after routine blood tests. Presenting symptoms include weight loss, fever, splenomegaly, purpura, night sweats, abdominal fullness and gout. The evolution of CML occurs *via* a biphasic or triphasic course. The majority of the cases (approximately 85%) are diagnosed during the asymptomatic chronic phase (CP) where the cells are mainly differentiated, minimally invasive and maintain their functionality [24]. If left untreated, the disease inevitably progresses after three to five years to an intermediate accelerated phase (AP) and to blast crisis (BC) [25]. Nevertheless, up to a quarter of patients develop directly the BC phase [26]. The typical features of AP and BC are differentiation arrest and accumulation of blast cells in the blood and BM (Table **1**) [27]. Especially during BC, there is a clonal expansion of immature population of blasts that may have developed additional cytogenetic abnormalities, and exhibit enhanced proliferation and reduced susceptibility to apoptosis. The BC transformation can be myeloid, lymphoid or both, with median survival measured in months [28].

Table 1. Staging system for CML according to the World Health Organisation (WHO) Classification 2008 (NA; not applicable).

Feature	CP	AP	BC
Blast % in blood or BM	<10%	10–19%	≥ 20%
Basophil % in blood	<20%	≥ 20%	NA
Thrombocytosis ($\times 10^9$/L)	≤1000 or responsive to therapy if >1000	>1000, unresponsive to therapy	NA
Thrombocytopenia ($\times 10^9$/L)	≥100 or related to therapy if <100	<100, unrelated to therapy	NA
Splenomegaly	Responsive to therapy	Persistent or increasing, unresponsive to therapy	NA
Extramedullary blast tumour (chloroma)	Absent	Absent	Present
New cytogenetic changes that develop after the initial BM karyotype	Absent	Present	NA

The progression of the disease is a complex and gradual process that represents the accumulation of genetic and epigenetic alterations; these may include enhanced

BCR-ABL activity following gene amplification, increased promoter activity or other less direct mechanisms, such as decreased miR-203 levels, inhibition of SHP-1 and inactivation of protein phosphatase 2A (PP2A) [29]. Interestingly, BCR-ABL may contribute to the progression process by activating mitogenic, anti-apoptotic and anti-differentiation mediators (*e.g.* MYC, JAK2, hnRNP-E2, Mdm 2, STAT5, BMI1 and BCL2), inhibiting tumour suppressors (*e.g.* p53 or CEBP α), or through aberrant splicing of modulators like GSK3ß and PYK2 [29].

3. DIAGNOSIS AND MONITORING

3.1. Morphological Identification

The typical laboratory findings in CP CML are leukocytosis, either high or low platelet count, a mild anemia and reduced activity of leukocyte alkaline phosphatase activity [30]. Blasts can be present in peripheral blood, such as eosinophils, mature or immature. Prognostic predictors of this disease are the proportion of basophils, circulating blasts and platelets in peripheral blood [31-33]. In addition, the hypercellularity is present in bone marrow, derived from CML patients, whereas myeloid maturation is present with the abundance of myelocytes [34, 35]. The number of megakaryocytes may be increased, and Gaucher-like cells can be present in 10% of cases [36-38] and in addition, it is possible to find different degree of fibrosis in bone marrow [39]. As the disease progresses, it is observed excess blasts, morphologic alterations of the myeloid cells, and increase in bone marrow stromal fibers [40]. Today, several tests are available to measure treatment effectiveness, such as full blood cell count (hematologic response), cytogenetic analysis (conventional or fluorescent *in situ* hybridization), and real-time quantitative polymerase chain reaction (RT-PCR) for BCR-ABL messenger RNA (molecular response). Today, guidelines are:

1) The evaluation of hematologic responses has to be considered every 2 weeks in order to obtain a complete hematologic response (CHR), and then every 90 days;

2) The evaluation of cytogenetic responses has to be considered every 6 months in order to obtain until a complete cytogenetic response (CCyR), then every years;

3) The evaluation of molecular responses has to be considered every 3 months or 1 month when there is the increase of BCR-ABL mRNA level. The BCR-ABL transcript level from peripheral blood (PB) has

to measure in order to evaluate the drugs response. However, the disease can progress in patients without BCR-ABL mRNA levels decrease after 6 months of treatment.

3.2. Cytogenetic Analysis

Today, cytogenetic and/or molecular testing is used in order to identify the specific genetic BRC-ABL abnormality. The Ph chromosome is in 95% of CML patients at onset of disease. When the CML cases present a masked translocations, it must be utilized molecular biology methods, such as fluorescence in situ hybridization (FISH) or RT-PCR in order to detect the BCR-ABL fusion gene [41, 42]. Occasionally, other chromosomal abnormalities can be detected in CML patients at onset *e.g.* trisomy 8, i(17q), an extra Ph chromosome and trisomy 19, including also monosomies of chromosomes 7, 17, and loss of Y; trisomies of chromosomes 17, 21; and translocation t(3;21)(q26.2;q22) [43]. The sensitivity of cytogenetic analysis decreases in detection of residual disease post therapy because of a limited of examined cells (only 3-4% Ph-positive cells). However, such technique is now considered standard tool in order to monitor CML patients.

3.3. Fluorescence *in situ* hybridization (FISH)

Such technique is used to reveal BCR-ABL rearrangements in CML. The BCR and ABL location genes in the genome in either metaphase and/or interphase cells can be detected with fluorescent DNA probes. The frequency of false positivity can be as high as 3-10%, using traditional FISH (also known as S-FISH or dual-FISH with two probes) [44-47]. To increase the sensitivity, it has been introduced a third probe that localizes the breakpoints in either BCR or ABL, and also four probes (Double FISH) to reduce the frequency of both false-positive and false-negative results respect to S-FISH [45-50]. Such method allows the BCR-ABL detection in about 95% of CML cases and the approximately 5% of cases with masked translocations [51, 52]. In addition, it is able to identify rare variant breakpoints occurring outside the regions covered by PCR primers [53-55]. FISH can also be made on interphase cells derived from both peripheral blood and bone marrow of patients [56-58].

3.4. Quantitative RT-PCR

PCR techniques, including RT-PCR, multiplex PCR and nested-PCR, is capable of detecting BCR-ABL oncogene in the CML with high sensitivity [59]. The decrease of BCR-ABL transcript levels is identified through QRT-PCR [60-64] in CP CML patients, treated with imatinib, and correlates with prognosis [65-67].

4. THE CASUAL RELATIONSHIP BETWEEN BCR-ABL AND CML

BCR-ABL has a an pivotal role in the leukemic change of hemopoietic cells and in the pathogenesis of CML [68]. In 1990, Elefanty *et al.* introduced BCR-ABL carrying retrovirus into BM cells, and reconstituted irradiated mice with these cells, resulting in a mild CML-like syndrome among several haemopoietic neoplasms [22, 69]. In the same year, introduction of BCR-ABL by retrovirus into murine haemopoietic stem cells was employed to reconstitute lethally irradiated mice, and a myeloproliferative syndrome similar to human CP CML was developed prominently among half of recipient mice [70]. Several years later, an efficient method was capable of inducing a CML-like disease in mice after transplantation of BM cells, transfected with BCR-ABL vector, and it was transplantable into secondary recipients [71]. In addition, BCR-ABL is associated with emergence of additional genomic abnormality and progression of CP CML into BC [72, 73]. Therefore, the leukemogenic nature of BCR-ABL makes it a therapeutic target in treating CML [28].

4.1. BCR

BCR is a 160 kDa protein with different domains (Fig. **1**) [74]. There is a coiled-coil oligomerization domain in the N-terminus of BCR [75], that enables BCR to form a homotetramer. Downstream of such domain, there is a Src homology (SH) 2 binding domain [76], by which it is able to interact with the same domain on c-ABL in a phosphotyrosine independent manner [76]. In addition, there is a serine/threonine kinase located in the N-terminus that is encoded by the first exon of BCR [77]. Located in the C-terminal of BCR is a guanosine triphosphate (GTPase) activating domain, which can increase the GTPase hydrolysis rate of a p21Rac GTP binding protein (G protein) [78]. Also in the C-terminal of BCR, there is a domain homologous to GEFs of p21Rho/Rac family of G proteins [79]. There is also a pleckstrin homology domain (PH domain) and a calcium dependent lipid binding (CaLB) domain in the BCR protein. Having different types of structural and functional properties, BCR protein may be involved in different signalling pathways and serve as the cross point of these pathways, but the understanding of the biological function of BCR protein is limited (Fig. **1**) [74].

4.2. ABL

c-ABL protein is a non-receptor tyrosine kinase (Fig. **1**) [24]. There are two splice variants of c-ABL, encoding two proteins with their difference in the first exon

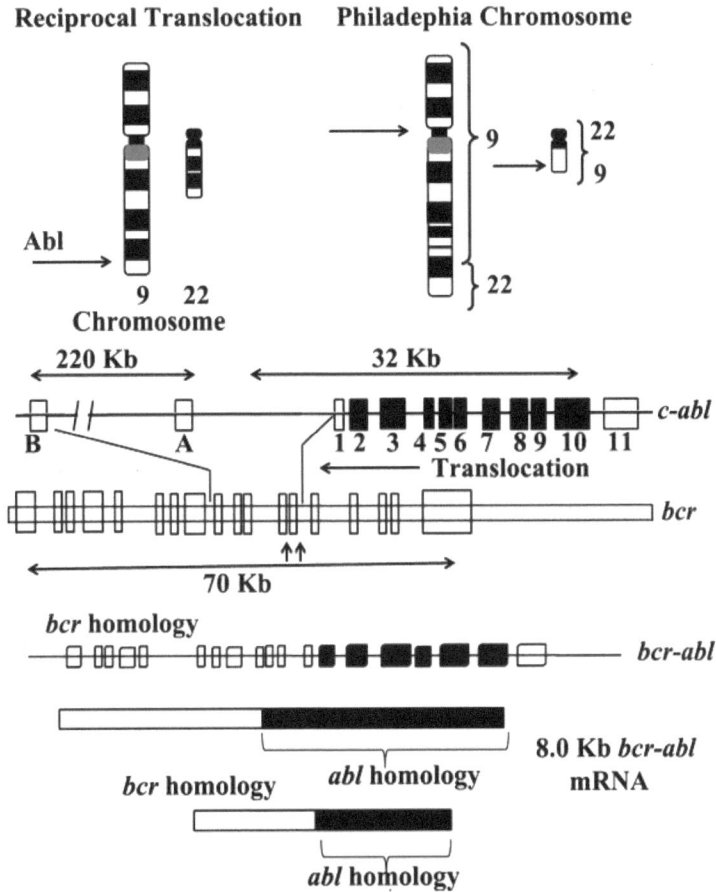

Figure 1: The translocation of t(9;22) in CML and functional domains of BCR-ABL protein.

(1a and 1b). c-ABL 1b is slightly longer than 1a and contains a myristoyl group at its N-terminus, and this group links c-ABL 1b with lipid membrane [18]. Located in the N-terminus of c-ABL are three SH domains: SH3, SH2 and SH1. The SH3 domain binds to proline-rich (PxxP) motifs [80, 81], and inhibits c-ABL tyrosine kinase activity [24]. Such protein can be activated by deletion of the SH3 domain [82]. The SH2 domain binds to phosphotyrosine residues in a sequence specific manner [80, 81]. A point mutations in the SH2 domain is able to inhibit the both phosphotyrosine binding and oncogenic ability of activated ABL [83]. The following SH1 domain carries a tyrosine kinase [80]. Downstream there are three proline-rich sequences, which can interact with SH3-containing proteins [81]. Towards the C-terminus of c-ABL are nuclear localization signals (NLSs) and nuclear export signal (NES), enabling it to travel between cell nucleus and cytoplasm [84, 85]. Located in the C-terminus are DNA and actin binding domains

[84, 86]. c-ABL is mainly located in the cytoplasm of human haemopoietic cells with a low-level presence in the nucleus [87], whereas it is largely present in the nucleus of fibroblasts [88]. The tyrosine kinase domain of c-ABL is regulated *in vivo*, and its activity is very low in normal context [76]. It can be activated by cell cycle progression, DNA damage and integrin mediated adhesion. c-ABL tyrosine kinase domain is increased during cell cycle progression to S phase [80]. Activated c-ABL causes the extensive phosphorylation of C-terminal domain of RNA polymerase II and activates transcription [80, 89]. In addition, it is induced by DNA damage and enhanced cell cycle arrest or apoptosis [90]. Furthermore, integrin-mediated adhesion can induce transient cytoplasmic translocation and activation of such protein in fibroblasts, which then activates mitogen activated protein kinase (MAPK) cascade [91, 92].

4.3. BCR-ABL

The p210 BCR-ABL protein forms dimers or tetramers *in vivo* that autophosphorylate each other (Fig. **1**) [93]. Fusion of BCR to c-ABL inhibits the latter's SH3 kinase regulatory domain, and actives the ABL tyrosine kinase activity in this protein [24]. Two domains of BCR are thought to be essential for ABL tyrosine kinase activation and oncogenic ability of the BCR-ABL protein: amino acids 1-63 and 176-242 [75]. The first domain is the coiled-coil oligomerization domain, and it mediates the deregulation of ABL tyrosine kinase and enhancement of filament (F)-actin binding ability of ABL [75]. Mutations in the oligomerization domain inhibit the BCR-ABL oncoprotein [75], whereas disruption of its tetramer by a competitive peptide of the coiled-coil oligomerization domain of BCR efficiently inhibited both its tyrosine kinase activity and its oncogenic function [94]. The second critical domain for kinase activation and transforming ability is the SH2 binding domain of BCR, which deregulates tyrosine kinase by binding to ABL SH2 domain [76]. On the other hand, ABL tyrosine kinase impairs BCR serine/threonine kinase by phosphorylating tyrosine (Y) 360 residues on BCR and BCR-ABL [95]. A deleted form of BCR which cannot be tyrosine phosphorylated, inhibits tyrosine kinase domain of active c-ABL and BCR-ABL, suggesting such phosphorylation of BCR suppresses its inhibitory activity [96].

5. BCR-ABL AND CONSTITUTIVE ACTIVATION OF PROLIFERATION AND SURVIVAL PATHWAYS

The BCR-ABL has a constitutively active tyrosine kinase domain and has a large variety of targets and activates a plethora of pathways that protect the cell from undergoing apoptosis and lead to transformation. At first the "discordant maturation hypothesis" was proposed, where it was assumed that the most mature proliferating cells in CP CML caused the expansion of Ph$^+$ cells. However, this was refuted by different investigations demonstrating that the myeloid expansion is a result of increased numbers of primitive CML progenitor cells [97]. BCR-ABL is localized exclusively to the cytoplasm and is able to constitutively phosphorylate tyrosine of a host substrate. Importantly, due to autophosphorylation, there is increased phospho-tyrosine (p-Tyr) on the BCR-ABL oncoprotein itself. A number of target substrates of BCR-ABL have been reported: (1) adaptor molecules such as CrkL and p62Dok (docking protein); (2) cell membrane and cytoskeleton related proteins such as talin and paxillin and (3) proteins with catalytic function such as Ras-GAP (GTPase activating proteins) and phospholipase Cg (PLCg) [28]. Tyrosine phosphorylation of these substrates actives various cytoplasmic and nuclear signalling cascades.

5.1. BCR-ABL and the Ras-Raf-MEK-ERK Pathway

Ras is a small guanine-nucleotide-binding protein (G-protein) that is active when bound to GTP, or inactive when GDP-bound. Ras is activated after the exchange of Ras-bound GDP for GTP *via* GEFs. In contrast, GAPs inactivate Ras by catalysing its intrinsic GTPase activity. In normal cells, activation of Ras by haemopoietic Growth Factors (GFs), such as Interleukin-3 (IL-3) leads to the recruitment of the serine/threonine Raf-1 kinase to the cell membrane. Activated Raf-1 can then phosphorylate mitogen-activated protein kinase (MAPK) kinase (MEK) on both serine and threonine residues. MEK is then able to phosphorylate and activate extracellular signal-regulated kinase (ERK), which can, in turn, phosphorylate transcription factors, such as c-Jun and c-Fos and thereby direct gene regulation [98]. In addition, elevated Ras activation occurs in CML cells [99]. BCR-ABL protein can interact with the adapter molecule Grb-2 after the autophosphorylation of Y177 [100]. Activated Grb-2 can then bind the positive regulator, Son of sevenless (SOS; a GEF) and then stabilise Ras in the active form. Grb-2 mutants, that lack SH3 domains, inhibit the RAS activation and suppress the BCR-ABL activity [101]. Two other BCR-ABL adaptor molecules, Shc (Src homology 2 domain) and CrkL, can also activate Ras [102, 103]. The importance of Ras in CML cell growth was demonstrated in BCR-ABL-positive cell line models, transfected with dominant-negative Ras vector, that inhibited malignant transformation [104] (Fig. **2**). Further investigations have shown that, in BCR-ABL-transformed haemopoietic cells, the Ras constitutively activation can induce the anti-apoptotic pathway [105, 106]. These data highlight an important role for the Ras-Raf-MEK-ERK pathway in the pathogenesis of BCR-ABL positive leukaemias.

5.2. BCR-ABL and the JAK-STAT Pathway

The Janus kinase (JAK)/signal transducers and activators of transcription (STAT) pathway are able to transmit signals from a different cytokines and growth factors into the cell nucleus in order to induce transcription of their target genes. This pathway contributes significantly to the integrity of organisms. Its activation is involved in different cellular processes such as growth, immune response, proliferation and hematopoiesis. The JAK-STAT system acts as the second messenger system through a receptor, JAK and STAT components. After the activation of receptor *via* different chemical signals, it triggers, in turn, the kinase function, that autophosphorylates itself. Indeed, JAK is able to phosphorylate STAT proteins, that forms a STAT dimer. This dimer becomes an active transcription factor, that goes into nucleus and links to specific promoters in the DNA. In mammals, there are seven STAT genes, that bind to different promoters, which regulate the expression of several genes. This pathway is evolutionarily conserved mechanism including worms, mammals and human. Although this pathway is not clear, it is known that disrupted or dysregulated JAK-STAT signaling can cause cancers and other disease [107]. STAT5 plays a major role in leukemia initiation, maintenance, progression as well as imatinib resistance. STAT5 activation causes the increase of the anti-apoptotic proteins Mcl-1 and Bcl-XL expression as well as Cyclin D1 expression which is essential to the progression from G1 to S phase of the cell cycle [108-110]. The mechanism of STAT activation by BCR-ABL is complex, involving both direct phosphorylation as well as intermediate kinases of the SRC and JAK families. JAK2 is an important non-receptor kinase that is activated by point mutations in many myeloproliferative diseases. It has also been correlated with BCR-ABL mediated leukemogenesis, as pharmacological inhibition of JAK2 decreases the viability of both imatinib-sensitive and resistant CML cells. However a more recent report has shown that inhibition of JAK2 was correlated with "off-target" effects of JAK2 inhibitors such as TG101348, TG101209 and AG490 on BCR-ABL directly [111]. Samanta *et al.* showed that activation of JAK2 by BCR-ABL induced phosphorylation of GAB2, thus linking JAK2 to the phosphatidylinositol-3-kinase (PI3K) and Ras pathways [112, 113]. Although BCR-ABL phosphorylates JAK2 on Tyr1007, cells, expressing kinase-inactive JAK2, do not exhibit reduced levels of active STAT5 [114]. More recently, Hantschel *et al.* showed that STAT5 activation is uncoupled from JAK2 activation in BCR-ABL positive cells and that STAT5 is directly phosphorylated by BCR-ABL [111].

5.3. BCR-ABL and Nuclear Factor κB (NF-κB) pathway

Rel/NF-κB (nuclear factor kappa-light-chain-enhancer of activated B cells) proteins are a family of transcription factors that play a pivotal role in immune responses, developmental processes, cellular growth, and apoptosis. In addition, these are constitutively active in different disease, including cancer and neurodegenerative diseases. Different studies have demonstrated that mutations or amplifications in Rel/NF-κB/IkB genes occurred in several human cancers. In addition, this signaling pathway can be dysregulated in different types of cancer, abrogating growth suppression and apoptosis. BCR-ABL activates NF-κB in a tyrosine kinase-dependent fashion [115]. However, it has also been suggested that Ras activation is involved in transactivation of NF-κB [116]. BCR-ABL is able to activate PI3K and INK that stimulate NF-κB transcriptional activity. Thus, it is still possible that multiple co-operating pathways contribute to the activation of NF-κB in CML. Once activated, NF-κB is able to mediate many key genes transcription, necessary for BCR-ABL transformation as well as many surface molecules to regulate cell adhesion and cell-cell interactions. Its activation happens possibly through crosstalk from the PI3K/AKT signaling pathway in both CML and AML [117-119]. Furthermore, inhibition of the NF-κB complex by blocking of various proteins in the signaling pathway suggests that it may be a possible therapeutic target in CML [119-121]. Interleukin 1 (IL-1) is one cytokine known to activate NF-κB in several types of immune cells [122]. Several studies have showed that IL1 provides a selective advantage for the CML stem cells [123, 124]. The mechanisms underlying the growth advantage of primitive CML cells respect to Hematopoietic Stem Cells (HSCs) after IL-1 stimulation is not yet known. However, has been observed that expression of the IL1 receptor accessory protein (IL1RAP) was upregulated in primitive $CD34^+CD38$-CML cells compared to the corresponding HSCs. The upregulation of this receptor seems to be sufficient to induce features of a myeloproliferative disease. Other studies have found a detectable level of NF-κB activity in chronic and blast phases of disease [124-126].

5.4. BCR-ABL and Myc Pathway

Myc is a proto-oncogene that is able to promote cell cycle progression. Its up-regulation has been found in different human malignancies and that is why c-myc can be a diagnostic marker for some cancer types. Myc is activated by BCR-ABL through the SH2 domain: its overexpression restores transformation in mutants carrying SH2 deletion, but dominant negative upregulation suppresses transformation (Fig. **2**) [127]. In this pathway, mediated by Ras/Raf, cyclin-dependent kinases (cdks), and E2F transcription factors in v-ABL-expressing cells, stimulates the Myc promoter. Another study shows similar results for BCR-ABL transformed murine myeloid cells [128]. Cell-context dependent-Myc expression can activate a proliferative or an apoptotic signalling [129]. Therefore, this dual action is compensated in CML cells by different mechanisms, such as the PI3K pathway.

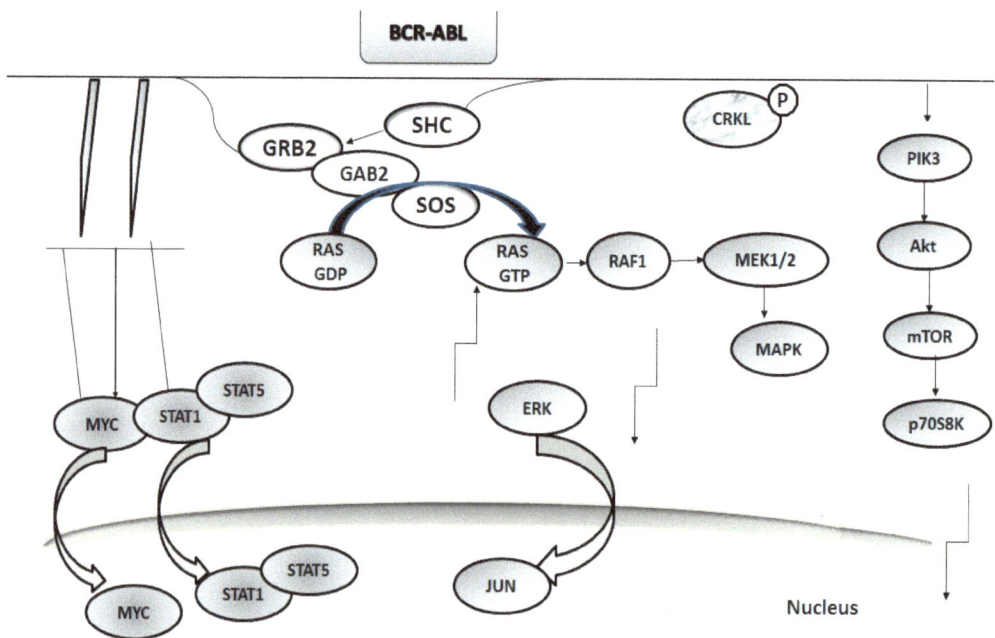

Figure 2: Schematic Diagram of Signaling pathways of BCR-ABL (Reproduced and modified from [24]).

5.5. BCR-ABL and the PI3K/AKT Pathway

PI3K is a heterodimer containing an 85-kDa (p85) regulatory subunit, one SH3 domain, two SH2 domains and a 110-kDa (p110) catalytic subunit [130] and it interacts with ligand-activated GFs and oncogene PTKs [131] by the SH2 and SH3 domains of p85 [132]. The PI3K signalling pathway plays a crucial role in many cellular processes, such as survival, proliferation, differentiation, mobility and metabolism in a number of cell types [133, 134]. Activated PI3K can produce PI-(4,5)-bisphosphate (PIP2) which may be converted to PI-(3,4,5)-triphosphate (PIP3), an important second messenger [135]. PIP3 and other PI3K lipid products on the inner cell membrane generate binding sites for other proteins that contain pleckstrin homology domains. PI3K signalling is negatively regulated by phosphatases, such as PTEN [136]. PI3K signalling is deregulated in different cancers and can also contribute to cellular transformation induced by BCR-ABL (Fig. **2**). The author, Skorski showed that PI3K activity was regulated by BCR-ABL and required for the growth of CML cells, using antisense oligonucleotides against PI3K expression [137]. Furthermore, wortmannin (a specific inhibitor of PI3K), was able to suppress the BCR-ABL positive cells proliferation, but not normal haemopoietic cells [138]. These data are supported by the fact that BCR-ABL-transformed cells have increased PI3K class 1A activity and an accumulation of PIP3 [139]. A YXXM motif is contained within the ABL sequence of BCR-ABL, which when phosphorylated corresponds to the optimal binding sequence for the SH2 domains of the p85 regulatory subunit of PI3K [140]. However, the BCR-ABL-PI3K interaction does not appear to be direct as mutation of the YXXM motif and does not attenuate PI3K activity. These data demonstrate that PI3K signalling pathway is activated by other tyrosine-phosphorylated proteins through the recruitment of BCR-ABL [140].

5.6. The Adapter Protein Gab2

The majority of evidence indicates that the main pathway for PI3K activation in BCR-ABL$^+$ cells occurs *via* the Y177 autophosphorylation site on the BCR portion of the fusion protein (Fig. **2**). The Y177 site is important for Grb2 binding *via* its SH2 and generates a docking site for Gab2 through its SH3 domain [141]. BCR-ABL can then phosphorylate Gab2 on tyrosines within its YXXM motif which, in turn, works as binding sites for the SH2 domains of PI3K regulatory subunits. Mutations within the Y177 site resulted in decreased transformation by BCR-ABL. Similarly, loss of Gab2 inhibited myeloid transformation both *in vitro* and *in vivo* [141].

5.7. The Adapter Protein CrkL

A further possible Gab2-independent mechanism of PI3K activation involves the 39-kDa protein CrkL. Such protein was originally identified by ten Hoeve *et al.* [142] and is phosphorylated directly by BCR-ABL protein in BCR-ABL positive cells [102]. Crk was encoded by oncoprotein as v-Crk (viral Crk, CT10 regulator of kinase). It is ubiquitously expressed and contains one SH2 domain and two SH3 domains [143]. The association among Crk and other focal adhesion proteins such as paxillin is involved in different cellular process such as the regulation of cellular motility and integrin-mediated cell adhesion [144]. In CML, CrkL interacts and is phosphorylated on tyrosine 207 (Fig. **2**) [145] directly by the BCR-ABL oncoprotein [102, 142, 146]. Tyrosine phosphorylated CrkL then binds to c-CBL with its SH2 domain and to PI3K, c-ABL and BCR-ABL through its

SH3 domain [147, 148]. The exact function of CrkL phosphorylation in leukaemic cell is not yet clear, however, it seems likely that CrkL, binding to other tyrosine phosphorylated proteins, can trigger other pathways, involved in leukaemic transformation.

5.8. AKT a Major Downstream Signalling Effector of PI3K

AKT, a 57-kDa serine/threonine kinase, is activated in a several human malignancies and has essential role in BCR-ABL-mediated transformation of hematopoietic cell lines and primary tumors from CML patients (Fig. **2**). The AKT family has three highly conserved isoforms: AKT1/α, AKT2/ß, and AKT3/µ [149]. The BCR-ABL oncoprotein binds with PI3 kinase that ultimately leads to the activation of the kinase AKT. This activation occurs with the adaptor protein complexes, such as GRB2/GAB2 and CrKL/CBL which respectively bind to BCR-ABL residues 244, 245 (*i.e.* tyrosine 177) or domains 246 and provide docking sites for PI3 kinase [150-152]. The PI3 kinase/AKT pathway activation is important key for BCR-ABL induced leukaemogenesis that is inhibited in murine BM cells *in vitro* and in immunodeficient mice, using a dominant negative AKT mutant. In particular, recent research has shown that the AKT3/µ isoforms is over-expressed in p210 BCR-ABL positive myeloid precursor cells and in certain CML cell lines. The proliferation and colony formation of such cells are inhibited, using small molecule isozyme-selective non-ATP competitive inhibitors of AKT. Finally, pharmacological inhibition of all three AKT family members overcomes imatinib-resistance of cells containing the T315I BCR-ABL mutant [153]. In addition, PI3 kinase/AKT pathway seems to be linked at least to the increased expression and transcriptional activation function of Myc [137]. Moreover, AKT activation can phosphorylate different survival proteins such as the GSK3β that controls Mcl-1 [154] and other downstream proteins, decreasing the apoptotic rate.

5.9. Forkhead Box, Subgroup O

The forkhead box, (subgroup O (FoxO) family of transcription factors) are a subclass of the Fox forkhead transcription factors that play several roles in proliferation, cellular metabolism and survival process [155]. The mammalian members of this family include FoxO1, FoxO3a, FoxO4 and FoxO6 and each protein presents three evolutionarily conserved AKT phosphorylation sites. Recent investigations have demonstrated the importance of FoxO function in haematological malignancies. BCR-ABL transformation was shown to inhibit FoxO3a, *via* the constitutive activation of PI3K/AKT signalling and the subsequent cytoplasmic sequestration of FoxO3a. The transduction of an active

FoxO3a triple mutant in CML cells leads on the induction of apoptosis [156, 157]. Furthermore, FoxO3a suppression by BCR-ABL affected the expression of the cell cycle regulatory gene, cyclin D2 [158]. These data suggest that inhibition may represent a potentially important mechanism in CML tumorigenesis.

5.10. Bcl-2-associated Death Promoter

The Bad protein, pro-apoptotic member of the Bcl-2 gene family, is involved in apoptotic pathway. Different studies have found that BCR-ABL can promote the phosphorylation and inhibition of Bad *in vitro*. However, within the same study, it was also noted that BCR-ABL-mediated phosphorylation of Bad was not associated with cell survival, since the cells survive in the absence of Bad phosphorylation [159]. This result, thereby, suggests that the transforming ability of BCR-ABL is due to activation of complex network of signalling cascades.

5.11. Murine Double Minute 2

Mdm2 is an E3 ubiquitin ligase which, when phosphorylated by AKT, negatively regulates the tumour suppressor, p53 [160]. Several studies suggest that BCR-ABL may promote cell survival by p53 downregulation *via* increased expression of Mdm2 [161].

5.12. Glycogen Synthase Kinase 3β

GSK3β is a serine/threonine kinase that is a downstream substrate of AKT and becomes inactivated following phosphorylation [162]. Two downstream targets of GSK3β, include cyclin D1 and β-catenin. Activated GSK3β, phosphorylates these proteins and targets them from proteasome-mediated degradation [163]. Recent studies have demonstrated that HSCs which were deficient in β-catenin, demonstrated a reduction in long-term self-renewal as measured by serial transplantation assays. Furthermore, BCR-ABL-transduced cells deficient in β-catenin, failed to develop CML [164].

5.13. Tuberous Sclerosis-2 and the Mammalian Target of Rapamycin Pathway

The serine/threonine kinase, mammalian target of rapamycin (mTOR) functions as both a nutrient sensor and an important downstream substrate of PI3K/AKT signalling, following GFs or oncoprotein stimulation [165]. The activation of mTOR results in the regulation of processes, such as cell growth, cell-cycle progression, protein synthesis and autophagy [166, 167]. mTOR can form two multi-protein complexes, mTOR complex 1 (mTORC1) and mTOR complex 2 (mTORC2) [166]. mTORC1 is composed of mTOR Raptor, mammalian MLST8, the non-core components PRAS40 and DEPTOR. It has been demonstrated to be involved in cellular proliferation, autophagy and translation in response to nutrients. mTORC2 is composed of mTOR, RICTOR, MLST8, and mSIN1. mTORC2 regulates the cytoskeleton through the activation of F-actin stress fibers, paxillin, RhoA, Rac1, and PKCα. In addition, the serine residue S473 of AKT is phosphorylated by mTORC2, thus affecting metabolism and survival. After this reaction, PDK1 phosphorylates, in turn, the threonine T308, leading to full AKT activation. mTORC2 is less well understood than the first mTOR-containing complex; however studies suggest that it is involved in the PI3K/AKT pathway as it directly phosphorylates AKT [168]. The mTOR pathway has proven to be a good target for the drug treatment of several malignancies which display excessive PI3K/AKT signaling, including breast and ovarian cancers [169]. The exact mechanism by which BCR-ABL activates mTOR in CML cells has yet to be defined but appears to be PI3K/AKT-dependent (Fig. **2**) [158]. PI3K/AKT-signaling regulates mTOR indirectly through activation of tuberous sclerosis complex 2 (TSC2). TSC2 forms a heterodimer with tuberous sclerosis complex 1 (TSC1) which functions as a check on mTOR-dependent signaling in the absence of GFs/oncogene signals. When the activated AKT phosphorylates the TSC2, the TSC1 could not interact with TSC2 and its ability to inhibit the small G-protein Rheb is blocked. Rheb is then able to activate mTOR once released from this inhibition [170]. The mTOR pathway controls cell growth through regulators of translation, such as 4EBP1 and the S6K1 and S6K2 [171]. Phosphorylation of 4EBP1 by mTOR blocks its ability to inhibit eIF4E. The eIF4E protein is then released and is free to bind the 5'-cap structure of mRNAs to allow an increase in cap-dependent translation [172]. The S6K activation of the S6 protein of 40S ribosomes promotes cell growth and proliferation by an unknown mechanism [173].

6. TYROSINE KINASE INHIBITORS (TKIs) IN THE CML THERAPY

6.1. Treatment of CML in pre-TKIs Era

The history of CML reveals many attempts to treat the disease with highly toxic and unselective agents. The earliest attempts were conducted with arsenic at the end of 19th century; this treatment resulted in shrinkage of the spleen and improvement in the leukocyte count and anemia, but the response lasted only few months. Radiotherapy, mainly given to the spleen, was introduced at the beginning of 20th century. This therapy mode induced a rapid decrease in spleen size and leukocyte count, and the response lasted from weeks to months and occasionally years [174]. The third major therapy mode was busulfan, which came into the clinic in the 1950s and was the treatment of choice for the next 35 years. Busulfan was found to be more effective and convenient than radiotherapy, and increased the survival of CML patients [175]. It was replaced by hydroxyurea and Interferon-α (IFN-α), which manifested less toxic effects. Hydroxyurea was more effective than busulfan in terms of prolonging survival [176], and turned out to be an excellent debulking agent by its effective normalizing of blood counts. It is commonly used as an adjuvant prior to more specific therapy modes [177]. Due to a lack of cytogenetic responses, hydroxyurea never gained position as a first line therapy. Usage of IFN-α began in the early 1980s; it increased the survival of CML patients when compared to busulfan and hydoxyurea, and was most effective as a combination therapy with low-dose cytarabine [178]. Allogeneic hematopoietic stem transplantation (HSCT) was tested experimentally as a treatment mode as early as 1979 [179] and provided a cure for the disease. In the 1980s, it became more popular HLA-matching techniques developed [180]. In a follow-up study conducted in 2005, the 15-year (patients collected 1982-92) overall survival rate was 53% with allogeneic HSCT patients [181]. Today, there are different TKIs in CML therapy, as showed in Table **2**.

Table 2. FDA-approved small molecule kinase inhibitors for clinical use. ([1]Gastro-intestinal stromal tumors=GIST and [2]Hypereosinophilic syndrome=HES).

Inhibitors	Main Kinase Target	Indication	FDA Approval Year
Imatinib	BCR-ABL, c-Kit, PDGFR	CML, Ph-positive ALL, GIST[1], HES[2]	2001
Dasatanib	BCR-ABL, SRC family	CML, Ph-positive ALL	2006
Nilotinib	BCR-ABL, c-Kit, PDGFR	CML CML	2007
Bosutinib	BCR-ABL, SRC family	CML resistant to other therapy	2012
Ponatinib	Abl, FLT3, FGFR, VEGFR, KIT, PDGFR, RET, EPH, Src family	CML resistant to other therapy (T315I), Ph-positive ALL.	2013

6.2. Imatinib

The specific BCR-ABL inhibitor, imatinib mesylate, was initially known by the names GCP 57148 or STI571 [182]. It was an efficient inhibitor of ABL tyrosine kinases (BCR-ABL and the normal counterpart) [183] as well as other kinases, such as PDGFR-α/β and c-Kit [184]. The first indication for imatinib as a possible treatment for CML was indicated in preclinical studies, where the drug reduced 92-98% the BCR-ABL positive colonies in colony-forming assays, performed using both peripheral blood (PB) and BM of primary patient samples [183]. In addition, a decreased BCR-ABL phosphorylation and an induction of cell death were observed in both BCR-ABL expressing cell lines and primary patient samples [185]. The efficacy of imatinib was further confirmed *in vivo* mice experiments in which dose dependent inhibition of tumor growth [183], blockage of BCR-ABL phosphorylation, and increased tumor-free survival were observed with imatinib therapy [186]. The mechanism of binding of imatinib to BCR-ABL was revealed with computational docking studies and X-ray crystallography [182, 187] and it interacts with the catalytically inactive form of the ABL KD. In this inactive conformation, the activation loop is unphosphorylated and in a closed conformation and the conserved DFG motif in the N-terminal part of the loop is pointed outward. The active conformation of the activation loop would instead be open, enabling substrate binding. The inactive conformation provides specificity for imatinib, since this conformation differs from that of other kinases. The side chain of T315 also forms a hydrogen bond with the secondary amino group in imatinib. T315 is replaced with methionine in many other kinases, rendering the possibility to form a hydrogen bond with imatinib. In total, there are six different hydrogen bonds through van der Waals interactions between imatinib and ABL [195]. Such binding prevents autophosphorylation of BCR-ABL, thereby avoiding phosphorylation of downstream signaling molecules essential for leukemogenesis, which eventually leads to inhibition of proliferation and apoptosis [183, 185].

6.3 Clinical Efficacy of Imatinib Therapy

Despite the existence of other potential kinase targets, CML was chosen as an indication for the first clinical phase I studies conducted with imatinib. Such drug was rapidly shown to be superior to other therapy modes, and was chosen as a first line therapy mode for CML in 2001. A phase II clinical trial was conducted in 532 IFN-α-failure CML patients; imatinib was judged as a safe and efficient therapy mode for this type of patient group after a 6-year follow-up [188]. The next phase 3 trial, the International Randomized Study of Interferon and ST1571 (IRIS) study, enrolled 1106 newly-diagnosed CML patients, who were treated either imatinib or IFN-α combined with cytarabine. Imatinib was shown to be superior to the IFN-α/cytarabine in terms of better tolerability, quality of life, and therapy response. The patients showed different response during imatinib and IFN-α/cytarabine treatment: the cytogenetic response rates were 84% and 35%, respectively. For this reason, the majority of patients, treated with IFN-α/cytarabine, were switched to the imatinib therapy [189-191]. Long-term follow-up of these patients at 6 years has confirmed the efficacy of imatinib therapy, since the event-free survival was 83% and general survival was 88% [192].

6.4 Second Generation Inhibitors: Nilotinib and Dasatinib

Dasatinib (formerly BMS-354825) has a broad kinase inhibition profile compared to imatinib, with main targets consisting of BCR-ABL, ABL, C-Kit, PDGFR, and SRC family kinases (SRC, FGR, FYN, LYN, and YES). In total, dasatinib inhibits more than 30 different kinases. The potency of dasatinib for unmutated BCR-ABL is over 300 times higher than that of imatinib and it was able to inhibit many mutated BCR-ABL isoforms that cause imatinib resistance [193]. However, some point mutations are inaccessible by dasatinib: *e.g.*, T315I/A, F317 L/V/C, and V299L [194]. The main indications for dasatinib therapy have been imatinib resistant CML and Ph-positive ALL [195, 196], but dasatinib has been approved also for the treatment of newly diagnosed CML patients. The SRC/ABL Tyrosine Kinase Inhibition Activity Research Trial C (START-C) evaluated the efficacy of dasatinib on CML patients who were intolerant or resistant to imatinib. The response rates at 15 months were: complete hematologic response (CHR), 91%; major cytogenetic response (MCyR), 59%; and complete cytogenetic response (CCyR), 49%; with 96% overall survival. A phase 3 DASatinib versus Imatinib Study In treatment-Naive CML patients (DASISION) study when compared dasatinib versus imatinib in newly diagnosed chronic phase CML patients [197] have shown faster and deeper responses in favor of dasatinib after 2-year follow-up study [197, 198].

Nilotinib (formerly AMN107) suppresses BCR-ABL activity with 20 times higher potency than imatinib and has the ability to inhibit imatinib-resistant mutations, except for T315I [199, 200]. Other nilotinib-resistant mutations seen in patients are E255K/V, Y253H, and F359V/C/I [201]. It is an effective treatment for CML with imatinib resistance and tolerance [202, 203], although in imatinib-resistant Ph$^+$ ALL patients, nilotinib showed only limited efficacy [203]. Patients with nilotinib-resistant mutations had less favorable responses to nilotinib [204]. A phase 3 study, called Evaluating Nilotinib Efficacy and Safety in Clinical Trials–Newly Diagnoses Patients (ENESTnd), when compared imatinib and nilotinib in newly diagnosed CML patients, showed that the responses with nilotinib were faster and deeper when compared to imatinib responses [205].

6.5. Other Tyrosine Kinase Inhibitors that Target BCR-ABL

Bosutinib (previously SKI-606) is a second-generation TKI with a target spectrum consisting of BCR-ABL, SRC, TEC, and STE20 family kinases. It does not inhibit c-Kit and PDGF receptors. However, as seen with other TKIs, the T315I and V299L mutations were inaccessible with bosutinib [206]. This drug demonstrated efficacy with imatinib resistant and intolerant CML patients [207]. Patients who had failed either with dasatinib or nilotinib, or both, also benefited from bosutinib therapy. Such therapy induced responses with patients carrying dasatinib-resistant F317L mutation and nilotinib resistant F359C/I/V mutations [208]. Bosutinib was recently approved for the therapy of CML patients who do not respond to other therapies.

Bafetinib (INNO-406) is a selective dual BCR-ABL/Lyn inhibitor, but it has no efficacy against the T315I mutation. Furthermore, it shows efficacy with patients resistant to multiple TKIs in a phase 1 study, although further studies are required to evaluate its efficacy [209].

Third generation TKIs are emerging in the field of CML therapy, to overcome the resistance seen with previous inhibitors. In their development, emphasis is on the T315I mutation and several candidates are on preclinical and phase 1 trials [210].

One of the most promising TKIs able to inhibit the T315I mutation is ponatinib (AP24534). In addition, it inhibits SRC, VEGFR, FGFR and PDGFR family kinases [211]. Clinical trials with ponatinib are ongoing and initial results from a phase 2 trial, called Ponatinib P-positive ALL and CML Evaluation (PACE), involves CML and Ph-positive ALL patients with dasatinib and/or nilotinib intolerance or resistance (including T315I). Results are promising: after a 6 month follow-up, 38/61 of chronic phase CML patients with the T315I mutation had entered major CyR [212]. However, this drug was correlated with a nearly 12% incidence of cardiovascular events such as serious arterial thrombotic events in adults, resulting in withdrawal from the market [213, 214]. For example, the upcoming a new trial in UK (SPIRIT 3) will evaluate both the selective use of dose-optimized ponatinib in patient who are not responding optimally to first-treatment, and the cardiovascular risk of CML patients in order to find the right balance between efficacy and risk for each patient.

7. MECHANISMS OF RESISTANCE TO TKIs

A proportion of patients did not respond to imatinib, which can be either due to primary resistance (also referred to as "refractoriness") characterized by the absence of an initial response, or to acquired resistance, where the patient loses the response. Such resistance can be further subclassified into hematologic, cytogenetic, and molecular resistance based on clinical and laboratory criteria. Initial response rates are lower in patients with advanced-stage disease and responses tend be short-lived in this patients group compared to the responses seen in chronic phase patients [209]. Different reports supported the idea that TKIs resistance is a consequence of the interaction of multiple factors including treatment compliance, genetic changes, BCR-ABL kinase domain mutations, or their combinations [215]. Therefore, such resistance can be divided in BCR-ABL–dependent and –independent mechanisms.

7.1. BCR-ABL-Dependent Mechanisms

7.1.1. Mutations in the BCR-ABL Kinase Domain as a Mechanism of Resistance to Imatinib

The point mutations in the region coding for tyrosine kinase domain of BCR-ABL can cause acquired resistance to imatinib. Different mutations have been detected in ABL domain and they can alter the three-dimensional structure of BCR-ABL interfering with the binding of imatinib or eliminate important molecules for imatinib binding [216]. So far, 100 different point mutations have been discovered, including T315I, Y253H and F255K, among others [217-221]. However, 15 amino acid substitutions are present in 85% of the mutations, whereas only seven mutated residues namely G250, Y253, E255, T315, M351, F359 and H396 are found in 66% of the reported cases [219]. Other mutations are clustered into four regions in the kinase domain: the first region, the P-loop, (amino acids 248-256), interacts with the ATP phosphate group and also binds with imatinib. The second cluster has a crucial role in autoregulation of both kinase activity and imatinib binding. The fourth cluster is located in the activation loop starting (amino acids 381-407) and the mutations in this cluster do not allow the binding with imatinib [221].

7.1.2 Mutations in the BCR-ABL Kinase Domain as a Mechanism of Resistance to Second and Third-Generation TKIs

Kinase domain point mutations have been found in different *in vitro* mutagenesis assays. Nilotinib resistance is mainly correlated with mutations in the P-loop or T315I, while dasatinib resistance is associated with mutations that are often localized at contact residues (L248, V299, T315, F317). Some of these mutations are sensitive to imatinib or nilotinib, such as V299L, T315A, F317L/V/I, but T315I is resistance against all of the drugs [204, 211, 212, 222]. Clinical studies have also demonstrated that dasatinib-reduced response occurred in imatinib-resistant cells with T315I and V299L mutations and different substitutions at amino acid F317. Of note, L299V and T315I mutations induced a reduced response to bosutinib. Furthermore, another article revealed different mutations (particular Y253 and E255 mutations) after ponatinib exposure [211]. However, clinical resistance has also emerged with the use of second and third generation tyrosine kinase inhibitors after imatinib failure.

7.1.3. Overexpression of BCR-ABL in TKIs Resistance

Imatinib resistance is caused by increased expression of the BCR-ABL kinase from genomic amplification or increased mRNA levels [223-226]. It is demonstrated that increased BCR-ABL expression level restored oncogenic signaling in presence of imanitib [227]. It is also possible that up-regulation of BCR–ABL may be an early phenomenon, that occurred before the emergence of a dominant clone with a mutant kinase domain [228]. Significantly, the genomic amplification is likely due to the genomic instability. On the contrary, Virgili *et al.* [229] found that loss of the remaining normal ABL allele in CML patients with deletion in 9q34, who did not achieve CCyR during treatment, yields a novel mechanism of imatinib resistance. These results demonstrated that these gene and protein expression alterations could explain all resistant mechanisms. These data indicate a complicated relationship among acquired resistance-related genes.

7.2. BCR-ABL-Independent Resistance Mechanisms to TKIs

The mechanisms for increased efflux of imanitib and/or decreased drug influx, in addition to drug sequestration, involving P-glycoprotein member (P-170 glycoprotein) of ABC transporter influences the imanitib resistance. Thus, such resistance seems to be a multifactorial phenomenon including various mechanisms such as augmented drug efflux and overexpression of the P-170 glycoprotein that activates drug efflux [230]. In addition, several articles have shown different hOCT-1 gene expression levels between cytogenetic responders and non-responders [231-233]. The intracellular concentration of dasatinib is not altered by hOCT1 compared to imatinib: patients with a low OCT1 expression or activity thus it may be candidates for first-line treatment with dasatinib [234]. Furthermore, the increased serum protein α1 acid glycoprotein expression, that sequesters imanitib [235], low serum drug concentration, and alternative signaling pathway activation through Ras/Raf/MEK kinase, STAT, ERK2, are involved in this process [236]. Another report shows that elevated mRNA levels of prostaglandin-endoperoxide synthase 1/cyclooxygenase 1 reduces the imanitib concentration since it encodes an enzyme that metabolizes such drug. Finally, the LYN kinase is markedly upregulated in cell lines and patient samples who are resistant to imatinib [207]. Knowing-down of LYN with short interfering RNA or a SRC kinase inhibitor is able to trigger apoptosis in resistant cells identifying a potential targeted therapeutic options [237]. Another studies have elucidated the molecular mechanism by which Sphingosine Kinase 1 (SphK1) induces the acquisition of resistance to the anticancer agent imatinib in CML [238, 239]. SphK1 contributes to the resistance of CML cells to this anticancer agent as demonstrated by the finding that siRNA knock-down of SphK1 expression sensitises resistant CML cells to imatinib whereas enforced expression of SphK1 prevents apoptosis in response to this drug in sensitive CML cells [240, 241]. Salas and colleagues [239] demonstrated that sphingosine 1-phosphate (S1P)/sphingosine 1-phosphate receptor 2 (S1P2) signalling regulates BCR-ABL stability in CML cells by inhibiting the activity of the PP2A. This prevents the dephosphorylation of BCR-ABL and therefore its proteasomal degradation, resulting in the accumulation of the oncoprotein. Thus, inhibition of the SK1/S1P/S1P2 pathway sensitises CML cells to imatinib by restoring the PP2A-dependent dephosphorylation and subsequent degradation of BCR-ABL [134]. Furthermore, the increase in imatinib-induced apoptosis is also shown in CD34$^+$ progenitor cells isolated from the peripheral blood of a CML patient who exerted clinical drug resistance due to the expression of the BCR-ABL mutant with T315I conversion and in IL-3-dependent murine myeloid 32D cells stably expressing wt-, Y253H-, or T315I-BCR-ABL, when compared to their controls *in situ* [251]. Interestingly, FTY720 (Fingolimod, Gilenya) that is a sphingosine analog and is also a potent PP2A–activating drug (PAD), was previously reported to induce apoptosis in CML cells *via* direct activation of PP2A [241].

7.3. p-CrkL as Equivalent Marker of BCR-ABL Kinase Inhibition by TKIs

Phosphorylated CrkL (p-CrkL) is considered commonly alternate marker for BCR-ABL tyrosine kinase activity [242]. It was firstly identified in 1994 that a 39 kDa protein was consistently tyrosine phosphorylated in peripheral blood neutrophils from all 18 chronic phase CML patients screened, but not in normal controls [102]. It was also found that the same protein was tyrosine phosphorylated in BCR-ABL positive cell lines, such as K562 and BV173 [102]. Later, this 39 kDa protein was identified as CrkL after protein purification [102]. The CrkL gene had been discovered in 1993; it is located on chromosome 22 and centromeric to the BCR gene [243]. Its protein is very similar to that of a viral oncogene v-Crk which is from avian sarcoma virus CT10, and CrkL is 60% homologous to Crk that is the human homolog of v-Crk [243]. CrkL protein contains a SH2 domain and two tandem SH3 domains without having any catalytic domain [243]. It was revealed that CrkL interacted with ABL directly [102]. It was also found that CrkL was tyrosine phosphorylated by Abl and BCR-ABL in transfected COS-1 cells and complexed with both of them in K562 and transfected COS-1 cells [244]. Subsequently, it was identified that BCR-ABL phosphorylated CrkL on tyrosine 207 [245]. p-CrkL has been used as a marker to determinate the effect of imatinib on inhibition of ABL kinase activity in patient samples by Western blotting. A method was developed to detect the level of p-CrkL in CD34$^+$ CML progenitor cells using intracellular flow cytometry. This method takes much less time than Western blotting, but more importantly requires as few as 104 cells for each test, while Western blotting requires more than 105 cells, making the flow cytometry method a preferred choice with rare stem cell populations [246]. A further validation of the flow cytometry method was reported in 2009, since it showed that p-CrkL and total tyrosine phosphorylation are reduced to the same levels in K562 cells with 48 hours of imatinib treatment [246]. However, a direct comparison of p-CrkL and BCR-ABL tyrosine kinase activity has not been made, especially at short time-points (≤24h).

8. ALTERNATIVE STRATEGY TO OVERCOME RESISTANCE IN CML TREATMENT

8.1. Allosteric Inhibition

Alternative strategy to overcome imatinib resistance is to target BCR-ABL kinase activity through an allosteric site at a distance from the active site. In general, allosteric kinase inhibitors are anticipated to target regulatory mechanisms unique to a given kinase and therefore may exhibit improved selectivity and reduced off-target effects compared to active inhibitors [247]. Using a high-throughput cytotoxicity assay, Gray and coworkers discovered GNF2, the first allosteric inhibitor of BCR-ABL [248]. Biochemical and structural studies showed that GNF2 binds the myristic acid binding pocket of ABL and stabilizes the inactive conformation of the kinase. Surprisingly, GNF2 alone was not active against the imatinib-resistant mutant T315I or several other prominent imatinib-resistant mutants. However, the combination of GNF2 with nilotinib was effective at inhibiting the T315I mutant. *In vivo* studies, using a murine CML bone marrow transplantation model, showed that the co-treatment of GNF5 (an analog of GNF2 with improved pharmacokinetics) and nilotinib reduced spleen size, white blood cell counts and STAT5 phosphorylation while increasing overall survival. In addition, Hydrogen exchange mass spectrometry (HXMS) studies revealed conformational changes in the ATP binding site upon GNF5 binding that directly support allosteric communication between the myristic acid binding pocket and the active site, which are separated by more than 30 Å in the crystal structure. Hence, this 46 new class of drug defines and potentiates the development of allosteric inhibitors that could be clinically tested [249, 250]. Besides allosteric inhibition by GNF2, peptides that target the BCR-derived region of the protein and induced modification of the kinase activity expanded the understanding of allosteric inhibitors. Specifically these peptides target the N-terminal coiled-coil region of BCR-ABL and inhibit oligomerization, reduce kinase activity and increases sensitivity to both imatinib and GNF2 [251]. In a more recent study Feng *et al.* have shown that adenoviral transduction of SH2-DED (dead effector domain) in CML cells, resulted in its binding to pTyr177 through the SH2 domain and activation of caspase-8 through the DED domain. As a result, BCR-ABL positive leukemia cells exhibited reduction in cell proliferation and enhancement of apoptosis [252]. Whether or not peptide-based approaches such as these will ever be translated to the clinic remains unclear.

8.2. JAK Inhibitors

In CML, the JAK2/STAT5 pathway is directly activated by the BCR-ABL protein and STAT5 was believed to be dispensable in CML, given that BCR-ABL transduction-transplantation experiments, using BM from mice with a truncated form of STAT5, was still able to induce a CML-like disease [253-256]. Normally, STAT5 is activated through phosphorylation of tyrosine residues by the upstream protein JAK2. Thus, it can be reasoned that, because of a direct activation of STAT5 by BCR-ABL [257, 258], JAK2 is redundant in CML. However, it has been argued that inhibition of JAK2 reduces the transforming effects of BCR-ABL and that JAK2 either cooperates with BCR-ABL to activate STAT5 or mediates cross-activation of other signaling pathways [259]. One of the main proofs supporting involvement of JAK2 in CML, is that inhibition of JAK2 with a small-molecule inhibitor was shown to induce apoptosis in CML cells [260]. Over the last ten years, the role of JAK2 has been considered in haematological malignancies, since different JAK2 mutations have been discovered in BCR-ABL negative myeloproliferative diseases [261, 262]. Such result has increased the development of several JAK2 pharmacological inhibitors, which now are showing a good encouraging results [263-265]. However, JAK2 is part of the BCR-ABL signaling network pathway and is activated in CML cells [266]. JAK2, with the point mutation, is also involved in CML maintenance [113, 267-269]. However, these studies have yielded conflicting results on the relationship between BCR-ABL and JAK2. A recent report has shown that co-administration of ruxolitinib (JAK2 inhibitor) and nilotinib results in enhanced reduction of viability and proliferation of CML CD34$^+$ cells, including the quiescent fraction, *in vitro* and reduced engraftment of CML CD34$^+$ cells *in vivo* compared to nilotinib alone [270]. These data are in according with studies which described a high decrease survival of primary CML-CD34$^+$ progenitors, after a concomitant treatment with dasatinib and nonclinically JAK2 inhibitors [271-274] but are contrary to those, reported by Hantschel *et al.* [275], showing that BCR-ABL is still able to transform murine BM cells in which JAK2 has been deleted, both *in vitro* and *in vivo*. Additionally, JAK2 deletion seems to accelerate CML development in mouse models by preferentially causing elimination of normal HSCs [276]. This divergence might arise from the fact that JAK2 became indispensable only when BCR-ABL kinase does not function as been reported [277]. In addition, it is important to underlie that JAK2-STAT5 signaling module uncoupled in the different murine leukemia models in the absence of exogenous GFs [275]. Indeed, the combined treatment with ruxolitinib and nilotinib partially inhibited the JAK2 kinase activity as shown by the reduced phosphorylation levels of JAK2 and STAT5 only in presence of GFs [271]. The same results have been obtained in the report by Lin *et al.* where co-treatment of BMS-911543 (JAK2 selective inhibitor) and dasatinib significantly eradicated CML-LSCs [278]. Based on current studies, a therapeutic strategy targeting JAK2 in chronic phase CML patients needs more studies in clinical trial. Thus, several JAK2 inhibitors are tested in clinical trials in Europe and USA (INCB018424, TG101348, CYT387, CEP-701, AZD1480, SB1518, and XL019) and others (NCB016562, NVP-BSK805, and R723) are in preclinical development.

8.3 Allogeneic Hematopoietic Stem Cell Transplantation (HSCT)

Today, allogeneic HSCT is considered as second-line therapy after tyrosine kinase inhibitors therapy failure [279]. HSCs from bone marrow, peripheral blood, or umbilical cord blood are used for transplantation when bone marrow is completely destroyed by disease (cancer) or therapy (chemotherapy, or radiation therapy). This procedure has significant morbidity and toxicity and for this reason, it is currently performed as a last-resort for incurable hematological malignancies. These indications are reported for 95% of the 50,000 stem-cell transplantation procedures performed each year worldwide. Ongoing advances such as more accurate molecular HLA typing of unrelated donors, however, are improving results and broadening the potential use of transplantation [280]. In addition, the probability of relapse for patients transplanted in their first chronic phase is 10-30%, and it can be increased in patients transplanted in advanced phases of the disease [281].

8.4. The Approach towards Patients Resistant to Tyrosine Kinase Inhibitors

Different treatments can be effective in imatinib resistant patients: a simple increase in the imatinib dose up to the use of new generation of TKIs or alternatively, allogeneic stem cell transplantation. Whereas some studies could show that higher doses of imatinib might improve the response, other studies could not found these results [282, 283]. In this scenario, second or third-generation TKIs use seems to be much more promising [284-287]. Hence, the increased dose is not a good therapeutic for patients with moderate tolerability to imatinib. The choice of new generation TKIs for the second-line therapy in resistant or intolerant patients, has to regard the BCR-ABL mutation status. Actually, preliminary results suggest that a treatment failure for dasatinib and bosutinib might be correlated with mutations at position V299 and V317. For nilotinib, mutations in the glycine-rich loop could constitute a problem (Y253, E255, F311 and F359). Unusually, these TKIs cannot be utilized in patients with the T315I mutation. In this context, the ponatinib could be a versatile approach, but it is necessary to better elucidate its mechanism. In summary, those patients as well as patients after treatment failure to several TKIs and/or in advanced stages should be considered for allogeneic stem cell transplantation. In addition, other new TKIs are in preclinical and clinical development such Bafetinib, an ABL and LYN kinase inhibitor, very similar to imatinib and DCC-2036. Such new multi-kinase inhibitor that binds the switch control pocket of ABL, has shown to exert high activity against the T315I mutation in preclinical studies [288]. Other studies on the alkaloid Homoharringtonine (omacetaxine), which is a global inhibitor of protein synthesis, showed clinical effectiveness even in patients treated with multiple TKIs [289].

8.5 The Stop Imatinib (STIM) Study

Stop Imatinib (STIM) study was conducted on 100 CML patients with CML who had a complete molecular remission (CMR) for at least 2 year after imanitib treatment. After stopping imatinib treatment, 41% of CML patients with a stable CMR for at least 2 years were able to sustain the CMR for 12 months, while 38% continued in complete molecular remission for up to 2 years after withdraw of imatinib. An update of the STIM trial at 48 months of follow-up, conducted on 80 patients with CM-CML has demonstrated that 39% of patients have maintenance of CMR at 24 and 36 months [290]. The author also used CMR to analyze molecular relapse, and the corresponding rate was 56% (45 patients) after a median of 4 months off therapy. Eight patients (18%) lost CMR after 6 months, 2 of which were late molecular relapses at months 35 and 40. Of the 51 patients who remained in MMR, 28 had positive BCR-ABL transcript: 12 patients had positive BCR-ABL transcripts below the MMR threshold; while 16 patients had less than 2 consecutive positive values below the MMR threshold. Thirty-one patients in the cohort restarted treatment (mainly imatinib) after the loss of MMR. In all patients who were retreated, MMR was regained at a median follow-up of 17 months, 23 patients regained CMR, and 8 patients are in MMR while they were undergone to treatment. The cumulative incidence of a second CMR was 84% at 24 months, and there was no loss of hematologic response during the time off treatment [291]. The Australian group has recently conducted a prospective clinical trial (TWISTER study), in which imatinib treatment was stopped in 40 CP-CML patients with undetectable minimal residual disease for at least 2 years. The relapses happened during 4 months of stopping imatinib. Twenty one patients treated with interferon before imatinib, showed slower achievement of undetectable minimal residual disease associated with relapse risk. Highly sensitive patient-specific BCR-ABL DNA PCR showed the original CML clone in all patients with stable undetectable minimal residual disease, even several years after imatinib withdrawal [292]. Another study have identified 50 patients with imatinib withdrawal for at least 6 months and analyzed 43 of them, aiming at characterizing the clinical outcomes and profiles of CP-CML patients who could discontinue imatinib. The complete molecular response rate was 47% after imatinib withdrawal. The authors concluded that imatinib dose, prior to interferon treatment, and the intensity of the molecular response has effect in long-term sustained complete molecular response after imatinib withdrawal [293]. Finally, a Korean study and several sporadic cases have also reported on discontinuation of imatinib and currently there are several ongoing clinical trials examining the alternative of stopping treatment in patients treated with the newer TKIs [294, 295].

9. PERSONALIZED MEDICINE OF CML

Personalizing treatment for CML patients depends amongst others on the phase of their disease – *i.e.* chronic, accelerated, or blast phase. According to the recently published European Leukemia Net (ELN) recommendations, the first-line treatment of chronic phase (CP)-CML can be using one of the three TKIs following this indication: imatinib 400 mg daily, nilotinib 300 mg twice daily and dasatinib 100 mg daily. Conversely, accelerated phase (AP) and blastic phase (BP) in newly diagnosed, TKI-naïve patients should be treated with imatinib 800 mg daily or dasatinib 140 mg daily. In another trial (ENESTnd), 846 patients were randomly enrolled to receive imatinib 400 mg once per day, nilotinib 300 mg twice per day, or nilotinib 400 mg twice per day [296, 297]. After one year, the patients, treated with nilotinib, had both the higher cumulative rate of CCyR (15%) and higher probability of major molecular response than in imatinib-treated patients. Yet, this superior rate of responses is not correlated with improvements of overall survival. In another trial, called the DASISION, 519 CML patients were randomly selected to receive either dasatinib (100 mg) or imatinib (400 mg) once daily as first-line therapy. A higher proportion of CCyR was induced by dasatinib after 12 months, but this effect was stopped after 24 months. Comorbidities play a role in personalizing treatment and choosing the right TKI according to the individual side effect profile. Each agent is associated with a specific safety profile that may contraindicate its use for certain patients [298]. Recently, several groups have suggested that BCR-ABL level measurement at 3 months was associated strongly correlates with CCyR achievement, molecular response as well as with overall survival and progression free survival during standard imatinib treatment. In a seminal paper, Marin *et al.* [309] considered 282 patients with CP-CML treated with imatinib (400 mg daily as first-line therapy) and then with dasatinib or nilotinib after imatinib failure. Transcript levels were present in 9.84% of patients after 3 months and had significantly lower probabilities of overall survival, progression-free survival, cumulative rate of CCyR, than those with lower transcript levels. Transcript levels at 3 months were the strongest predictor for the different outcomes, allowing early clinical intervention [299, 300]. Similarly, the German group analyzed 1303 newly diagnosed imatinib treated patients. The persistence of BCR-ABL mRNA levels at 3 months of >10%, 1–10% and 61% separated patients into high-risk, intermediate-risk and low-risk groups with a 5-year overall survival of 87%, 94% and 97%, respectively. In addition, such group also identified high-risk and-low risk patients by cytogenetic analysis at 3 months. They found that cytogenetic and molecular landmarks were of clinical significance also at 6 months [301]. The MD Anderson group reported their data from clinical trials with imatinib, nilotinib and dasatinib as initial therapy from 2000 to 2011. Their data also show that cytogenetic response at 3 months significantly discriminated for 3-year overall survival [302]. In summary, personalized cancer management includes testing for disease-causing mutations, assaying for minimal residual disease and applying targeted therapy.

10. EPIGENETIC MECHANISMS IN CML DEVELOPMENT

10.1 DNA Methylation

Epigenetic mechanisms are able to change the gene expression without involving DNA sequence. These changes are caused by different factors such as age, chronic stress and different types of diseases. Three systems are affected by this process: DNA methylation, histone modification and noncoding RNAs (ncRNAs). Generally, hypermethylation of 5' region of the genes alters their expression and causes the loss of their expression, whereas hypomethylation is found in different types of cancer and it is likely correlated with cancer progression [303]. Typically, there is hypermethylation of tumor suppressor genes and hypomethylation of oncogenes. Such process plays a vital role in the development of CML. The methylation at Pa promoter of ABL is a marker of CML pathogenesis [304, 305]. Indeed, the range of methylation frequency varies from 26% [306] to 77% [307], 78% [305] and 81% [308] in chronic phase CP-CML. A recent study has reported the high methylation at Pa promoter in BM samples derived from CP-CML patients when compared to those derived from healthy donors [309]. In BM samples of patients derived from AP-CML, each progenitor cell presents both methylated and unmethylated alleles, and then, in this case, the methylation can depend on an allele-specific process [304]. The important cell cycle regulatory gene p15(INK4b) has been shown to be inactivated in myelodysplastic syndrome [310, 311]. It has been found that its promoter is hypomethylated in CML patients [312, 313], whereas the gene is hypermethylated in 18% and 24% of patient samples [307, 314]. Little is known about the expression and epigenetic modification of this gene in CML. Ras association domain-containing protein 1 (RASSF1A) gene encodes a protein similar to the RAS effector proteins. Deleted or mutated expression of this gene has been correlated with the CML pathogenesis, suggesting its tumor suppressor function. Furthermore, the hypermethylation of its CpG-island promoter region was correlated with complete inactivation of this gene [315]. However, the aberrant methylation of additional genes seems to be correlated with both progression of CML and potential resistance to TKIs treatment. The AP-2 alpha protein is a sequence-specific DNA-binding protein that interacts with cellular enhancer elements to regulate transcription of select genes and recruiting transcription machinery. Early B cell factor 2 (Ebf2), a transcription factor, mediates the maintenance of hematopoietic progenitors, in part by regulating Wnt signaling. The hypermethylation of such genes has been observed in BC compared to CP in the study of 55 CP CML and 8 BC CML samples [316]. A recent study has showed that ATG16L1 to be a bona fide autophagy protein [317]. Another autophagy gene DAPK-1, a positive regulator of apoptosis, is frequently inactivated in human cancers by methylation [318]. Both genes may play role in CML development: ATG16L2 gene methylation occurred in 69% of CML patients and it is linked to a lower probability of achieving a major molecular response compared with the unmethylated cases at baseline [319]. The aberrant methylation pattern of DAPK-1 was found in 51% of CML cases and it was correlated with both gender the CML. Recently, genome wide methylation studies in K562 cells have underlined that five gene were linked to AP or BC progression in CML: OSCP1, TFAP2E, PGRA and PGRB were hypermethylated in patients with myeloproliferative disease while CDKN2B (p15) was deleted in K562 cells. Moreover, several methylated genes such calcitonin, HIC1, ER, PDLIM4, HOXA4, HOXA5, DDIT3 were associated with different stages of disease [319-325].

10.2 The Inhibition of DNA Methylation in CML

5-Azacytidine (chemical analogue of cytidine) is present in DNA and RNA. Azacitidine and its deoxy derivative, decitabine are two drugs that are used in the treatment of myelodysplastic diseases [326]. In a randomized trial of Myelodysplastic syndromes (MDs), around 16% of people, receiving azacitidine beside to supportive treatment, presented a complete or partial response—blood cell counts and bone marrow morphology returning to normal—and 2/3 patients did not require blood transfusions [327-329]. Kantarijan *et al.* [339] reported that 130 CML patients, treated with decitabine at different concentrations, had hematological response rate that varied in a range of 28% in BC, 55% in AP, and 63% in CP. The 25 CML patients in phase II trial were treated with decitabine since they have previously developed imatinib resistance [313]. Such drug was used with imatinib in 18 and 10 patients with CML in AP and BC, respectively. The overall hematologic response rate was 30% in BC and 50% in AP, obtaining the similar result in a previous phase I trial, where decitabine was administered alone. The combination therapy of hydralazine and magnesium valproate may be valid treatment in patients with myelodysplastic syndromes [330]. These drugs were used to revert the imatinib resistance in patients with CML [331]. Eight patients have received hydralazin, magnesium valproate in combination with imatinib at the time of progression. Seven patients obtained hematologic and cytogenetic responses or at least stabilization of disease after this treatment.

10.3 Histone Deacetylase Inhibitors (HDACIs) therapy in CML

Different studies have shed light on the important role of histone modifications in diverse cellular processes such as DNA replication, transcription, and cellular differentiation. Histone acetylation and deacetylation are reactions where the lysine residues at the N-terminus of histone proteins are acetylated and deacetylated as part of gene regulation. Histone acetylation and deacetylation are mediated by interplay of enzymes that activate and deactivate the paired processes. Histone acetylation, by histone acetyltransferase (HAT), acetylates the lysine residues to decrease compaction, in part due to the weakened charge attraction between the DNA and histone proteins; this increases gene activation and transcription. While histone deacetylase (HDAC) works to remove that acetyl group from the lysine and increase compaction and decrease transcription. Changes of the balance between histone acetylation and deacetylation is considered a key factor in cancer development, and an overexpression is often observed in cancer cells. However, it has been found their recruitment following oncogenic translocation protein complexes in different types of lymphomas and leukemias [332, 333]. Different HDAC inhibitors are able to restore normal acetylation of histone proteins and transcription factors [334]. At least 12 different HDACIs are developed for clinical trials in patients with hematologic and leukemias, lymphomas, and multiple myeloma [335]. From several studies, it has been possible to conclude that HDACIs, including vorinostat, depsipeptide, LBH-589, PDX-101, and several others, have antiproliferative activity in both hematologic malignancies and solid tumors [335, 336]. Vorinostat (SAHA) has been the first HDACI to be approved by the Food and Drug Administration for clinical use in cutaneous T-cell lymphoma patients [337]. The treatment with SAHA down-regulated the mRNA and protein level of BCR-ABL in K562 and LAMA-84 cells. Moreover, the BCR-ABL expression levels strongly decreased after co-administration with SAHA and imatinib. In particular, such drug triggered apotosis in $CD34^+$ cells derived from BC CML patients that become imatinib resistant during the treatment [338]. Furthermore, the study of Fiskus *et al.* [349], where vorinostat and dasatinib co-treatment is responsible for both reduction of the BCR-ABL protein level and induction of cell death in imatinib-sensitive and imatinib-resistant cells as well as in primary CML cells derived from patients in advanced stages of the disease. These data suggest that these inhibitors such as SAHA can cause induce acetylation and inhibition Heat Shock Protein 90 (a chaperone for BCR-ABL), and hence stimulates degradation of client proteins BCR-ABL [339, 340].

11. THE MicroRNAs ROLE IN THE PATHOGENESIS AND DRUG RESISTANCE IN CML

The discovery of small RNAs like MicroRNAs (miRNAs) has revolutionized the concept of genes and gene regulation. It is clear that miRNAs play a key role in virtually all biological processes and that their dysregulation is associated with disease. The first miRNA, discovered in the nematode *Caernohabditis elegans* development, was found to be responsible for silencing the lin-14 gene [341, 342]. The second important miRNA let-7 was also identified in *Caernohabditi elegans* [343], and then in human beings and in animal [344, 345]. Subsequently, hundreds of miRNAs and their biological functions have been identified [346]. They depict a pivotal roles in the cellular growth, proliferation, and differentiation *via* the negative regulation of over one-third of all human genes at the translational level [347]. The miRNAs may have either an oncogenic or a tumor-suppressive function [348]. The first investigation discovered that miR-15 and miR-16-1 had a deleted region, 13q14, in CLL [349] and they are as tumor suppressors, and their expression inversely correlates with anti-apoptotic BCL2 expression [350]. Another early findings have detected over 50% of miRNA genes are located within regions of loss of heterozygosity, amplification, fragile sites, viral integration sites, and other cancer-related genomic regions [351, 352]. Using qRT-PCR, a report found one upregulated miRNA (miR-96) and four downregulated (miR-151, miR-150, miR-125a, and miR 10a) in $CD34^+$ cells derived from CML patients when compared with the same cells derived from healthy donors. Furthermore, an upregulation of miR-150 and miR-151 was reported in Mo7e leukemic cell line after the imanitib treatment, indicating a link between these miRNAs and BCR-ABL downregulation [353]. The same investigation suggests that miR-10a downregulation is independent of BCR-ABL in CML patients [353]. Additionally, transcription factor upstream stimulatory factor 2 (USF2) has been considered as a miR-10a target and it appears to be upregulated in CML patients using gene array studies [353]. The miR-17-92 cluster is responsible for upregulation of BCR-ABL fusion gene [354]. This cluster is conserved in the chromosome 13 open reading frame 25 (c13orf25) genomic region and its transcription is dependent on by c-Myc [354]. The cluster plays likely an oncogenic role in CML as overexpression induced proliferation *in vitro*. Interestingly, such miRNA-17-92 exhibited an up-regulation in CML-CP, but not in blast crisis CML-BC, whereas this phase showed increased BCR-ABL activity [355]. Different members of the miR-17-92 cluster regulate the transcription of different genes, that can induce or inhibit cell death process, such as E2F1, PTEN, and BIM [356]. These experiments point out a specific function for this miRNA during two stage of the disease (CML-CP and CML-BC). Furthermore, has been found that an aberrant activity of RNA binding proteins (RBPs) is linked to increased BCR-ABL activity during blast crisis [356]. The CML-BC progenitors do not undergo neutrophil differentiation when C/EBPα expression is suppressed by the translation inhibitory activity of the RNA-binding protein hnRNP-E2. In fact, a BCR-ABL upregulation is necessary to enhance hnRNP-E2 expression, which depends on phosphorylation of hnRNP-E2 serines 173, 189, and 272 and threonine 213 by the BCR-ABL-activated ERK1/2. Additionally, myeloid differentiation is inhibited through interaction with C/EBPα [357, 358]. A miR-328 downregulation has been shown in CML-BC secondary to BCR-ABL activity both *in vitro* and *in vivo* using micro-array, northern blot, and qRT-PCR methods [355]. Postulating that miRNA and RBPs may interact, this study demonstrated that hnRNP-E2 mediated the loss of miR-328 in CML-BC *via* an inverse correlation between the two forms, suggesting hnRNP-E2 may regulate miR-328 activity. Therefore, the interaction between them occurs in a seed sequence independent manner. Re-expression of miR-328 is able to restores differentiation and block blast survival by simultaneously interacting with the hnRNP E2 (PCBP2), and the mRNA encoding the survival factor PIM1 [355]. Another article describes that the RBP, hnRNP-A1, is also upregulated in CML-BC and is linked to pri-miR-17-92 at the same time [359]. This study provides support for the idea that the lack of miR-17-92 expression can play a pivotal role during BC stage in CML, as described previously. Another hallmark of CML is miR-150 whose a reduced expression is found in CD34$^+$ cells [360] derived from CML patient at diagnosis and also in total leukocytes of PB derived from CML patients in AP and BC [361]. Moreover, such miRNA was significantly increased after short therapy with imatinib [362]. Different researches showed that the amount of miR-146a was also decreased at CML diagnosis [360-362], whereas it was normalized after imatinib treatment [361, 362]. Another recent study on 17 patients has been demonstrated that the miR-451 expression level increased after therapy with imatinib. In addition, miR-451 level and BCR-ABL transcript level are inversely correlated in some CML patients. However, this study suggests that expression of miR-451, being heterogeneous among the patients, can be regulated by other mediators [363]. A recent work of Rokah *et al.* [364] has characterized miRNAs expression profile of CML cell lines and patients compared to its normal counterpart derived from healthy donors, using miRNA microarrays and miRNA real-time PCR. The expression levels of miR-31, miR-155, and miR-564 were decreased in CML and influenced by BCR-ABL activity. Besides, the predicted targets and affected pathways of the deregulated miRNAs were aberrantly expressed in K562 cell lines in comparison to blood cells of healthy individuals. In addition, MiR-130a expression is also regulated by BCR-ABL in K562 cells. SiRNA knockdown of BCR-ABL in K562, decreased miR-130a and miR-130b and increased the expression of their putative target, the growth negative regulator CCN3 [365]. Recent evidence indicates the miRNAs expression could be linked to the onset of resistance to TKIs. In this context, a recent work has demonstrated the relationship among expression change in specific miRNAs and resistance to imatinib or responsiveness to imatinib after the treatment in CML patients. In peripheral blood mononuclear cells (PBMCs) derived from patients newly diagnosed with CML and treated with imatinib for two weeks, it had been observed an increase in the expression of miR-150 and miR-146a and a decrease of miR-142-3p and miR-199b-5p [366]. Lopotova` *et al.* [367] and Scholl *et al.* [368] have reported that CML patients with imatinib-resistant showed lower levels of miRNA-451 compared with responders. In line with this report, Liu *et al.* [369] described a reciprocal regulatory correlation between c-Myc and miRNA-144/451. In particular, Myc is upregulated in imatinib-resistant K562 cells, where it inhibited miRNA-144/451 transcription. Finally, Ohyashiki *et al.* have identified downregulation of miR-148b in patients in the STOP-IM group and in a subset of the IM group [370].

12. LEUKEMIA STEM CELLS (LSCs) IN CML

Cancer stem cells (CSCs) were first identified in patients with AML [371-373]. A CD34$^+$CD38-stem cell-like population was isolated from a human AML. Tumor was developed when such cells were injected in immunodeficient mouse host. In addition, the results suggested that normal primitive cells were targets for leukemic transformation [371]. The frequency of such cells derived from peripheral blood of those AML patients was one engraftment unit in 250,000 [374]. Weissman and colleagues suggested that: 1) such cells may generate tumors through their self-renewal process and differentiation into different cell types; 2) such cells can be distinct population in tumors and provoke relapse and metastasis by generating a new tumors and 3) such cells can be to be serially transplanted [375]. The study by Pérez-Caro and colleagues describes a mouse model of CML [376]: bone marrow cells from wild type mice were transduced with a retrovirus expressing BCR-ABL under control of Sca-1 promoter, specifically expressed in the haematopoietic stem cells. In this case, the mice developed CML-like disease and such results suggested that LSCs were located in BCR-ABL-expressing HSC population. In order to confirm this theory, bone marrow cells, isolated from primary CML mice, were transplanted into secondary recipient mice, that developed the leukemia and then died [377]. It is important to note that LSCs in CML are not a static population: evidence demonstrates that the both expanded progenitor pool (CD34$^+$Lin-) and unexpanded population of HSCs are found in BM from patients with CML-BC [378]. Such data suggested that LSCs for CML-BC could reside in more differentiated progenitor cells during this stage. Indeed, real-time PCR results showed more abundant BCR-ABL transcripts in myeloid progenitors than HSCs. Furthermore, a mouse model, expressing BCR-ABL in an established line of E2A-knockout mouse bone marrow cells also showed that BCR-ABL transformed GMPs functioned as LSCs [379]. Despite, Wnt/β-catenin pathway is generally triggered in HSCs but not in GMPs, a reactivation of β-catenin signaling occurred in GMPs derived from CML patients in blast crisis and from mice in blast crisis [378, 379]. These data point out that GMPs act as LSCs in CML blast crisis [380]. Finally, these results provide further evidence that different properties of LSCs may cause the difference between these two disease phases.

12.1. Resistance of LSCs to BCR-ABL Kinase Inhibitors

Some data confirmed that primitive quiescent leukemic stem cells are resistant to imatinib. One study shows that non dividing Lin-CD34$^+$ stem cells, derived from the PB of CML-CP patients and cultured with and without GFs and imatinib, did not undergo apoptosis [377]. The imatinib treatment alone could not cure CML, since it is not able to kill the quiescent BCR-ABL-expressing LSCs, as confirmed by both *in vitro* and *in vivo* studies [377, 380]. In addition, inhibition of both SRC and BCR-ABL kinase by dasatinib does not induce the cell death in leukemic stem cells in B-ALL and CML mice, suggesting a BCR-ABL kinase-independent pathway in such cells [377]. This result is supported by the subsequent findings that quiescent human CD34$^+$CD38-CML cells are resistant to both dasatinib and imatinib treatment [381]. In addition, BCR-ABL-expressing HSCs were identified in the side population of bone marrow cells from the imatinib- or dasatinib-treated CML mice, and they are able to develop the disease into recipient mice [377]. This result specify that neither imatinib nor dasatinib are able to eliminate completely BCR-ABL-expressing HSCs, suggesting that involvement of other pathways. To confirm these data, the BCR-ABL-expressing HSCs in dasatinib-treated CML mice were compared with HSCs in placebo-treated mice. Despite the fact, dasatinib decreased only the numbers of BCR-ABL-expressing HSCs in CML mice [377]. In addition, LSCs pool in bone marrow of imatinib-treated CML mice increased with time during the treatment. The same number of bone marrow cells from primary CML mice, (enriched for Sca-1$^+$ cells) were transplanted into three groups of tertiary recipient mice after different time points of imatinib treatment. It was observed that the bone marrow cells derived from the secondary CML mice, after longer imanitib treatment, induced faster tertiary CML in recipient mice [382]. Overall, these findings suggest that neither dasatinib nor imatinib are capable of eradicating completely LSCs.

13. CRITICAL MOLECULAR PATHWAYS IN LSCs

BCR-ABL may to play a crucial role in survival of LSCs since it activates a complex molecular network and interacts with different downstream signaling pathways. Indeed, the Alox5 significantly affects the functions of LCS but not of normal hematopoietic stem cells.

13.1. Wnt/β-catenin Pathway

Recent reports have shown that the Wnt/β-catenin pathway promotes the self-renewal capacity of stem cells such as hematopoietic stem cells [383]. The activation of β-catenin is observed in myeloid progenitors derived from in CML-BC patients in blast crisis [378]. Furthermore, in a CML mouse model the absence of β-catenin down-regulated ability of BCR-ABL to support both LSCs long-term renewal and CML development [377], as shown in the serial replating and transplantation assays.

13.2. The Hedgehog (Hh) Signaling Pathway

The Hh signaling pathway is one of the key regulators of Drosophila development [384]. In mammals, three Hedgehog homologues, DHH, IHH, and SHH have been found. Recently, it has been demonstrated that Hedgehog signaling plays an important role in regulating adult stem cells [385]. The SHH interacts with PTCH1 receptor in order to remove the SMO inhibition. The SMO transmits the signal *via* Gli transcription factors: the activators Gli1 and Gli2 and the repressor Gli3. These activated factors reach the nucleus and control the transcription of hedgehog target. It was observed the increased transcript levels of Gli1 and Ptch1 in CD34$^+$ cells derived from CML patient in both chronic phase and blast crisis [386]. Additionally, two studies showed that the BCR–ABL-transduced haematopoietic progenitors into irradiated mice developed CML in 94% of recipients, whereas the same cells with SMO deletion (SMO$^{-/-}$) provoked CML in only 47% of recipients [386-388]. By contrast, SMO overexpression caused an increased percentage of LSCs [386-388]. Thus genetic loss and gain of function experiments show that Hh and SMO pathway controls the frequency and maintenance of CML stem cells. Such underlying mechanism can be regulated by Numb, that is upregulated in LSCs following SMO deletion and such overexpression suppresses the propagation of LSCs *in vitro* [387].

13.3 Alox5 Pathway

The Alox5 gene encodes Arachidonate 5-lipoxygenase, (also known as 5-lipoxygenase, 5-LOX or 5-LO), that belongs to the lipoxygenase gene family. Different selective 5-LO inhibitors are used in different diseases [389, 390] and are able to induce apoptosis of CML cells *in vitro* [391, 392]. A recent study points out that Alox5 is differentially expressed in CD34$^+$ CML cells. Furthermore, the Alox5 gene was chosen for functional study of gene expression profiles between normal HSCs and LSCs by DNA microarray analysis. Such study has found that Alox5 is upregulated by BCR-ABL, but imatinib is not able to downregulate its expression level [393]. Furthermore, recipients of BCR-ABL transduced bone marrow cells from Alox5-/- donor mice did not develop the disease [393].

13.4. PTEN Pathway

The tumor suppressor gene PTEN expression is deleted in several cancers. By gene expression profiling of CD34$^+$ subsets from the bone marrow derived from both CML patients and healthy donors, it was found that it is differentially expressed [394]. In a recent study, the authors have generated a CML mouse model carrying PTEN deletion, using PTEN conditional knockout mice (PTEN$^{fl/fl}$). Bone marrow cells of PTEN$^{fl/fl}$ mice were transfected with the BCR-ABL-iCre-GFP retrovirus or BCR-ABL-GFP retrovirus as a control, and then transplanted into recipient mice. The CML development occurred much faster in mice that received bone marrow cells with PTEN deletion [395]. Additionally, a same number of

LSCs was sorted from CML mice, that have received both type of transfected bone marrow cells, and then transplanted into another recipient mice. The survival of CML was significantly longer in mice that received bone marrow cells with PTEN deletion [395], indicating that it is a potent tumor suppressor in BCR-ABL–induced CML cells.

13.5. Forkhead Box O (FOXO)

Forkhead box O (FOXO) transcription factors are a superfamily of evolutionarily conserved and are key components in tumour suppression by mediating the expression of genes involved in stress resistance, DNA damage repair, cell cycle arrest and apoptosis. A marked decrease of the HSCs compartment (including the short- and long-term HSCs populations) has been shown using FoxO1/3/4 triple conditional knockout mice model. Such decrease is associated with increased cell division and terminal differentiation at the expense of self-renewal [396]. Since FoxO3a acts downstream of the PTEN/PI3K/AKT pathway, it can have a key roles in the maintenance of CML LSCs [397]. In quiescent wild-type HSCs, FOXO proteins are localized in the nucleus and their transcriptional activity causes cell cycle arrest. FOXOs deletion stimulates an over-production of ROS in the proportion of cycling HSCs, leading to eventually HSC depletion [396]. A transduction/transplantation mouse model that reproduces CML like myeloproliferative disease has shown that FOXO3a can sustain leukemic stem cells [398]. Furthermore, FoxO3a localization in CML-LSCs is controlled by TGF-β. The combined treatment with TGF-β and BCR-ABL inhibition and FoxO3a knowdown, causes the LSCs eradication and CML regression [398]. In a recent study, Pellicano *et al.* shows that TKIs induced a dephosphorylation of FOXO1 and 3a and their relocalization from cytoplasm (inactive) to nucleus (active) in CD34$^+$ CML cells. Such process mediated also the expression of key FOXO target genes, such as Cyclin D1, ATM, CDKN1C, and BCL6 and induced G1 arrest. The same phenomenon was found in a CML transgenic mouse model, treated for six days with dasatanib. In addition, a regulation of FOXO target genes, including p57/CDKN1C and BCL6, occurs in CD34$^+$ cells derived from CML patients, treated with imatinib. Although these data suggest that FOXO activation can play a role in TKI treatment, more investigation would be required to better understand this mechanism.

13.6. MSR1 Pathway

MSR1 is a member of a family termed scavenger receptors and is mostly expressed in macrophages and dendritic cells and it binds to a broad range of ligands including Acetylated-low density lipoprotein (Ac-LDL), bacterial surface components and apoptotic cells [399]. MSR1 could mediate receptor internalization and cell adhesion and plays important roles in host cell interactions, macrophage adhesion and phagocytosis of apoptotic cells [400]. The binding MSR1 will activate a series of signal pathways, including Src kinase pathway and PI3K kinase-AKT pathway, to promote the cell growth, metabolism, survival and glucose homeostasis [401]. The Src kinase Lyn is associated with the cytoplasmic domain of MSR1, and it will rapidly activate Lyn [402]. PI3K-AKT-GSK3β pathway was also activated rapidly during the MSR1 mediated cell adhesion [401]. Recently, MSR1 was reported to be associated with prostate cancer in men of both African American and European genetically although the mechanism of MSR1 regulating cancer development is still unclear [403]. In CML development, MSR1 down-regulation is mediated by BCR-ABL and it is regained by Alox5 deletion, leading an acceleration of CML development. Furthermore, this deletion activated cell cycle progression and inhibited apoptosis of LSCs. MSR1 affects CML development through the PI3K-AKT pathway and β-catenin. The enhancement of MSR1 function may be of significance in the development of novel therapeutic strategies targeting CML.

13.7. SIRT1 Pathway

Sirtuin 1 (SIRT1) is the most conserved mammalian NAD$^+$-dependent protein deacetylase that is becoming a key molecule in different cancer and it can work as either a tumor suppressor or tumor promoter depending on the pathways and type of cancers. It deacetylate proteins that contribute to different cellular process [404, 405]. In CML, BCR-ABL through STAT5 upregulates SIRT1 that is able to induce leukemogenesis [404, 405]. In addition, such upregulation is evident both in CD34$^+$chronic CML progenitor cells and in later stages of CML [331]. In particular, SIRT1 deletion significantly inhibits BCR-ABL expression in HSCs and CML progression. Such result suggests that SIRT1 can acquire oncogenic properties in bone marrow stem cell [404]. Co-administration of tenovin-6 (TV-6) (small-molecule inhibitor of SIRT1) and imatinib increased apoptosis in CML mice model and human CML progenitor cells, but did not improve survival [405]. Consistently, in a another study, it has found that combining TV-6 and imatinib increased apoptosis of leukemic stem cells, also in cells harboring T315I mutation compared to either agent alone [404]. Therefore, these controversial results need more investigation.

13.8. PML Pathway

PML functions as tumour suppressor in the haemopoietic system. In CML, it positively regulates self-renewal in CML-initiating cells, while its deletion leads to complete elimination of the leukaemic stem cell pool and decreases disease progression. In addition, PML expression level is associated with poor overall survival in CML patients [406]. Arsenic trioxide is able to reduce PML expression and induces murine CML stem cells to enter the cell cycle, so the cells tend to respond more easily to therapy. Accordingly, arsenic associated with cytosine arabinoside activates apoptosis in the leukemic stem cells.

13.9. Genomic Instability Survival Pathways of LSCs

BCR-ABL tyrosine kinase is associated with elevated levels of Reactive Oxygen Species (ROS) and oxidative DNA damage in both murine and human BCR-ABL-transformed cell lines in comparison to non-transformed cell lines [407, 408]; similarly, both ROS and oxidative DNA damage are elevated in primary CML-CP and CML-BP cells in comparison to counterparts from healthy donors [184]. Most recently, Nieborowska-Skorska *et al.* published that ROS and oxidative DNA damage are elevated in human $CD34^+$ progenitor cells and $CD34^+CD38$-stem cell-enriched populations of CML-CP patient cells in comparison to counterparts from healthy donors [409]. To explore underlying mechanisms, *i.e.* Bolton-Gillespie *et al.* studied the effect of low and high BCR-ABL expression on ROS and oxidative DNA damage in BCR-ABL-transduced human $CD34^+$ cells. They had detected elevated ROS levels and oxidative DNA damage in high BCR-ABL-expressing $CD34^+$ cells compared to low BCR-ABL-expressing cells [410]. Furthermore, Nieborowska-Skorska *et al.* have further demonstrated that Rac2-PAK serine/threonine pathway changes mitochondrial membrane potential and electron flow *via* the mitochondrial respiratory chain complex III thereby stimulating high production of ROS in TKI-naïve and TKI-treated LSCs, using CML-CP primary cells and an inducible model of murine CML-CP. In this process, AKT seemed to play a pivotal role in generation of ROS-induced oxidative DNA damage [411]. These data support the hypothesis that genomic instability may originate from LSCs, but do not exclude the potential role of Leukemia Progenitor Cells (LPCs), and may have important clinical implications for CML treatment since additional genetic aberrations that encode primary resistance may protect LSCs, including the quiescent subpopulation, from eradication by TKIs, and the continuous accumulation of genetic errors may trigger disease relapse and progression. Furthermore, these results suggest that the therapy could be addressed in genomic instability mechanisms for CML treatment.

13.10. BCR-ABL-Independent Survival Pathways of LSCs

Recently, parallel studies have demonstrated that CML stem cells can survival without BCR-ABL activity [267, 268]. Indeed, BCR-ABL activity is suppressed by imatinib in all stem (CD34$^+$CD38-, CD133$^+$) and progenitor (CD34$^+$CD38$^+$) cells and in quiescent and cycling progenitors derived from newly diagnosed CML patients. The expansion of such cells is reduced when they are treated *in vitro* with imatinib for a short time [267]. Moreover, CML stem cells, derived by the CML transgenic mouse model, with inducible expression of BCR-ABL, can recur *in vivo* and start again leukemia in secondary recipients. In addition, BCR-ABL deletion in the human CD34$^+$ CML cells, treated for several days in physiologic growth factors, leads to partial inhibition of p-CrkL and p-STAT5, suppression of proliferation and colony forming cells, without reduction of cellular intake. The input cells are reduced (50%) after dasatanib treatment, whereas they increase mostly after the complete growth factor deletion plus dasatinib treatment. Moreover, the surviving fraction, enriched for primitive leukemic cells, was able both to grow up in a long-term culture-initiating cell assay and to expand after dasatinib elimination and addition of growth factors [268]. The results indicate that primary human and murine CML stem cells are not addicted to the oncogene kinase activity. Therefore, such data show that CML stem cells survival is Bcr-Abl kinase independent mechanisms.

14. THERAPEUTIC STRATEGIES FOR DELETION OF LSCs IN CML

14.1. Change in Cellular Properties of LSCs

The stimulation of primitive quiescent CML cells into cell cycle can be a potential strategy to kill such cells using BCR-ABL kinase inhibitors. The combined treatment with G-CSF (able to promote cell cycle entry) and imatinib could be an alternative approach, although such treatment did have achieved expected outcome [412, 413]. Finally, Pandolfi and colleagues have found that PML deficiency in non-proliferating LSCs causes LSCs depletion [406].

14.2. Targeting Critical Signaling Pathways in LSCs

Zileuton. Recently, a new drug has been developed to kill the LSCs in AML using an *in silico* screen of public gene expression database [414]. As described above, pivotal role Alox5 has been found in survival of LSCs but not normal hematopoietic stem cells using in CML mice [393]. Zileuton, active inhibitor of 5-lipoxygenase, suppresses leukotrienes (LTB$_4$, LTC$_4$, LTD$_4$ and LTE$_4$) formation. Such drug has been tested alone or in combination with imatinib in CML mice and the results have showed that it was more effective than imatinib. The combined treatment leads prolonged survival associated with less severe leukemia. Myeloid leukemia cells gradually reduced during double treatment in both peripheral blood that bone marrow derived from CML mice. In particular, normal myeloid cells did not decrease during zileuton treatment in the peripheral blood derived from the same animals. In addition, it is observed a probable arrest of LT (long-term)-LSCs differentiation since the ratio between the percentage of LT-LSCs and ST (short-term)-LSCs/multiple progenitor cells (MPP) augmented during the treatment. Beside, this treatment did not influence differentiation of LT (long-term)-HSCs in the same animals, suggesting that there is not suppression of normal HSCs.

BMS-214662. BMS-214662 is a farnesyltransferase inhibitor formed of a benzodiazepine core with a thienyl sulfonylurea moiety in the 4-position and a methyl-imidazole group on the 1-position that exhibits broad spectrum cytotoxicity against different tumors [415]. The both proliferating and quiescent primitive CD34$^+$CD38- CML stem cells underwent apoptosis during the treatment with such drug alone or imatinib or dasatinib combination. In particular, the different combinations had little or no effect on normal CD34$^+$ HSCs [416]. Such drug triggered the apoptotic pathways involving Bax, ROS, Cytochrome c, and Caspase-9/3 in association with protein kinase Cβ (PKCβ), E2F1 and Cyclin A-associated Cyclin-dependent kinase 2. In addition, the double treatment with the PKC modulators, bryostatin-1 or hispidin, markedly induced apoptosis the CML CD34$^+$ and CD34$^+$CD38- cells, suggesting the potential role of BMS-214662 for eradication of this disease [416-418].

FTY720. FTY720 is also a protein phosphatase 2A (PP2A)–activating drug (PAD). Such protein is a tumor suppressor and it is deleted in numerous cancers. PP2A is a tumor suppressor. A recent study has shown that SET (a nucleus/cytoplasm-localized phosphoprotein) expression is upregulated by BCR-ABL in CML cell lines and CD34$^+$ cells derived from CML patients in CP and BC. This upregulation, in turn, causes a decreased activity of PP2A, since SET is its potent inhibitor. The BRC-ABL decrease occurred in CML cell lines and CD34$^+$ cells derived from CML patients, after restoring PP2A activity [342]. For this reason, FTY720 was tested *in vitro*, and it triggered apoptosis in myeloid and lymphoid cell lines, imatinib/dasatinib-sensitive and -resistant myeloid and lymphoid cell lines and CD34$^+$ cells derived from both PB and BM of CML-BC patients but not of normal CD34$^+$ bone marrow cells derived of healthy donors. The CD34$^+$cells from CML patients, treated with both myeloid cytokines and FTY720, showed the same PP2A activity levels when compared to normal CD34$^+$ cells [241].

Bortezomib. Bortezomib, proteasome inhibitor, possesses antitumor activity in both solid and haematological malignancies. The proteasome has a pivotal role in the cellular homeostasis and it eliminates several regulatory proteins, including transcription factors, signaling molecules, and cell cycle inhibitors. Its inhibition results in cell cycle arrest or programmed cell death. It has been found that cancer cells are more sensitive to proteasome inhibition when compared to normal cells hence these inhibitors can be considered a novel strategy for cancer therapy. The proteasome induced protein degradation *via* 3 catalytic specificities: chymotrypsin-like (CT-L), trypsin-like (T-L), and post-glutamyl hydrolytic (PG) [419, 420]. Such drug is a reversible and specific inhibitor of CT-L activity and several clinical trials are testing its action in different types of cancers, including in multiple myeloma, mantle cell, follicular non-Hodgkins lymphoma, peripheral T-cell lymphoma and chronic lymphocytic leukemia [421]. Proteasomal activity was up-regulated in CML cells [344], and the BCR-ABL-positive cell lines sensitive or resistant to imatinib inhibited the growth after the deletion of this activity [422]. This drug inhibits the colony formation of $CD34^+$ $BCR-ABL^+$ progenitor cells derived from CML patients in *vitro*. It has been also observed that it activated cell death and its inhibition blocked of $CD34^+$ 38-long-term culture-initiating (LTC-IC) cells [346]. Furthermore, it impairs the function of CML-LSCs by decreasing levels of engraftment of patient-derived $CD34^+$ CML cells in mice [423]. Interestingly, bortezomib induced apoptosis of different BCR-ABL mutants, including T315I, H396P and M351T, suggesting that imatinib resistance does not affect the bortezomib sensitivity [423, 424].

Interferon-α is a pharmaceutical drug that is obtained from T lymphocytes of human blood in response to foreign antigens. It is an antiproliferative cytokine therapeutically applied to treat cancer expansion. It has been used in CML but its mechanism of action is not yet understood [425]. A report showed that 12 CML patients were treated with imatinib and with discontinued therapy. Six of them had again the relapse with detectable expression BCR-ABL mRNA, while others were in molecular remission. The previous treatment of these latter patients was made with IFN-α [426]. These data indicate that IFN-α could be eliminate the LSCs in CML. Other studies are necessary in order to gain more insight into this mechanisms.

14.3. Targeting CXCR4 Signaling in CML-LSCs

C-X-C chemokine receptor type 4 (CXCR-4) is encoded by the CXCR4 gene [427, 428]. It is a human chemokine receptors, that is triggered after the binding with the chemokine CXCL12 (also known as Stromal Cell-Derived Factor-1, SDF-1), with potent chemotactic activity for lymphocytes. Structure-function studies have shown that this interaction occurs in a 2-site binding: the extracellular N-terminal domain of the chemokine receptor (site I) initially binds to a ligand, that, in turn, interacts with the chemokine receptor at a different extracellular site (site II) to induce an activation signal [429]. CXCR4's ligand or SDF-1 plays a pivotal role in hematopoietic stem cell homing to the bone marrow and in hematopoietic stem cell quiescence. CXCR4 and its ligand CXCL12 interaction appears to play a critical for leukemic cells. Stroma-derived CXCL12 promotes the proliferation of HSCs and inhibits their terminal differentiation [430]. In CML, stromal interactions are deregulated and therefore, they develop malignant phenotype and prevent the TKIs-induced apoptosis in CML cells through various molecular mechanisms. The up-regulation of BCR-ABL down-regulates CXCR4 expression level, and this is correlated with the cell migration defects in CML. Several reports propose that TKIs re-establish CXCR4 expression and induce CML cells migration towards bone marrow micro-environment niches, which acquire stroma-mediated chemo-resistance of CML progenitor cells [431-438]. Elucidating the role of the BM milieu and the CXCR4/CXCL12 axis in the response to TKIs and in disease progression may provide a rational basis for the development of new drugs. Using leukemia MO7e cells, Geay *et al.* have shown that BCR-ABL suppresses the chemotactic response to CXCL12 and increases CXCR4 expression [436]. In another study, Ptasznik and colleagues have demonstrated that BCR-ABL alters CXCL12-mediated adhesion through beta2 integrin [439]. Such mechanisms involve the Src-related kinase Lyn that is activated by BCR-ABL in CML. In this fashion, such protein is unresponsive to CXCL12-mediated signaling and allows the egress of immature cells from the BM [435]. CXCR4 know-down with the small molecule inhibitor such as AMD3100 (plerixafor) is effective in combination with TKIs both *in vitro* and *in vivo*. Thus, pre-treatment with AMD3100 *in vitro* increased the imatinib-induced apoptosis of stroma-protected the BV173 lymphoid BCR-ABL-positive cells in culture, and significantly reduced the repopulating CML content in BM and spleen of SCID mice [440]. Combination of AMD3100 and nilotinib replaced the sensitivity of stroma-protected human K562 and KU812F CML cells *in vitro*. In addition, it activates nilotinib-induced reduction of tumor in a murine model of CML [441]. In agreement with previously reported data demonstrating an up-regulation of CXCR4 in K562 and KBM-5 cells upon imatinib treatment, another article shows a similar increase in CXCR4 surface levels in LAMA-84 cells [442]. Previously, Abraham *et al.* have established a role for the CXCR4 antagonist BKT140 in mobilization of hematopoietic stem cells from the BM [443]. Co-administration of BKT140 and imatinib triggered cell apoptosis and amended the BMSC-mediated resistance to imatinib [442]. Recent works have demonstrated that imatinib is able to upregulated BCL6 in K562 and LAMA-84 cells, and such increase mediated resistance to TKIs and enables the survival of CML stem cells and Ph-positive ALL cells [444, 445]. In accordance with these reports, it is observed that imatinib treatment in the presence of BMSCs further elevated BCL6 mRNA levels. Importantly, BKT140 treatment effectively down-regulates imatinib-induced increase of BCL6 levels in the absence as well as the presence of BMCSs [442]. Although the mechanisms is not yet clear, all data indicates the importance of CXCR4/CXCL12 axis in this process and further substantiate that CXCR4 is rational target for combinational therapy in CML.

15. LSCs AND THE HYPOXIC MICROENVIRONMENT

Hypoxia, low levels of oxygen, is a known characteristic of solid tumors and leukemia resulting in the increase in drug resistance. Mammalian bone marrow provides a relatively hypoxic niche essential for maintaining the self-renewal and survival of primitive HSCs [446]. The hypoxic niche is crucial for bone marrow function because *in vitro* culture of hematopoietic progenitors under hypoxic conditions displays a decrease in cellular proliferation accompanied by increase of the cells at the G0 phase of the cell cycle [447]. Although the molecular mechanisms of hypoxia in HSCs survival remains largely unknown, accumulating evidence suggests that hypoxia-inducible factor 1 (HIF1) can mediate the effect of hypoxia on HSCs [425]. In addition, it has been shown that cyclin-dependent kinase inhibitors that are regulated by HIF1α are associated with the function of hypoxia in HSCs [448]. A recent study on HSCs in an animal model has shown that regulation of the HIF1α level is crucial for the HSCs survival [449, 450]. HIF1 belongs to the family of basic helix-loop-helix (bHLH) transcription factors, and is a heterodimer formed of a constitutively expressed HIF1β subunit and a HIF1α subunit [451]. The expression of HIF1α is mediated by different signals. Under normoxic conditions, the HIF1α degradation occurs through the oxygen-hydroxylation of proline residues 402 and 564 by a proline hydroxylase (PHD). The HIFα hydroxylated form is identified by the Von Hippel-Lindau protein (pVHL) through E3 ubiquitin-protein ligase [452, 453]. In order to target LSCs that are responsible for relapse in patients, a hypoxia-selective cytotoxic compound that is produced naturally by the *Micromonospora* bacteria strain, Rakicidin A [454], was tested in CML cells with SC-like features. CML cells continuously cultured at 1% O_2 and the sub-clones (hypoxia adapted cells) that survived these hypoxic conditions were treated with Rakicidin A. These cells showed shrinkage, nuclear condensation and fragmentation which indicated that these cells are in apoptosis. Hypoxia adapted cells are less viable when treated with up to over 7.5 μM Rakicidin A compared to parental cells that are not hypoxia adapted. Apoptosis by Rakicidin A was induced *via* caspase dependent and mitochondria dependent pathways, as proved by caspase-3 inhibition, and the resulting delay in apoptosis [455]. The effect of combination therapy of Rakicidin A and imatinib was tested in hypoxia adapted cells and in non hypoxia adapted cells. Such combination was synergistic in hypoxia adapted cells, meaning cell death was induced in LSCs. This combined chemotherapy was rather antagonistic than synergistic in non-hypoxia adapted cells showing that Rakicidin A and imatinib specifically induces cell death in hypoxia adapted cells [455]. The ideal therapy will eliminate the LSCs while saving the HSCs. A way to achieve this might be to develop a drug that is inactive until it reaches a hypoxic area and then specifically targets LSCs. The HIF pathway may be a molecular target for the elimination of LSCs. Finally, it is still unknown whether HIF1α plays a role in LSCs in CML.

16. AUTOPHAGY IN BLOOD CANCERS

Although autophagy inhibition holds promise for treating hematologic malignancies based on the rationales described above, there is another side to the coin, reflecting the view of the role of autophagy as a tumor suppressive mechanism. Such role is supported by accumulating evidence and suggests that autophagy activation could also be a viable approach for treating cancer, at least in certain contexts and with well-defined measures. In addition, it is also possible, although highly controversial, that in well-defined circumstances, autophagy could turn into a true cell death program ('autophagic cell death') instead of being cytoprotective. On the other hand, autophagy has been considered as a survival mechanism above all within cancer cells. Cells in the core of the tumour or highly proliferating cancer cells that have surpassed the potential of their vascular niche, have to overcome adverse conditions such as hypoxia and limited access to nutrients and GFs [455, 456]. Unlike normal cells that have low basal autophagy levels, cancer cells seem to be "addicted" to elevated levels of autophagy under nutrients full conditions. In support of the cancer cell "autophagy addiction" hypothesis, studies demonstrate the autophagy dependence of RAS-driven cancers [457-459]. Several data have shown that autophagy is required for the maintenance of mitochondrial integrity, which is necessary for the survival of RAS-expressing cells during starvation [457]. Most importantly, tumourigenic ability of RAS was ablated in an autophagy deficient background [457-459]. Taken altogether, autophagy seems to have a dual role in cancer and it can act as a tumour suppressor by ensuring cellular integrity, however, in established tumours it seems to support the survival of cancer cells [460]. Therefore, modulation of autophagy for cancer therapy should be context-specific and take into account the type of tumour, the stage of disease and the nature of the treatment [460].

16.1 Inhibition of Autophagy in CML

The inhibition of autophagy can be an effective approach in order to promote imanitib-induced cell death in LSCs [461]. Autophagy inhibitors can be a novel approach in cancer treatment, when autophagy acts as a mechanism that helps cancer cells recover from the insult of anticancer treatments. TKIs stimulate autophagy in CML cells, likely through inhibition of the PI3K-AKT-mTORC1 axis [461-463]. In turn, autophagy promotes survival and leukemogenic potential of CML stem cells [464] which are insensitive to TKIs [465, 466]. In 2008, inhibition of BCR-ABL by bafetinib, a dual BCR-ABL/LYN TKI, or imatinib, was shown to induce protective autophagy [467, 468]. In 2009, a study provided for the first time rigorous evidence that TKIs induce autophagy in CML primary cells and its inhibition augmented TKIs effects [461]. However, inhibition of the imatinib-induced autophagy in CML and primary CML cells with CQ enhanced the effects of imatinib. Furthermore, CQ potentiated the effects of dasatanib and nilotinib. Importantly, induction of autophagy was able to suppress BCR-ABL activity, since imatinib treatment did not enhance autophagic activity in the cells carrying the resistant clone BCR-ABLT315I. Nevertheless, these data do not clarify if the observed effect on stem/progenitor cells is due to the inhibition of cellular autophagy or lysosomal activity, or other CQ-associated off-target effects. Sheng and colleagues recently have shed more light regarding the mechanism by which inhibition of BCR-ABL lead to autophagy induction [469]. Oncogenic BCR-ABL mimics the effects of GFs and activates the PI3K/AKT pathway, which inhibits FoxO4. This inhibition causes increased levels of the ATF5 and its target, mTOR. Subsequently, mTOR suppresses autophagy. Upon BCR-ABL inhibition, for example during TKIs treatment, decreases PI3K/AKT signaling and thus, FoxO4 suppresses ATF5 activity. In turn, such process decreased transcription of mTOR, whereas increased autophagic pathway. Yu *et al.* showed that imatinib induced autophagy through inhibition of miRNA-30a and up-regulation of beclin-1 and ATG5 expression [470]. In addition, the authors observed that the rapamycin (mTOR inhibitor) did not influence miR-30a levels, antagomir-30a and mTOR mRNA level, whereas miR-30a mimic did not change the rapamycin-induced LC3 formation. These findings suggest that BCR-ABL expression can inhibit autophagy both in a mTOR-dependent and -independent manner to support this idea, K562 cells were treated with OSI-027 (an mTOR inhibitor) and combination of OSI-027 and CQ, that caused increased apoptosis respect to OSI-027 alone [471]. Therefore, autophagy may be a key defensive mechanism after the treatment with mTOR inhibition and co-administration of mTOR and autophagy inhibitors in CML cells. Other data regarding the protective role of autophagy in CML were supported by the different studies [472-475]. Most importantly, in 2011, Altman and colleagues demonstrated that autophagy was necessary for BCR-ABL transformation and deletion of ATG3 in a mouse model prevented BCR-ABL-mediated leukaemogenesis [464]. However, there are studies supporting the induction of autophagy in combination with targeting of non-BCR-ABL related pathways for the treatment of CML [476, 477]. Resveratrol (RSV), a phytoalexin, has been proposed to induce apoptosis and autophagy in an AMPK-dependent manner, and inhibition of RSV-induced autophagy partially rescued RSV-mediated apoptosis within CML cells sensitive or resistant to imatinib [476]. Additionally, it has been proposed that treatment with arsenic trioxide, an autophagy inducer, has antileukaemic effects that are partially attributed to the autophagic degradation of BCR-ABL [478]. In a recent study, K562 cells and peripheral blood mononuclear cells from newly diagnosed CML patients have been treated with PP1, a Src kinase inhibitor, and LY294002, a specific PI3K tyrosine kinase inhibitor, after imatinib exposure. The combination of these two inhibitors with imatinib is able to trigger both apoptosis and autophagy in leukemia cells in association with the stress of the endoplasmic reticulum [479]. Another study has shown that a novel compound, Thiotanib activates autophagy in CML cells and, meanwhile, it causes growth inhibition, cell cycle arrest, and apoptosis of CML cells. Interestingly, it could suppress phosphorylation of AKT and mTOR, up-regulate the beclin-1 expression level and stop the formation of the Bcl-2 and Beclin-1 complex for triggering autophagy [480]. Take together, these data highlight a new approach for TKIs reforming and further provide an indication of the efficacy enhancement of TKIs in combination with autophagy inhibitors.

16.2. Clinical Experience with Autophagy Inhibitors in Blood Cancers

Both CQ and its derivative hydroxychloroquine (HCQ) inhibit autophagy at a late stage, by inhibiting lysosomal acidification and preventing the fusion between autophagosomes and lysosomes. CQ and HCQ are not exclusive autophagy inhibitors and affect other cellular pathways as well. However, since these drugs are FDA approved, they can be used in the clinic for the inhibition of autophagy, and a number of clinical trials are investigating their effects in combination with other anti-cancer agents [481]. Early stage pharmacological autophagy inhibitors target the PI3K-kinases, such as 3-methyladenine (3-MA), a PI3K-III inhibitor, and wortmannin (WM), a pan-PI3K inhibitor [482]. Nonetheless, PI3K inhibitors may also suppress class I PI3K kinases upstream of mTOR, and, partially, induce autophagy. Moreover, like CQ and HCQ, PI3K inhibitors are not targeting autophagy solely, and some of them are associated with high toxicity at clinically relevant doses [483]. Chemical autophagy inhibitors can block autophagy at the fusion stage, such as vinblastine and nocodazole [484], or at the degradation stage, such as ammonium chloride or lysosomal protease inhibitors E64d and pepstatin [485]. Another late autophagy stage inhibitor is bafilomycin A1, which inhibits lysosomal acidification [486]. Finally, the findings from the study of Bellodi and colleagues [487] provided the foundation for the beginning of the phase II clinical trial CHOICES (CQ and IM Combination to Eliminate Stem cells) clinical trial. CHOICES is a randomised trial investigating the effects of IM *versus* HCQ and IM for CML patients that have been on IM for >1 year and are in McyR with residual disease detectable by qRT-PCR.

CONCLUSION

Tyrosine kinase inhibitors are new promising therapy for different type of cancer. Moreover, fixed dosing is still used during the therapy, although several problems, correlated with intersubject variations, have not disappeared and may increase due the wide variations in absorption and bioavailability of such drugs. However, resistance in CML is recognized as a more narrow but evolving problem for which solutions are increasingly available. With a palette of medications available and data supporting intervention after early diagnostic suggestion of underlying resistance, it may be possible to avoid subsequent clinical consequences, particularly proliferation of disease or transformation to advanced stages of CML. Tools available for the unstable and proliferative case of transformed disease are limited, and the task remains to balance the degree of empiric or preventative intensification of therapy for the purpose of subverting resistance with any risks of more potent therapies. Furthermore, the pivotal purpose of personalized medicine is to create a right treatment plan that it is able to disrupt the cancer cells but not the normal cells of the body. Advances in the understanding of the disease pathobiology, leukemic stem cells signal transduction can offer novel options both for the treatment and survival benefits for CML patients. Novel targeted therapies are evolving, beyond cancer progression and across the identification of novel targets and development of other drugs to apply to clinical practice. Finally, increasing evidence suggests that stemness results from the incessant convergence of cell-intrinsic features (genetic mutations and epigenetic regulation), local signals (of a chemical, mechanical, and molecular nature), stochastic events, and population forces that continuously shape the stem cell pool. In this scenario, the future development of successful clinical strategies will be tightly linked to a deeper understanding of the dynamic, adaptable, and evolving nature of such cells.

ACKNOWLEDGEMENTS

Declared none.

CONFLICT OF INTEREST

The author(s) confirm that this chapter contents have no conflict of interest.

ABBREVIATIONS

4E-BP1	=	eIF-4E-Binding Protein 1
ABC	=	Transporters ATP-binding cassette transporters
ALL	=	Acute Lymphoblastic Leukemia
ALOX5	=	5-lipoxygenase gene
AMPK 5'	=	Adenosine Monophospate (AMP)-Activated Protein Kinase
AML	=	Acute Myeloid Leukemia
ATG16L1	=	ATG16 autophagy related 16-like 1
ATM	=	Ataxia Telangiectasia-Mutated Protein
ATF5	=	Activating Transcription Factor 5
ATG5	=	Autophagy protein 5
ATP	=	Adenosine Triphosphate
Bax	=	Bcl-2 associated X protein
Bad	=	Bcl-2-associated death promoter
Bcl-2	=	B-cell lymphoma 2

Bcl-XL	=	Basal cell lymphoma-extra large
BMI1	=	B lymphoma Mo-MLV insertion region 1 homolog
BMSCs	=	Bone Marrow-derived Stem Cells
CCN	=	Cysteine-Rich Protein
CEBP α	=	CCAAT/Enhancer Binding Protein α
CDKN2B (p15)	=	Cyclin-Dependent Kinase Inhibitor 2B (P15, Inhibits CDK4)
CDKN1C	=	Cyclin-Dependent Kinase Inhibitor Family 1 Member C (p57, Kip2)
CQ	=	Chloroquine
CrK	=	Avian Sarcoma Virus CT10 Oncogene Homolog
CrKL	=	CrK-Like Protein
DAKI-1	=	Death-Associated Protein Kinase 1
DDIT3	=	DNA-Damage-Inducible Transcript
Deptor	=	DEP domain containing mTOR-interacting protein
Dhh	=	Desert hedgehog
E64d	=	(2S,3S)-trans-epoxysuccinyl-L-leucylamido-2-methylbutane ethyl ester
eIF4E	=	Eukaryotic initiation factor 4E
ER	=	Estrogen Receptor Alpha
ERK	=	Extracellular Signal-Regulated Kinase
FACS	=	Fluorescence-Activated Cell Sorting
FGFR	=	Fibroblast Growth Factor Receptor
Fgr	=	Feline Gardner-Rasheed sarcoma viral oncogene homolog

FYN	=	FYN oncogene related to SRC, FGR, YES
Gab2	=	GRB2-associated binding protein 2
G-CSF	=	Granulocyte-Colony Stimulating Factor
GDP	=	Guanosine Diphosphate
GEFs	=	Guanine Nucleotide Exchange Factors
GFs	=	Growth Factors
GFP	=	Green Fluorescent Protein
Gli1A	=	Transcriptional Activators 1A
Gli2A	=	Transcriptional Activators 2A
Gli3A	=	Transcriptional Activators 3A
Gli2R	=	Transcriptional Activators 2R
Gli3R	=	Transcriptional Repressor 3R
GMPs	=	Granulocyte-Macrophage Progenitors
GNF2	=	Allosteric, Non-ATP-competitive tyrosine kinase inhibitor
G-proteins	=	Guanine-Nucleotide-Binding Proteins
Grb-2	=	Growth Factor Receptor-Bound 2
GSK3β	=	Glycogen Synthase Kinase 3β
GTP	=	Guanosine Triphosphate
hnRNP-E2	=	Heterogeneous ribonucleoproteins E2
Hck	=	Hemopoietic cell kinase
HIC1	=	Hypermethylated In Cancer 1

Hh-Gli	= Hedgehog-Glioma-associated oncogene homolog zinc finger protein
HLA	= Human Leukocyte Antigen
hnRNP	= Heterogeneous nuclear ribonucleoprotein
hOCT-1	= Human Organic Cation Transporter 1
HOXA4-5	= Homeobox A4-5
HSCs	= Hematopoietic Stem cells
HSCT	= Hematopoietic Stem Cell Transplantation
IFN	= Interferon
Ihh	= Indian hedgehog
IKK	= IκB Kinase Enzyme
INK	= c-Jun N-terminal Kinase
JAKs	= Janus Family of Non-Receptor Tyrosine Kinases
KIT	= v-Kit Hardy-Zuckerman 4 feline sarcoma viral oncogene homolog
Lck	= Lymphocyte-specific protein tyrosine kinase
LTB4	= Leukotriene B4
LTC4	= Leukotriene C4
LTD4	= Leukotriene D4
LTE4	= Leukotriene E4
LYN	= v-yes-1 Yamaguchi sarcoma viral related oncogene homolog
MAPK	= Mitogen-Activated Protein Kinase

Mcl-1 = Induced Myeloid Leukaemia Cell Differentiation Protein

Myc = v-myc Avian Myelocytomatosis Viral Oncogene Homolog

Mdm 2 = Murine double minute 2

mLST8 = Mammalian Lethal with SEC13 Protein 8

mTOR = Mammalian Target of Rapamyci

mTORC1/2 = mTOR complex 1/2

mSIN1 = Mammalian Stress-Activated Protein Kinase Interacting Protein 1

MSR1 = Macrophage Scavenger Receptor 1

NF-κB = Nuclear Factor kappa-light-chain-enhancer of activated B cells

NOD/SCID = Non-Obese Diabetic/Severe Combined Immunodeficient

NOXA = NADPH Oxidase Activator

OSCP1 = Organic Solute Carrier Partner 1

PAK = p21-activated *kinase family*

PDGFR = Platelet-derived Growth Factor Receptor

PDK1 3 = Phosphoinositide-dependent protein kinase 1

Ph = Philadelphia Chromosome

PI3K = Phosphatidylinositol-3-kinase

PIM = Provirus integration site for Moloney murine leukemia virus

PKC = Protein Kinase C

PML = Promyelocytic Leukemia Protein

PP2A　　　　　　　= Protein Phosphatase 2A

PRAS40　　　　　　= Proline Rich Akt substrate 40

PGRA and PGRB = Progesterone Receptor Isoforms

PP1　　　　　　　= 4-amino-5-(4-methylphenyl)-7-(t-butyl)pyrazolo[3,4-d]-pyrimidine

PTCH1　　　　　　= Patched-1

PTEN　　　　　　= Phosphatase and Tensin Homologue Deleted on Chromosome Ten

PUMA　　　　　　= p53-up-Regulated Modulator of Apoptosis

PYK2　　　　　　= Proline-rich Tyrosine Kinase 2

Rac　　　　　　　= Protein Kinase

RAF　　　　　　　= v-Raf Murine Sarcoma 3611 Viral Oncogene Homolog 1

RAPTOR　　　　　= Regulatory-Associated Companion of mTOR

RAS　　　　　　　= Rat Sarcoma

RBPs　　　　　　= RNA-Binding Proteins

Rel　　　　　　　= Avian Reticuloendotheliosis Proto-Oncogene

Rheb　　　　　　= Ras Homolog Enriched in brain

rictor　　　　　　= Rapamycin-insensitive companion of mTOR

RT-PCT　　　　　= Reverse transcription polymerase chain reaction

S1P　　　　　　　= Sphingosine-1-phosphate

S6K1-2 & S6K2 = p70 ribosomal protein S6 kinases 1 and 2

SHP-1　　　　　　= Src homology region 2 domain-containing phosphatase-1

S1P2   = Sphingosine 1-phosphate receptor 2

SMO   = Smoothened

SphK1   = Sphingosine Kinase 1

Src   = v-Src avian sarcoma (Schmidt-Ruppin A-2) viral oncogene homolog

Shh   = Sonic hedgehog

STAT   = Signal Transducer and Activator of Transcription

Ste20   = Sterile 20-like kinase (MST) 1/2 (homologs of Drosophila Hippo)

TFAP2E   = Transcription Factor AP-2 Epsilon (Activating Enhancer Binding Protein 2 Epsilon

TEK   = Tyrosine Endothelial Kinase

TGF-β = Transforming Growth Factor-β

VEGFR   = Vascular Endothelial Growth Factor Receptor

WNT   = Wingless type MMTV integration site family member 1

Yes   = Proto-oncogene Tyrosine-Protein Kinase Yes (v-yes-1 Yamaguchi Sarcoma Viral Oncogene Homolog 1)

REFERENCES

[1]   Geary CG. The story of chronic myeloid leukaemia. Br J Haematol 2000; 110(1): 2-11.
[2]   Tefferi A. The history of myeloproliferative disorders: before and after Dameshek. Leukemia 2008; 22(1): 3-13.
[3]   Dameshek W. Some speculations on the myeloproliferative syndromes. Blood 1951; 6(4): 372-5.
[4]   Nowell PC, Hungerford DA. Chromosome studies on normal and leukemic human leukocytes. J Natl Cancer Inst 1960; 25: 85-109.
[5]   Nowell PC, Hungerford DA. A minute chromosome in human chronic granulocytic leukemia science 1960; 132: 1497.
[6]   Nowell PC, Hungerford DA. Chromosome studies in human leukemia. II. Chronic granulocytic leukemia. J Natl Cancer Inst 1961; 27: 1013-35.
[7]   Caspersson T, Farber S, Foley GE, *et al.* Chemical differentiation along metaphase chromosomes. Exp Cell Res 1968; 49(1): 219-22.

[8]   Caspersson T, Zech L, Johansson C. Differential binding of alkylating fluorochromes in human chromosomes. Exp Cell Res 1970; 60(3): 315-9.

[9]   Caspersson T, Gahrton G, Lindsten J, Zech L. Identification of the Philadelphia chromosome as a number 22 by quinacrine mustard fluorescence analysis. Exp Cell Res 1970; 63(1): 238-40.

[10] Rowley JD. Letter: A new consistent chromosomal abnormality in chronic myelogenous leukaemia identified by quinacrine fluorescence and Giemsa staining. Nature 1973; 243(5405): 290-3.

[11] Prakash O, Yunis JJ. High resolution chromosomes of the t(9;22) positive leukemias. Cancer Genet Cytogenet 1984; 11(4): 361-7.

[12] Heisterkamp N, Groffen J, Stephenson JR, *et al.* Chromosomal localization of human cellular homologues of two viral oncogenes. Nature 1982; 299(5885): 747-9.

[13] Heisterkamp N, Stephenson JR, Groffen J, *et al.* Localization of the c-abl oncogene adjacent to a translocation break point in chronic myelocytic leukaemia. Nature 1983; 306(5940): 239-42.

[14] Gale RP, Canaani E. An 8-kilobase abl RNA transcript in chronic myelogenous leukemia. Proc Natl Acad Sci USA 1984; 81(18): 5648-52.

[15] Shtivelman E, Lifshitz B, Gale RP, Canaani E. Fused transcript of abl and bcr genes in chronic myelogenous leukaemia. Nature 1985; 315(6020): 550-4.

[16] Konopka JB, Watanabe SM, Witte ON. An alteration of the human c-abl protein in K562 leukemia cells unmasks associated tyrosine kinase activity. Cell 1984; 37(3): 1035-42.

[17] Ben-Neriah Y, Daley GQ, Mes-Masson AM, Witte ON, Baltimore D. The chronic myelogenous leukemia-specific P210 protein is the product of the bcr/abl hybrid gene. Science 1986; 233(4760): 212-4.

[18] de Klein A, van Kessel AG, Grosveld G, *et al.* A cellular oncogene is translocated to the Philadelphia chromosome in chronic myelocytic leukaemia. Nature 1982; 300(5894): 765-7.

[19] Groffen J, Stephenson JR, Heisterkamp N, de Klein A, Bartram CR, Grosveld G. Philadelphia chromosomal breakpoints are clustered within a limited region, bcr, on chromosome 22. Cell 1984; 36(1): 93-9.

[20] Heisterkamp N, Stam K, Groffen J, de Klein A, Grosveld G. Structural organization of the bcr gene and its role in the Ph' translocation. Nature 1985; 315(6022): 758-61.

[21] Daley GQ, Van Etten RA, Baltimore D. Induction of chronic myelogenous leukemia in mice by the P210bcr/abl gene of the Philadelphia chromosome. Science 1990; 247(4944): 824-30.

[22] Elefanty AG, Hariharan IK, Cory S. bcr-abl, the hallmark of chronic myeloid leukaemia in man, induces multiple haemopoietic neoplasms in mice. EMBO J 1990; 9(4): 1069-78.

[23] Kelliher MA, McLaughlin J, Witte ON, Rosenberg N. Induction of a chronic myelogenous leukemia- like syndrome in mice with v-abl and BCR/ABL. Proc Natl Acad Sci USA 1990; 87(17): 6649-53.

[24] Faderl S, Talpaz M, Estrov Z, O'Brien S, Kurzrock R, Kantarjian HM. The biology of chronic myeloid leukemia. N Engl J Med 1999; 341(3): 164-72.

[25] Kantarjian HM, Dixon D, Keating MJ, *et al.* Characteristics of accelerated disease in chronic myelogenous leukemia. Cancer 1988; 61(7): 1441-6.

[26] Kantarjian HM, Deisseroth A, Kurzrock R, Estrov Z, Talpaz M. Chronic myelogenous leukemia: a concise update. Blood 1993; 82(3): 691-703.

[27] Wong S, Witte ON. Modeling Philadelphia chromosome positive leukemias. Oncogene 2001; 20(40): 5644-59.

[28] Smith DL, Burthem J, Whetton AD. Molecular pathogenesis of chronic myeloid leukaemia. Expert Rev Mol Med 2003; 5(27): 1-27.

[29] Perrotti D, Jamieson C, Goldman J, Skorski T. Chronic myeloid leukemia: mechanisms of blastic transformation. J Clin Invest 2010; 120(7): 2254-64.

[30] Cramer E, Auclair C, Hakim J, *et al.* Metabolic activity of phagocytosing granulocytes in chronic granulocytic leukemia: ultrastructural observation of a degranulation defect. Blood 1977; 50(1): 93-106.

[31] Hasford J, Ansari H, Pfirrmann M, Hehlmann R. Analysis and validation of prognostic factors for CML. Bone Marrow Transplant 1996; 17 (Suppl. 3): S49-54.

[32] Sokal JE. Prognosis in chronic myeloid leukaemia: biology of the disease *vs.* treatment. Baillieres Clin Haematol 1987; 1(4): 907-29.

[33]   Spiers AS, Bain BJ, Turner JE. The peripheral blood in chronic granulocytic leukaemia. Study of 50 untreated Philadelphia-positive cases. Scand J Haematol 1977; 18(1): 25-38.

[34]   Lorand-Metze I, Vassallo J, Souza CA. Histological and cytological heterogeneity of bone marrow in Philadelphia-positive chronic myelogenous leukaemia at diagnosis. Br J Haematol 1987; 67(1): 45-9.

[35]   Silver RT. Morphology of The Blood and Marrow in Clinical Practice. Ann Intern Med. New York, NY: Grune & Stratton 1970; 84: pp. 874-5.

[36]   Dosik H, Rosner F, Sawitsky A. Acquired lipidosis: Gaucher-like cells and "blue cells" in chronic granulocytic leukemia. Semin Hematol 1972; 9(3): 309-16.

[37]   Anastasi J, Musvee T, Roulston D, Domer PH, Larson RA, Vardiman JW. Pseudo-Gaucher histiocytes identified up to 1 year after transplantation for CML are BCR/ABL-positive. Leukemia 1998; 12(2):   233-7.

[38]   Hayhoe FG, Flemans RJ, Cowling DC. Acquired lipidosis of marrow macrophages: birefringent blue crystals and Gaucher-like cells, sea-blue histiocytes, and grey-green crystals. J Clin Pathol 1979; 32(5): 420-8.

[39]   Castro-Malaspina H, Moore MA. Pathophysiological mechanisms operating in the development of myelofibrosis: role of megakaryocytes. Nouv Rev Fr Hematol 1982; 24(4): 221-6.

[40]   Cotta CV, Bueso-Ramos CE. New insights into the pathobiology and treatment of chronic myelogenous leukemia. Ann Diagn Pathol 2007; 11(1): 68-78.

[41]   Cortes JE, Talpaz M, Beran M, *et al.* Philadelphia chromosome-negative chronic myelogenous leukemia with rearrangement of the breakpoint cluster region. Long-term follow-up results. Cancer   1995; 75(2): 464-70.

[42]   Aurich J, Duchayne E, Huguet-Rigal F, *et al.* Clinical, morphological, cytogenetic and molecular aspects of a series of Ph-negative chronic myeloid leukemias. Hematol Cell Ther 1998; 40(4): 149-58.

[43]   Mark HF, Sokolic RA, Mark Y. Conventional cytogenetics and FISH in the detection of BCR/ABL fusion in chronic myeloid leukemia (CML). Exp Mol Pathol 2006; 81(1): 1-7.

[44]   Tkachuk DC, Westbrook CA, Andreeff M, *et al.* Detection of bcr-abl fusion in chronic myelogeneous leukemia by *in situ* hybridization. Science 1990; 250(4980): 559-62.

[45]   Dewald GW, Schad CR, Christensen ER, *et al.* The application of fluorescent *in situ* hybridization to detect Mbcr/abl fusion in variant Ph chromosomes in CML and ALL. Cancer Genet Cytogenet 1993; 71(1): 7-14.

[46]   Garcia-Isidoro M, Tabernero MD, Garcia JL, *et al.* Detection of the Mbcr/abl translocation in chronic myeloid leukemia by fluorescence *in situ* hybridization: comparison with conventional cytogenetics   and implications for minimal residual disease detection. Hum Pathol 1997; 28(2): 154-9.

[47]   Werner M, Ewig M, Nasarek A, *et al.* Value of fluorescence *in situ* hybridization for detecting the   bcr/abl gene fusion in interphase cells of routine bone marrow specimens. Diagn Mol Pathol 1997;   6(5): 282-7.

[48]   Sinclair PB, Green AR, Grace C, Nacheva EP. Improved sensitivity of BCR-ABL detection: a triple-probe three-color fluorescence *in situ* hybridization system. Blood 1997; 90(4): 1395-402.

[49]   Landstrom AP, Tefferi A. Fluorescent *in situ* hybridization in the diagnosis, prognosis, and treatment monitoring of chronic myeloid leukemia. Leuk Lymphoma 2006; 47(3): 397-402.

[50]   Buño I, Wyatt WA, Zinsmeister AR, Dietz-Band J, Silver RT, Dewald GW. A special fluorescent *in situ* hybridization technique to study peripheral blood and assess the effectiveness of interferon therapy in chronic myeloid leukemia. Blood 1998; 92(7): 2315-21.

[51]   Calabrese G, Stuppia L, Franchi PG, *et al.* Complex translocations of the Ph chromosome and Ph negative CML arise from similar mechanisms, as evidenced by FISH analysis. Cancer Genet Cytogenet 1994; 78(2): 153-9.

[52]   Nacheva E, Holloway T, Brown K, Bloxham D, Green AR. Philadelphia-negative chronic myeloid leukaemia: detection by FISH of BCR-ABL fusion gene localized either to chromosome 9 or chromosome 22. Br J Haematol 1994; 87(2): 409-12.

[53]   van der Plas DC, Soekarman D, van Gent AM, Grosveld G, Hagemeijer A. bcr-abl mRNA lacking abl exon a2 detected by polymerase chain reaction in a chronic myelogeneous leukemia patient. Leukemia 1991; 5(6): 457-61.

[54]   Iwata S, Mizutani S, Nakazawa S, Yata J. Heterogeneity of the breakpoint in the ABL gene in cases with BCR/ABL transcript lacking ABL exon a2. Leukemia 1994; 8(10): 1696-702.

[55]   Páldi-Haris P, Barta A, Lengyel L, *et al.* Molecular background of a new case of chronic myelogenous leukemia with bcr-abl chimera mRNA lacking the A2 exon. Leukemia 1994; 8(10): 1791.

[56]   Duba HC, Peter S, Hilbe W, *et al.* Monitoring of remission status by fluorescence *in situ* hybridisation in chronic myeloid leukaemia patients treated with interferon-alpha. Int J Oncol 1999; 14(1): 145-50.

[57]   Froncillo MC, Maffei L, Cantonetti M, *et al.* FISH analysis for CML monitoring? Ann Hematol 1996; 73(3): 113-9.

[58]   Tchirkov A, Giollant M, Tavernier F, *et al.* Interphase cytogenetics and competitive RT-PCR for residual disease monitoring in patients with chronic myeloid leukaemia during interferon-alpha therapy. Br J Haematol 1998; 101(3): 552-7.

[59]   Otazú IB, Zalcberg I, Tabak DG, Dobbin J, Seuánez HN. Detection of BCR-ABL transcripts by multiplex and nested PCR in different haematological disorders. Leuk Lymphoma 2000; 37(1-2): 205- 11.

[60]   Lion T, Izraeli S, Henn T, Gaiger A, Mor W, Gadner H. Monitoring of residual disease in chronic myelogenous leukemia by quantitative polymerase chain reaction. Leukemia 1992; 6(6): 495-9.

[61]   Malinge MC, Mahon FX, Delfau MH, *et al.* Quantitative determination of the hybrid Bcr-Abl RNA in patients with chronic myelogenous leukaemia under interferon therapy. Br J Haematol 1992; 82(4): 701-7.

[62]   Thompson JD, Brodsky I, Yunis JJ. Molecular quantification of residual disease in chronic myelogenous leukemia after bone marrow transplantation. Blood 1992; 79(6): 1629-35.

[63]   Mensink E, van de Locht A, Schattenberg A, *et al.* Quantitation of minimal residual disease in Philadelphia chromosome positive chronic myeloid leukaemia patients using real-time quantitative RT-PCR. Br J Haematol 1998; 102(3): 768-74.

[64]   Branford S, Hughes TP, Rudzki Z. Monitoring chronic myeloid leukaemia therapy by real-time quantitative PCR in blood is a reliable alternative to bone marrow cytogenetics. Br J Haematol 1999; 107(3): 587-99.

[65]   Branford S, Rudzki Z, Harper A, *et al.* Imatinib produces significantly superior molecular responses compared to interferon alfa plus cytarabine in patients with newly diagnosed chronic myeloid leukemia in chronic phase. Leukemia 2003; 17(12): 2401-9.

[66]   Merx K, Müller MC, Kreil S, *et al.* Early reduction of BCR-ABL mRNA transcript levels predicts cytogenetic response in chronic phase CML patients treated with imatinib after failure of interferon alpha. Leukemia 2002; 16(9): 1579-83.

[67]   Wang L, Pearson K, Ferguson JE, Clark RE. The early molecular response to imatinib predicts cytogenetic and clinical outcome in chronic myeloid leukaemia. Br J Haematol 2003; 120(6): 990-9.

[68]   Jørgensen HG, Holyoake TL. A comparison of normal and leukemic stem cell biology in Chronic Myeloid Leukemia. Hematol Oncol 2001; 19(3): 89-106.

[69]   Harris AW, Bath ML, Rosenbaum H, McNeall J, Adams JM, Cory S. Lymphoid tumorigenesis by v-abl and BCR-v-abl in transgenic mice. Curr Top Microbiol Immunol 1990; 166: 165-73.

[70]   Gishizky ML, Johnson-White J, Witte ON. Efficient transplantation of BCR-ABL-induced chronic myelogenous leukemia-like syndrome in mice. Proc Natl Acad Sci USA 1993; 90(8): 3755-9.

[71]   Pear WS, Miller JP, Xu L, *et al.* Efficient and rapid induction of a chronic myelogenous leukemia-like myeloproliferative disease in mice receiving P210 bcr/abl-transduced bone marrow. Blood 1998; 92(10): 3780-92.

[72]   Laneuville P, Sun G, Timm M, Vekemans M. Clonal evolution in a myeloid cell line transformed to interleukin-3 independent growth by retroviral transduction and expression of p210bcr/abl. Blood 1992; 80(7): 1788-97.

[73]   Salloukh HF, Laneuville P. Increase in mutant frequencies in mice expressing the BCR-ABL activated tyrosine kinase. Leukemia 2000; 14(8): 1401-4.

[74]   Reuther GW, Fu H, Cripe LD, Collier RJ, Pendergast AM. Association of the protein kinases c-Bcr and Bcr-Abl with proteins of the 14-3-3 family. Science 1994; 266(5182): 129-33.

[75]   McWhirter JR, Galasso DL, Wang JY. A coiled-coil oligomerization domain of Bcr is essential for the transforming function of Bcr-Abl oncoproteins. Mol Cell Biol 1993; 13(12): 7587-95.

[76]   Pendergast AM, Muller AJ, Havlik MH, Maru Y, Witte ON. BCR sequences essential for transformation by the BCR-ABL oncogene bind to the ABL SH2 regulatory domain in a non- phosphotyrosine-dependent manner. Cell 1991; 66(1): 161-71.

[77]   Maru Y, Witte ON. The BCR gene encodes a novel serine/threonine kinase activity within a single exon. Cell 1991; 67(3): 459-68.

[78]   Diekmann D, Brill S, Garrett MD, *et al.* Bcr encodes a GTPase-activating protein for p21rac. Nature 1991; 351(6325): 400-2.

[79]   Ron D, Zannini M, Lewis M, *et al.* A region of proto-dbl essential for its transforming activity shows sequence similarity to a yeast cell cycle gene, CDC24, and the human breakpoint cluster gene, bcr. New Biol 1991; 3(4): 372-9.

[80]   Raitano AB, Whang YE, Sawyers CL. Signal transduction by wild-type and leukemogenic Abl proteins. Biochim Biophys Acta 1997; 1333(3): F201-16.

[81]   Feller SM, Ren R, Hanafusa H, Baltimore D. SH2 and SH3 domains as molecular adhesives: the interactions of Crk and Abl. Trends Biochem Sci 1994; 19(11): 453-8.

[82]   Franz WM, Berger P, Wang JY. Deletion of an N-terminal regulatory domain of the c-abl tyrosine kinase activates its oncogenic potential. EMBO J 1989; 8(1): 137-47.

[83]   Mayer BJ, Jackson PK, Van Etten RA, Baltimore D. Point mutations in the abl SH2 domain coordinately impair phosphotyrosine binding *in vitro* and transforming activity *in vivo.* Mol Cell Biol 1992; 12(2): 609-18.

[84]   Deininger MW, Druker BJ. Specific targeted therapy of chronic myelogenous leukemia with imatinib. Pharmacol Rev 2003; 55(3): 401-23.

[85]   Taagepera S, McDonald D, Loeb JE, *et al.* Nuclear-cytoplasmic shuttling of C-ABL tyrosine kinase. Proc Natl Acad Sci USA 1998; 95(13): 7457-62.

[86]   McWhirter JR, Wang JY. An actin-binding function contributes to transformation by the Bcr-Abl oncoprotein of Philadelphia chromosome-positive human leukemias. EMBO J 1993; 12(4): 1533-46.

[87]   Wetzler M, Talpaz M, Van Etten RA, Hirsh-Ginsberg C, Beran M, Kurzrock R. Subcellular localization of Bcr, Abl, and Bcr-Abl proteins in normal and leukemic cells and correlation of expression with myeloid differentiation. J Clin Invest 1993; 92(4): 1925-39.

[88]   Van Etten RA, Jackson P, Baltimore D. The mouse type IV c-abl gene product is a nuclear protein, and activation of transforming ability is associated with cytoplasmic localization. Cell 1989; 58(4): 669-78.

[89]   Duyster J, Baskaran R, Wang JY. Src homology 2 domain as a specificity determinant in the c-Abl-mediated tyrosine phosphorylation of the RNA polymerase II carboxyl-terminal repeated domain. Proc Natl Acad Sci USA 1995; 92(5): 1555-9.

[90]   White E, Prives C. DNA damage enables p73. Nature 1999; 399(6738): 734-735, 737.

[91]   Lewis JM, Baskaran R, Taagepera S, Schwartz MA, Wang JY. Integrin regulation of c-Abl tyrosine kinase activity and cytoplasmic-nuclear transport. Proc Natl Acad Sci USA 1996; 93(26): 15174-9.

[92]   Renshaw MW, Lewis JM, Schwartz MA. The c-Abl tyrosine kinase contributes to the transient activation of MAP kinase in cells plated on fibronectin. Oncogene 2000; 19(28): 3216-9.

[93]   Ren R. Mechanisms of BCR-ABL in the pathogenesis of chronic myelogenous leukaemia. Nat Rev Cancer 2005; 5(3): 172-83.

[94]   Beissert T, Puccetti E, Bianchini A, *et al.* Targeting of the N-terminal coiled coil oligomerization interface of BCR interferes with the transformation potential of BCR-ABL and increases sensitivity to STI571. Blood 2003; 102(8): 2985-93.

[95]   Liu J, Wu Y, Ma GZ, *et al.* Inhibition of Bcr serine kinase by tyrosine phosphorylation. Mol Cell Biol 1996; 16(3): 998-1005.

[96]   Liu J, Wu Y, Arlinghaus RB. Sequences within the first exon of BCR inhibit the activated tyrosine kinases of c-Abl and the Bcr-Abl oncoprotein. Cancer Res 1996; 56(22): 5120-4.

[97]   Marley SB, Lewis JL, Scott MA, Goldman JM, Gordon MY. Evaluation of "discordant maturation' in chronic myeloid leukaemia using cultures of primitive progenitor cells and their production of clonogenic progeny (CFU-GM). Br J Haematol 1996; 95(2): 299-305.

[98]   Mandanas RA, Leibowitz DS, Gharehbaghi K, *et al.* Role of p21 RAS in p210 bcr-abl trans-formation of murine myeloid cells. Blood 1993; 82(6): 1838-47.

[99]   Skorski T, Kanakaraj P, Ku DH, *et al.* Negative regulation of p120GAP GTPase promoting activity by p210bcr/abl: implication for RAS-dependent Philadelphia chromosome positive cell growth. J Exp Med 1994; 179(6): 1855-65.

[100]   Pendergast AM, Quilliam LA, Cripe LD, *et al.* BCR-ABL-induced oncogenesis is mediated by direct interaction with the SH2 domain of the GRB-2 adaptor protein. Cell 1993; 75(1): 175-85.

[101]   Gishizky ML, Cortez D, Pendergast AM. Mutant forms of growth factor-binding protein-2 reverse BCR-ABL-induced transformation. Proc Natl Acad Sci USA 1995; 92(24): 10889-93.

[102]   Oda T, Heaney C, Hagopian JR, Okuda K, Griffin JD, Druker BJ. Crkl is the major tyrosine- phosphorylated protein in neutrophils from patients with chronic myelogenous leukemia. J Biol Chem 1994; 269(37): 22925-8.

[103]   Pelicci G, Lanfrancone L, Salcini AE, *et al.* Constitutive phosphorylation of Shc proteins in human tumors. Oncogene 1995; 11(5): 899-907.

[104]   Sawyers CL, McLaughlin J, Witte ON. Genetic requirement for Ras in the transformation of fibroblasts and hematopoietic cells by the Bcr-Abl oncogene. J Exp Med 1995; 181(1): 307-13.

[105]   Cortez D, Kadlec L, Pendergast AM. Structural and signaling requirements for BCR-ABL-mediated transformation and inhibition of apoptosis. Mol Cell Biol 1995; 15(10): 5531-41.

[106]   Goga A, McLaughlin J, Afar DE, Saffran DC, Witte ON. Alternative signals to RAS for hematopoietic transformation by the BCR-ABL oncogene. Cell 1995; 82(6): 981-8.

[107]   Aaronson DS, Horvath CM. A road map for those who don't know JAK-STAT. Science 2002; 296(5573): 1653-5.

[108]   de Groot RP, Raaijmakers JA, Lammers JW, Koenderman L. STAT5-Dependent CyclinD1 and Bcl-xL expression in Bcr-Abl-transformed cells. Mol Cell Biol Res Commun 2000; 3(5): 299-305.

[109]   Gesbert F, Griffin JD. Bcr/Abl activates transcription of the Bcl-X gene through STAT5. Blood 2000; 96(6): 2269-76.

[110]   Sillaber C, Gesbert F, Frank DA, Sattler M, Griffin JD. STAT5 activation contributes to growth and viability in Bcr/Abl-transformed cells. Blood 2000; 95(6): 2118-25.

[111]   Hantschel O, Warsch W, Eckelhart E, *et al.* BCR-ABL uncouples canonical JAK2-STAT5 signaling in chronic myeloid leukemia. Nat Chem Biol 2012; 8(3): 285-93.

[112]   Samanta AK, Lin H, Sun T, Kantarjian H, Arlinghaus RB. Janus kinase 2: a critical target in chronic myelogenous leukemia. Cancer Res 2006; 66(13): 6468-72.

[113]   Samanta A, Perazzona B, Chakraborty S, *et al.* Janus kinase 2 regulates Bcr-Abl signaling in chronic myeloid leukemia. Leukemia 2011; 25(3): 463-72.

[114]   Xie S, Wang Y, Liu J, *et al.* Involvement of Jak2 tyrosine phosphorylation in Bcr-Abl trans-formation. Oncogene 2001; 20(43): 6188-95.

[115]   Kirchner D, Duyster J, Ottmann O, Schmid RM, Bergmann L, Munzert G. Mechanisms of Bcr-Abl-mediated NF-kappaB/Rel activation. Exp Hematol 2003; 31(6): 504-11.

[116]   Reuther JY, Reuther GW, Cortez D, Pendergast AM, Baldwin AS Jr. A requirement for NF-kappaB activation in Bcr-Abl-mediated transformation. Genes Dev 1998; 12(7): 968-81.

[117]   Guzman ML, Neering SJ, Upchurch D, *et al.* Nuclear factor-kappaB is constitutively activated in primitive human acute myelogenous leukemia cells. Blood 2001; 98(8): 2301-7.

[118]   Karin M, Cao Y, Greten FR, Li ZW. NF-kappaB in cancer: from innocent bystander to major culprit. Nat Rev Cancer 2002; 2(4): 301-10.

[119]   Zhu X, Wang L, Zhang B, Li J, Dou X, Zhao RC. TGF-beta1-induced PI3K/Akt/NF-kappaB/MMP9 signalling pathway is activated in Philadelphia chromosome-positive chronic myeloid leukaemia hemangioblasts. J Biochem 2011; 149(4): 405-14.

[120]   Cilloni D, Messa F, Arruga F, *et al.* The NF-kappaB pathway blockade by the IKK inhibitor PS1145 can overcome imatinib resistance. Leukemia 2006; 20(1): 61-7.

[121]   Lounnas N, Frelin C, Gonthier N, *et al.* NF-kappaB inhibition triggers death of imatinib-sensitive and imatinib-resistant chronic myeloid leukemia cells including T315I Bcr-Abl mutants. Int J Cancer 2009; 125(2): 308-17.

[122]   Fitzgerald KA, O'Neill LA. The role of the interleukin-1/Toll-like receptor superfamily in inflammation and host defence. Microbes Infect 2000; 2(8): 933-43.

[123]   Zhang B, Ho YW, Huang Q, *et al.* Altered microenvironmental regulation of leukemic and normal stem cells in chronic myelogenous leukemia. Cancer Cell 2012; 21(4): 577-92.

[124]   Järås M, Johnels P, Hansen N, *et al.* Isolation and killing of candidate chronic myeloid leukemia stem cells by antibody targeting of IL-1 receptor accessory protein. Proc Natl Acad Sci USA 2010; 107(37): 16280-5.

[125]   Hsieh MY, Van Etten RA. IKK-dependent activation of NF-κB contributes to myeloid and lymphoid leukemogenesis by BCR-ABL1. Blood 2014; 123(15): 2401-11.

[126]   Cilloni D, Saglio G. Molecular pathways: BCR-ABL. Clin Cancer Res 2012; 18(4): 930-7.

[127]   Sawyers CL, Callahan W, Witte ON. Dominant negative MYC blocks transformation by ABL oncogenes. Cell 1992; 70(6): 901-10.

[128]   Stewart MJ, Litz-Jackson S, Burgess GS, Williamson EA, Leibowitz DS, Boswell HS. Role for E2F1 in p210 BCR-ABL downstream regulation of c-myc transcription initiation. Studies in murine myeloid cells. Leukemia 1995; 9(9): 1499-507.

[129]   Bissonnette RP, Echeverri F, Mahboubi A, Green DR. Apoptotic cell death induced by c-myc is inhibited by bcl-2. Nature 1992; 359(6395): 552-4.

[130]   Carpenter CL, Duckworth BC, Auger KR, Cohen B, Schaffhausen BS, Cantley LC. Purification and characterization of phosphoinositide 3-kinase from rat liver. J Biol Chem 1990; 265(32): 19704-11.

[131]   Cantley LC, Auger KR, Carpenter C, *et al.* Oncogenes and signal transduction. Cell 1991; 64(2): 281-302.

[132]   Skolnik EY, Margolis B, Mohammadi M, *et al.* Cloning of PI3 kinase-associated p85 utilizing a novel method for expression/cloning of target proteins for receptor tyrosine kinases. Cell 1991; 65(1): 83-90.

[133]   Kapeller R, Cantley LC. Phosphatidylinositol 3-kinase. BioEssays 1994; 16(8): 565-76.

[134]   Shepherd PR, Reaves BJ, Davidson HW. Phosphoinositide 3-kinases and membrane traffic. Trends Cell Biol 1996; 6(3): 92-7.

[135]   Divecha N, Irvine RF. Phospholipid signaling. Cell 1995; 80(2): 269-78.

[136]   Vanhaesebroeck B, Leevers SJ, Ahmadi K, *et al.* Synthesis and function of 3-phosphorylated inositol lipids. Annu Rev Biochem 2001; 70: 535-602.

[137]   Skorski T, Bellacosa A, Nieborowska-Skorska M, *et al.* Transformation of hematopoietic cells by BCR/ABL requires activation of a PI-3k/Akt-dependent pathway. EMBO J 1997; 16(20): 6151-61.

[138]   Skorski T, Kanakaraj P, Nieborowska-Skorska M, *et al.* Phosphatidylinositol-3 kinase activity is regulated by BCR/ABL and is required for the growth of Philadelphia chromosome-positive cells. Blood 1995; 86(2): 726-36.

[139]   Varticovski L, Daley GQ, Jackson P, Baltimore D, Cantley LC. Activation of phosphatidylinositol 3-kinase in cells expressing abl oncogene variants. Mol Cell Biol 1991; 11(2): 1107-13.

[140]   Jain SK, Susa M, Keeler ML, Carlesso N, Druker B, Varticovski L. PI 3-kinase activation in BCR/abl- transformed hematopoietic cells does not require interaction of p85 SH2 domains with p210 BCR/abl. Blood 1996; 88(5): 1542-50.

[141]   Sattler M, Mohi MG, Pride YB, *et al.* Critical role for Gab2 in transformation by BCR/ABL. Cancer Cell 2002; 1(5): 479-92.

[142]   ten Hoeve J, Arlinghaus RB, Guo JQ, Heisterkamp N, Groffen J. Tyrosine phosphorylation of CRKL in Philadelphia+ leukemia. Blood 1994; 84(6): 1731-6.

[143]   Reichman CT, Mayer BJ, Keshav S, Hanafusa H. The product of the cellular crk gene consists primarily of SH2 and SH3 regions. Cell Growth Differ 1992; 3(7): 451-60.

[144]   Salgia R, Uemura N, Okuda K, *et al.* CRKL links p210BCR/ABL with paxillin in chronic myelogenous leukemia cells. J Biol Chem 1995; 270(49): 29145-50.

[145]   de Jong R, ten Hoeve J, Heisterkamp N, Groffen J. Tyrosine 207 in CRKL is the BCR/ABL phosphorylation site. Oncogene 1997; 14(5): 507-13.

[146] Nichols GL, Raines MA, Vera JC, Lacomis L, Tempst P, Golde DW. Identification of CRKL as the constitutively phosphorylated 39-kD tyrosine phosphoprotein in chronic myelogenous leukemia cells. Blood 1994; 84(9): 2912-8.

[147] Chin H, Saito T, Arai A, *et al.* Erythropoietin and IL-3 induce tyrosine phosphorylation of CrkL and its association with Shc, SHP-2, and Cbl in hematopoietic cells. Biochem Biophys Res Commun 1997; 239(2): 412-7.

[148] Sattler M, Salgia R, Okuda K, *et al.* The proto-oncogene product p120CBL and the adaptor proteins CRKL and c-CRK link c-ABL, p190BCR/ABL and p210BCR/ABL to the phosphatidylinositol-3' kinase pathway. Oncogene 1996; 12(4): 839-46.

[149] Brazil DP, Yang ZZ, Hemmings BA. Advances in protein kinase B signalling: AKTion on multiple fronts. Trends Biochem Sci 2004; 29(5): 233-42.

[150] Ren SY, Bolton E, Mohi MG, Morrione A, Neel BG, Skorski T. Phosphatidylinositol 3-kinase p85alpha subunit-dependent interaction with BCR/ABL-related fusion tyrosine kinases: molecular   mechanisms and biological consequences. Mol Cell Biol 2005; 25(18): 8001-8.

[151] Chu S, Li L, Singh H, Bhatia R. BCR-tyrosine 177 plays an essential role in Ras and Akt activation and in human hematopoietic progenitor transformation in chronic myelogenous leukemia. Cancer Res 2007; 67(14): 7045-53.

[152] Sattler M, Verma S, Pride YB, Salgia R, Rohrschneider LR, Griffin JD. SHIP1, an SH2 domain   containing polyinositol-5-phosphatase, regulates migration through two critical tyrosine residues and   forms a novel signaling complex with DOK1 and CRKL. J Biol Chem 2001; 276(4): 2451-8.

[153] Watkins A. The Role of Akt Isoforms BCR-ABL induced Chronic Myeloid Leukemia A Dissertation Submitted for The Degree of Doctor of Philosophy Thomas Jefferson University Umi Number: 3524076 2012.

[154] Maurer U, Charvet C, Wagman AS, Dejardin E, Green DR. Glycogen synthase kinase-3 regulates mitochondrial outer membrane permeabilization and apoptosis by destabilization of MCL-1. Mol Cell 2006; 21(6): 749-60.

[155] Accili D, Arden KC. FoxOs at the crossroads of cellular metabolism, differentiation, and transformation. Cell 2004; 117(4): 421-6.

[156] Ghaffari S, Jagani Z, Kitidis C, Lodish HF, Khosravi-Far R. Cytokines and BCR-ABL mediate   suppression of TRAIL-induced apoptosis through inhibition of forkhead FoxO3a transcription factor.   Proc Natl Acad Sci USA 2003; 100(11): 6523-8.

[157] Essafi A, Fernández de Mattos S, Hassen YA, *et al.* Direct transcriptional regulation of Bim by FoxO3a mediates STI571-induced apoptosis in Bcr-Abl-expressing cells. Oncogene 2005; 24(14): 2317-29.

[158] Fernández de Mattos S, Essafi A, Soeiro I, *et al.* FoxO3a and BCR-ABL regulate cyclin D2 transcription through a STAT5/BCL6-dependent mechanism. Mol Cell Biol 2004; 24(22): 10058-71.

[159] Neshat MS, Raitano AB, Wang HG, Reed JC, Sawyers CL. The survival function of the Bcr-Abl oncogene is mediated by Bad-dependent and -independent pathways: roles for phosphatidylinositol 3- kinase and Raf. Mol Cell Biol 2000; 20(4): 1179-86.

[160] Mayo LD, Donner DB. A phosphatidylinositol 3-kinase/Akt pathway promotes translocation of Mdm2 from the cytoplasm to the nucleus. Proc Natl Acad Sci USA 2001; 98(20): 11598-603.

[161] Goetz AW, van der Kuip H, Maya R, Oren M, Aulitzky WE. Requirement for Mdm2 in the survival effects of Bcr-Abl and interleukin 3 in hematopoietic cells. Cancer Res 2001; 61(20): 7635-41.

[162] Pap M, Cooper GM. Role of glycogen synthase kinase-3 in the phosphatidylinositol 3-Kinase/Akt cell survival pathway. J Biol Chem 1998; 273(32): 19929-32.

[163] Diehl JA, Cheng M, Roussel MF, Sherr CJ. Glycogen synthase kinase-3beta regulates cyclin D1 proteolysis and subcellular localization. Genes Dev 1998; 12(22): 3499-511.

[164] Zhao C, Blum J, Chen A, *et al.* Loss of beta-catenin impairs the renewal of normal and CML stem cells *in vivo.* Cancer Cell 2007; 12(6): 528-41.

[165] Richardson CJ, Schalm SS, Blenis J. PI3-kinase and TOR: PIKTORing cell growth. Semin Cell Dev Biol 2004; 15(2): 147-59.

[166] Bhaskar PT, Hay N. The two TORCs and Akt. Dev Cell 2007; 12(4): 487-502.

[167] Wullschleger S, Loewith R, Hall MN. TOR signaling in growth and metabolism. Cell 2006; 124(3): 471-84.

[168] Sarbassov DD, Guertin DA, Ali SM, Sabatini DM. Phosphorylation and regulation of Akt/PKB by the rictor-mTOR complex. Science 2005; 307(5712): 1098-101.

[169] Sawyers CL. Will mTOR inhibitors make it as cancer drugs? Cancer Cell 2003; 4(5): 343-8.

[170] Ly C, Arechiga AF, Melo JV, Walsh CM, Ong ST. Bcr-Abl kinase modulates the translation regulators ribosomal protein S6 and 4E-BP1 in chronic myelogenous leukemia cells *via* the mammalian target of rapamycin. Cancer Res 2003; 63(18): 5716-22.

[171] Inoki K, Corradetti MN, Guan KL. Dysregulation of the TSC-mTOR pathway in human disease. Nat Genet 2005; 37(1): 19-24.

[172] Hay N, Sonenberg N. Upstream and downstream of mTOR. Genes Dev 2004; 18(16): 1926-45.

[173] Pende M, Um SH, Mieulet V, *et al.* S6K1(-/-)/S6K2(-/-) mice exhibit perinatal lethality and rapamycin-sensitive 5'-terminal oligopyrimidine mRNA translation and reveal a mitogen-activated   protein kinase-dependent S6 kinase pathway. Mol Cell Biol 2004; 24(8): 3112-24.

[174] Hoffman WJ, Craver LF. Chronic Myelogenous Leukemia: Value of Irradiation and Its Effect on the Duration of Life. JAMA 1931; 97: 836-8.

[175] Galton DA. Myleran in chronic myeloid leukaemia; results of treatment. Lancet 1953; 264(6753): 208-13.

[176] Hehlmann R, Heimpel H, Hasford J, *et al.* Randomized comparison of busulfan and hydroxyurea in chronic myelogenous leukemia: prolongation of survival by hydroxyurea. Blood 1993; 82(2): 398-407.

[177] Garcia-Manero G, Talpaz M, Kantarjian HM. Current therapy of chronic myelogenous leukemia. Intern Med 2002; 41(4): 254-64.

[178] Silver RT, Woolf SH, Hehlmann R, *et al.* An evidence-based analysis of the effect of busulfan, hydroxyurea, interferon, and allogeneic bone marrow transplantation in treating the chronic phase of   chronic myeloid leukemia: developed for the American Society of Hematology. Blood 1999; 94(5):   1517-36.

[179] Fefer A, Cheever MA, Thomas ED, *et al.* Disappearance of Ph1-positive cells in four patients with chronic granulocytic leukemia after chemotherapy, irradiation and marrow transplantation from an   identical twin. N Engl J Med 1979; 300(7): 333-7.

[180] Pavlu J, Szydlo RM, Goldman JM, Apperley JF. Three decades of transplantation for chronic myeloid leukemia: what have we learned? Blood 2011; 117(3): 755-63.

[181] Robin M, Guardiola P, Devergie A, *et al.* A 10-year median follow-up study after allogeneic stem cell transplantation for chronic myeloid leukemia in chronic phase from HLA-identical sibling donors. Leukemia 2005; 19(9): 1613-20.

[182] Capdeville R, Buchdunger E, Zimmermann J, Matter A. Glivec (STI571, imatinib), a rationally developed, targeted anticancer drug. Nat Rev Drug Discov 2002; 1(7): 493-502.

[183] Druker BJ, Tamura S, Buchdunger E, *et al.* Effects of a selective inhibitor of the Abl tyrosine kinase on the growth of Bcr-Abl positive cells. Nat Med 1996; 2(5): 561-6.

[184] Hantschel O, Rix U, Superti-Furga G. Target spectrum of the BCR-ABL inhibitors imatinib, nilotinib and dasatinib. Leuk Lymphoma 2008; 49(4): 615-9.

[185] Gambacorti-Passerini C, le Coutre P, Mologni L, *et al.* Inhibition of the ABL kinase activity blocks the proliferation of BCR/ABL+ leukemic cells and induces apoptosis. Blood Cells Mol Dis 1997; 23(3): 380-94.

[186] le Coutre P, Mologni L, Cleris L, *et al. In vivo* eradication of human BCR/ABL-positive leukemia cells with an ABL kinase inhibitor. J Natl Cancer Inst 1999; 91(2): 163-8.

[187] Schindler T, Bornmann W, Pellicena P, Miller WT, Clarkson B, Kuriyan J. Structural mechanism for STI-571 inhibition of abelson tyrosine kinase. Science 2000; 289(5486): 1938-42.

[188] Hochhaus A, Druker B, Sawyers C, *et al.* Favorable long-term follow-up results over 6 years for response, survival, and safety with imatinib mesylate therapy in chronic-phase chronic myeloid   leukemia after failure of interferon-alpha treatment. Blood 2008; 111(3): 1039-43.

[189] Hahn EA, Glendenning GA, Sorensen MV, *et al.* Quality of life in patients with newly diagnosed   chronic phase chronic myeloid leukemia on imatinib versus interferon alfa plus low-dose cytarabine: results from the IRIS Study. J Clin Oncol 2003; 21(11): 2138-46.

[190] Hughes TP, Kaeda J, Branford S, *et al.* International Randomised Study of Interferon *versus* STI571 (IRIS) Study Group. Frequency of major molecular responses to imatinib or interferon alfa

plus cytarabine in newly diagnosed chronic myeloid leukemia. N Engl J Med 2003; 349(15): 1423-32.

[191]   O'Brien SG, Guilhot F, Larson RA, *et al.* Imatinib compared with interferon and low-dose cytarabine for newly diagnosed chronic-phase chronic myeloid leukemia. N Engl J Med 2003; 348(11): 994- 1004.

[192]   Hochhaus A, O'Brien SG, Guilhot F, *et al.* Six-year follow-up of patients receiving imatinib for the first-line treatment of chronic myeloid leukemia. Leukemia 2009; 23(6): 1054-61.

[193]   Shah NP, Tran C, Lee FY, Chen P, Norris D, Sawyers CL. Overriding imatinib resistance with a novel ABL kinase inhibitor. Science 2004; 305(5682): 399-401.

[194]   Deguchi Y, Kimura S, Ashihara E, *et al.* Comparison of imatinib, dasatinib, nilotinib and INNO-406 in imatinib-resistant cell lines. Leuk Res 2008; 32(6): 980-3.

[195]   Hochhaus A, Baccarani M, Deininger M, *et al.* Dasatinib induces durable cytogenetic responses in patients with chronic myelogenous leukemia in chronic phase with resistance or intolerance to imatinib. Leukemia 2008; 22(6): 1200-6.

[196]   Kantarjian H, Pasquini R, Hamerschlak N, *et al.* Dasatinib or high-dose imatinib for chronic-phase chronic myeloid leukemia after failure of first-line imatinib: a randomized phase 2 trial. Blood 2007; 109(12): 5143-50.

[197]   Kantarjian H, Shah NP, Hochhaus A, *et al.* Dasatinib *versus* imatinib in newly diagnosed chronic- phase chronic myeloid leukemia. N Engl J Med 2010; 362(24): 2260-70.

[198]   Kantarjian HM, Shah NP, Cortes JE, *et al.* Dasatinib or imatinib in newly diagnosed chronic-phase chronic myeloid leukemia: 2-year follow-up from a randomized phase 3 trial (DASISION). Blood 2012; 119(5): 1123-9.

[199]   Weisberg E, Manley PW, Breitenstein W, *et al.* Characterization of AMN107, a selective inhibitor of native and mutant Bcr-Abl. Cancer Cell 2005; 7(2): 129-41.

[200]   Golemovic M, Verstovsek S, Giles F, *et al.* AMN107, a novel aminopyrimidine inhibitor of Bcr-Abl, has *in vitro* activity against imatinib-resistant chronic myeloid leukemia. Clin Cancer Res 2005; 11(13): 4941-7.

[201]   Soverini S, Hochhaus A, Nicolini FE, *et al.* BCR-ABL kinase domain mutation analysis in chronic myeloid leukemia patients treated with tyrosine kinase inhibitors: recommendations from an expert   panel on behalf of European LeukemiaNet. Blood 2011; 118(5): 1208-15.

[202]   Kantarjian H, Giles F, Wunderle L, *et al.* Nilotinib in imatinib-resistant CML and Philadelphia chromosome-positive ALL. N Engl J Med 2006; 354(24): 2542-51.

[203]   Kantarjian HM, Giles FJ, Bhalla KN, *et al.* Nilotinib is effective in patients with chronic myeloid   leukemia in chronic phase after imatinib resistance or intolerance: 24-month follow-up results. Blood 2011; 117(4): 1141-5.

[204]   Hughes T, Saglio G, Branford S, *et al.* Impact of baseline BCR-ABL mutations on response to nilotinib in patients with chronic myeloid leukemia in chronic phase. J Clin Oncol 2009; 27(25): 4204- 10.

[205]   Larson RA, Hochhaus A, Hughes TP, *et al.* Nilotinib *vs* imatinib in patients with newly diagnosed Philadelphia chromosome-positive chronic myeloid leukemia in chronic phase: ENESTnd 3-year   follow-up. Leukemia 2012; 26(10): 2197-203.

[206]   Remsing Rix LL, Rix U, Colinge J, *et al.* Global target profile of the kinase inhibitor bosutinib in primary chronic myeloid leukemia cells. Leukemia 2009; 23(3): 477-85.

[207]   Cortes JE, Kantarjian HM, Brümmendorf TH, *et al.* Safety and efficacy of bosutinib (SKI-606) in   chronic phase Philadelphia chromosome-positive chronic myeloid leukemia patients with resistance or intolerance to imatinib. Blood 2011; 118(17): 4567-76.

[208]   Khoury HJ, Cortes JE, Kantarjian HM, *et al.* Bosutinib is active in chronic phase chronic myeloid   leukemia after imatinib and dasatinib and/or nilotinib therapy failure. Blood 2012; 119(15): 3403-12.

[209]   Kantarjian H, le Coutre P, Cortes J, *et al.* Phase 1 study of INNO-406, a dual Abl/Lyn kinase inhibitor, in Philadelphia chromosome-positive leukemias after imatinib resistance or intolerance. Cancer 2010; 116(11): 2665-72.

[210]   Agrawal M, Garg RJ, Cortes J, Quintás-Cardama A. Tyrosine kinase inhibitors: the first decade. Curr Hematol Malig Rep 2010; 5(2): 70-80.

[211] O'Hare T, Shakespeare WC, Zhu X, *et al.* AP24534, a pan-BCR-ABL inhibitor for chronic myeloid leukemia, potently inhibits the T315I mutant and overcomes mutation-based resistance. Cancer Cell 2009; 16(5): 401-12.

[212] Cortes JE, Kim DW, Pinilla-Ibarz J, *et al.* PACE: A pivotal phase II trial of ponatinib in patients with CML and Ph+ALL resistant or intolerant to dasatinib or nilotinib, or with the T315I mutation. ASCO abstracts 2012; 6503.

[213] FDA asks manufacturer of the leukemia drug Iclusig (ponatinib) to suspend marketing and sales. U.S. Food and Drug Administration http://www. fda.gov/Drugs/DrugSafety/ucm373040.htm 2013.

[214] Lipshultz SE, Diamond MB, Franco VI, *et al.* Managing chemotherapy-related cardiotoxicity in survivors of childhood cancers. Paediatr Drugs 2014; 16(5): 373-89.

[215] Waller CF. Imatinib mesylate. Recent Results Cancer Res 2014; 201: 1-25.

[216] Shah NP, Nicoll JM, Nagar B, *et al.* Multiple BCR-ABL kinase domain mutations confer polyclonal resistance to the tyrosine kinase inhibitor imatinib (STI571) in chronic phase and blast crisis chronic myeloid leukemia. Cancer Cell 2002; 2(2): 117-25.

[217] Gorre ME, Mohammed M, Ellwood K, *et al.* Clinical resistance to STI-571 cancer therapy caused by BCR-ABL gene mutation or amplification. Science 2001; 293(5531): 876-80.

[218] Sawyers CL, Hochhaus A, Feldman E, *et al.* Imatinib induces hematologic and cytogenetic responses in patients with chronic myelogenous leukemia in myeloid blast crisis: results of a phase II study. Blood 2002; 99(10): 3530-9.

[219] Apperley JF, Part I. Part I: mechanisms of resistance to imatinib in chronic myeloid leukaemia. Lancet Oncol 2007; 8(11): 1018-29.

[220] Khorashad JS, Kelley TW, Szankasi P, *et al.* BCR-ABL1 compound mutations in tyrosine kinase inhibitor-resistant CML: frequency and clonal relationships. Blood 2013; 121(3): 489-98.

[221] Melo JV, Chuah C. Resistance to imatinib mesylate in chronic myeloid leukaemia. Cancer Lett 2007; 249(2): 121-32.

[222] Mahon FX, Hayette S, Lagarde V, *et al.* Evidence that resistance to nilotinib may be due to BCR- ABL, Pgp, or Src kinase overexpression. Cancer Res 2008; 68(23): 9809-16.

[223] le Coutre P, Tassi E, Varella-Garcia M, *et al.* Induction of resistance to the Abelson inhibitor STI571 in human leukemic cells through gene amplification. Blood 2000; 95(5): 1758-66.

[224] Mahon FX, Deininger MW, Schultheis B, *et al.* Selection and characterization of BCR-ABL positive cell lines with differential sensitivity to the tyrosine kinase inhibitor STI571: diverse mechanisms of resistance. Blood 2000; 96(3): 1070-9.

[225] Ursan ID, Jiang R, Pickard EM, Lee TA, Ng D, Pickard AS. Emergence of BCR-ABL kinase domain mutations associated with newly diagnosed chronic myeloid leukemia: a meta-analysis of clinical trials of tyrosine kinase inhibitors. J Manag Care Spec Pharm 2015; 21(2): 114-22.

[226] Jabbour E, Kantarjian H, Jones D, *et al.* Frequency and clinical significance of BCR-ABL mutations in patients imatinib mesylate. Leukemia 2006; 20(10): 1767-73.

[227] Volpe G, Panuzzo C, Ulisciani S, Cilloni D. Imatinib resistance in CML. Cancer Lett 2009; 274(1): 1- 9.

[228] Breccia M, Alimena G. Resistance to imatinib in chronic myeloid leukemia and therapeutic approaches to circumvent the problem. Cardiovasc Hematol Disord Drug Targets 2009; 9(1): 21-8.

[229] Virgili A, Koptyra M, Dasgupta Y, *et al.* Imatinib sensitivity in BCR-ABL1-positive chronic myeloid leukemia cells is regulated by the remaining normal ABL1 allele. Cancer Res 2011; 71(16): 5381-6.

[230] Shukla S, Ohnuma S, Ambudkar SV. Improving cancer chemotherapy with modulators of ABC drug transporters. Curr Drug Targets 2011; 12(5): 621-30.

[231] White DL, Saunders VA, Dang P, *et al.* OCT-1-mediated influx is a key determinant of the intracellular uptake of imatinib but not nilotinib (AMN107): reduced OCT-1 activity is the cause of low *in vitro* sensitivity to imatinib. Blood 2006; 108(2): 697-704.

[232] Wang L, Giannoudis A, Lane S, Williamson P, Pirmohamed M, Clark RE. Expression of the uptake drug transporter hOCT1 is an important clinical determinant of the response to imatinib in chronic myeloid leukemia. Clin Pharmacol Ther 2008; 83(2): 258-64.

[233] Crossman LC, Druker BJ, Deininger MW, Pirmohamed M, Wang L, Clark RE. hOCT 1 and resistance to imatinib. Blood 2005; 106(3): 1133-4.

[234] Hiwase DK, Saunders V, Hewett D, *et al.* Dasatinib cellular uptake and efflux in chronic myeloid leukemia cells: therapeutic implications. Clin Cancer Res 2008; 14(12): 3881-8.

[235] Bhamidipati PK, Kantarjian H, Cortes J, Cornelison AM, Jabbour E. Management of imatinib-resistant patients with chronic myeloid leukemia. Ther Adv Hematol 2013; 4(2): 103-17.

[236] Bixby D, Talpaz M. Mechanisms of resistance to tyrosine kinase inhibitors in chronic myeloid leukemia and recent therapeutic strategies to overcome resistance. Hematology Am Soc Hematol Educ Program 2009; 461-76.

[237] Dai Y, Rahmani M, Corey SJ, Dent P, Grant S. A Bcr/Abl-independent, Lyn-dependent form of imatinib mesylate (STI-571) resistance is associated with altered expression of Bcl-2. J Biol Chem 2004; 279(33): 34227-39.

[238] Marfe G, Di Stefano C, Gambacurta A, *et al.* Sphingosine kinase 1 overexpression is regulated by signaling through PI3K, AKT2, and mTOR in imatinib-resistant chronic myeloid leukemia cells. Exp Hematol 2011; 39(6): 653-665.e6.

[239] Salas A, Ponnusamy S, Senkal CE, *et al.* Sphingosine kinase-1 and sphingosine 1-phosphate receptor 2 mediate Bcr-Abl1 stability and drug resistance by modulation of protein phosphatase 2A. Blood 2011; 117(22): 5941-52.

[240] Baran Y, Salas A, Senkal CE, *et al.* Alterations of ceramide/sphingosine 1-phosphate rheostat involved in the regulation of resistance to imatinib-induced apoptosis in K562 human chronic myeloid leukemia cells. J Biol Chem 2007; 282(15): 10922-34.

[241] Neviani P, Santhanam R, Oaks JJ, *et al.* FTY720, a new alternative for treating blast crisis chronic myelogenous leukemia and Philadelphia chromosome-positive acute lymphocytic leukemia. J Clin Invest 2007; 117(9): 2408-21.

[242] Hamilton A, Alhashimi F, Myssina S, Jorgensen HG, Holyoake TL. Optimization of methods for the detection of BCR-ABL activity in Philadelphia-positive cells. Exp Hematol 2009; 37(3): 395-401.

[243] ten Hoeve J, Morris C, Heisterkamp N, Groffen J. Isolation and chromosomal localization of CRKL, a human crk-like gene. Oncogene 1993; 8(9): 2469-74.

[244] ten Hoeve J, Arlinghaus RB, Guo JQ, Heisterkamp N, Groffen J. Tyrosine phosphorylation of CRKL in Philadelphia+ leukemia. Blood 1994; 84(6): 1731-6.

[245] de Jong R, ten Hoeve J, Heisterkamp N, Groffen J. Tyrosine 207 in CRKL is the BCR/ABL phosphorylation site. Oncogene 1997; 14(5): 507-13.

[246] Hamilton A, Elrick L, Myssina S, *et al.* BCR-ABL activity and its response to drugs can be determined in CD34+ CML stem cells by CrkL phosphorylation status using flow cytometry. Leukemia 2006; 20(6): 1035-9.

[247] Reddy EP, Aggarwal AK. The ins and outs of bcr-abl inhibition. Genes Cancer 2012; 3(5-6): 447-54.

[248] Adrián FJ, Ding Q, Sim T, *et al.* Allosteric inhibitors of Bcr-abl-dependent cell proliferation. Nat Chem Biol 2006; 2(2): 95-102.

[249] Iacob RE, Zhang J, Gray NS, Engen JR. Allosteric interactions between the myristate- and ATP-site of the Abl kinase. PLoS One 2011; 6(1): e15929.

[250] Zhang J, Adrián FJ, Jahnke W, *et al.* Targeting Bcr-Abl by combining allosteric with ATP-binding-site inhibitors. Nature 2010; 463(7280): 501-6.

[251] Beissert T, Hundertmark A, Kaburova V, *et al.* Targeting of the N-terminal coiled coil oligomerization interface by a helix-2 peptide inhibits unmutated and imatinib-resistant BCR/ABL. Int J Cancer 2008; 122(12): 2744-52.

[252] Peng Z, Yuan Y, Li YJ, *et al.* Targeting BCR tyrosine177 site with novel SH2-DED causes selective leukemia cell death *in vitro* and *in vivo*. Int J Biochem Cell Biol 2012; 44(6): 861-8.

[253] Sexl V, Piekorz R, Moriggl R, *et al.* Stat5a/b contribute to interleukin 7-induced B-cell precursor expansion, but abl- and bcr/abl-induced transformation are independent of stat5. Blood 2000; 96(6): 2277-83.

[254] Hoelbl A, Kovacic B, Kerenyi MA, *et al.* Clarifying the role of Stat5 in lymphoid development and Abelson-induced transformation. Blood 2006; 107(12): 4898-906.

[255] Hoelbl A, Schuster C, Kovacic B, *et al.* Stat5 is indispensable for the maintenance of bcr/abl-positive leukaemia. EMBO Mol Med 2010; 2(3): 98-110.

[256] Ye D, Wolff N, Li L, Zhang S, Ilaria RL Jr. STAT5 signaling is required for the efficient induction and maintenance of CML in mice. Blood 2006; 107(12): 4917-25.

[257]   Wilson-Rawls J, Liu J, Laneuville P, Arlinghaus RB. P210 Bcr-Abl interacts with the interleukin-3 beta c subunit and constitutively activates Jak2. Leukemia 1997; 11 (Suppl. 3): 428-31.

[258]   Chai SK, Nichols GL, Rothman P. Constitutive activation of JAKs and STATs in BCR-Abl-expressing cell lines and peripheral blood cells derived from leukemic patients. J Immunol 1997; 159(10): 4720-8.

[259]   Warsch W, Walz C, Sexl V. JAK of all trades: JAK2-STAT5 as novel therapeutic targets in BCR- ABL1+ chronic myeloid leukemia. Blood 2013; 122(13): 2167-75.

[260]   Samanta AK, Chakraborty SN, Wang Y, *et al.* Jak2 inhibition deactivates Lyn kinase through the SET-PP2A-SHP1 pathway, causing apoptosis in drug-resistant cells from chronic myelogenous leukemia patients. Oncogene 2009; 28(14): 1669-81.

[261]   Baxter EJ, Scott LM, Campbell PJ, *et al.* Acquired mutation of the tyrosine kinase JAK2 in human myeloproliferative disorders. Lancet 2005; 365(9464): 1054-61.

[262]   Kralovics R, Passamonti F, Buser AS, *et al.* A gain-of-function mutation of JAK2 in myeloproliferative disorders. N Engl J Med 2005; 352(17): 1779-90.

[263]   Verstovsek S, Mesa RA, Gotlib J, *et al.* A double-blind, placebo-controlled trial of ruxolitinib for myelofibrosis. N Engl J Med 2012; 366(9): 799-807.

[264]   Harrison C, Kiladjian JJ, Al-Ali HK, *et al.* JAK inhibition with ruxolitinib *versus* best available therapy for myelofibrosis. N Engl J Med 2012; 366(9): 787-98.

[265]   Gallipoli P. The role of jak2 inhibitors in the management of cml: can we rest the case. Sci Proc 2014; 1: e383.

[266]   Zhang H, Li S. Molecular mechanisms for survival regulation of chronic myeloid leukemia stem cells. Protein Cell 2013; 4(3): 186-96.

[267]   Corbin AS, Agarwal A, Loriaux M, Cortes J, Deininger MW, Druker BJ. Human chronic myeloid leukemia stem cells are insensitive to imatinib despite inhibition of BCR-ABL activity. J Clin Invest 2011; 121(1): 396-409.

[268]   Hamilton A, Helgason GV, Schemionek M, *et al.* Chronic myeloid leukemia stem cells are not dependent on Bcr-Abl kinase activity for their survival. Blood 2012; 119(6): 1501-10.

[269]   Muxí PJ, Oliver AC. Jak-2 positive myeloproliferative neoplasms. Curr Treat Options Oncol 2014; 15(2): 147-56.

[270]   Gallipoli P, Cook A, Rhodes S, *et al.* JAK2/STAT5 inhibition by nilotinib with ruxolitinib contributes to the elimination of CML CD34+ cells *in vitro* and *in vivo*. Blood 2014; 124(9): 1492-501.

[271]   Hiwase DK, White DL, Powell JA, *et al.* Blocking cytokine signaling along with intense Bcr-Abl kinase inhibition induces apoptosis in primary CML progenitors. Leukemia 2010; 24(4): 771-8.

[272]   Traer E, MacKenzie R, Snead J, *et al.* Blockade of JAK2-mediated extrinsic survival signals restores sensitivity of CML cells to ABL inhibitors. Leukemia 2012; 26(5): 1140-3.

[273]   Nair RR, Tolentino JH, Argilagos RF, Zhang L, Pinilla-Ibarz J, Hazlehurst LA. Potentiation of Nilotinib-mediated cell death in the context of the bone marrow microenvironment requires a promiscuous JAK inhibitor in CML. Leuk Res 2012; 36(6): 756-63.

[274]   Quintarelli C, De Angelis B, Errichiello S, *et al.* Selective strong synergism of Ruxolitinib and second generation tyrosine kinase inhibitors to overcome bone marrow stroma related drug resistance in chronic myelogenous leukemia. Leuk Res 2014; 38(2): 236-42.

[275]   Tong H, Ren Y, Zhang F, Jin J. Homoharringtonine affects the JAK2-STAT5 signal pathway through alteration of protein tyrosine kinase phosphorylation in acute myeloid leukemia cells. Eur J Haematol 2008; 81(4): 259-66.

[276]   Grundschober E, Hoelbl-Kovacic A, Bhagwat N, *et al.* Acceleration of Bcr-Abl+ leukemia induced by deletion of JAK2. Leukemia 2014; 28(9): 1918-22.

[277]   Hurtz C, Hatzi K, Cerchietti L, *et al.* BCL6-mediated repression of p53 is critical for leukemia stem cell survival in chronic myeloid leukemia. J Exp Med 2011; 208(11): 2163-74.

[278]   Lin H, Chen M, Rothe K, Lorenzi MV, Woolfson A, Jiang X. Selective JAK2/ABL dual inhibition therapy effectively eliminates TKI-insensitive CML stem/progenitor cells. Oncotarget 2014; 5(18): 8637-50.

[279]   Al-Mawali A, Gillis D, Lewis I. The role of multiparameter flow cytometry for detection of minimal residual disease in acute myeloid leukemia. Am J Clin Pathol 2009; 131(1): 16-26.

[280] Goldman JM. How I treat chronic myeloid leukemia in the imatinib era. Blood 2007; 110(8): 2828-37.

[281] Dazzi F, Fozza C. Disease relapse after haematopoietic stem cell transplantation: risk factors and treatment. Best Pract Res Clin Haematol 2007; 20(2): 311-27.

[282] Hehlmann R, Lauseker M, Jung-Munkwitz S, *et al.* Tolerability-adapted imatinib 800 mg/d versus 400 mg/d versus 400 mg/d plus interferon-α in newly diagnosed chronic myeloid leukemia. J Clin Oncol 2011; 29(12): 1634-42.

[283] Baccarani M, Rosti G, Castagnetti F, *et al.* Comparison of imatinib 400 mg and 800 mg daily in the front-line treatment of high-risk, Philadelphia-positive chronic myeloid leukemia: a European LeukemiaNet Study. Blood 2009; 113(19): 4497-504.

[284] Farnsworth P, Ward D, Reddy V. Persistent complete molecular remission after nilotinib and graft-*versus*-leukemia effect in an acute lymphoblastic leukemia patient with cytogenetic relapse after allogeneic stem cell transplantation. Exp Hematol Oncol 2012; 1(1): 29.

[285] Tanaka H, Hirase C, Matsumura I. Molecular mechanisms in the resistance of CML stem cells to tyrosine kinase inhibitors and novel targets for achieving a cure. Rinsho Ketsueki 2015; 56(2): 139-49.

[286] Saglio G, Kim DW, Issaragrisil S, *et al.* Nilotinib *versus* imatinib for newly diagnosed chronic myeloid leukemia. N Engl J Med 2010; 362(24): 2251-9.

[287] Kantarjian H, Shah NP, Hochhaus A, *et al.* Dasatinib *versus* imatinib in newly diagnosed chronic- phase chronic myeloid leukemia. N Engl J Med 2010; 362(24): 2260-70.

[288] Eide CA, Adrian LT, Tyner JW, *et al.* The ABL switch control inhibitor DCC-2036 is active against the chronic myeloid leukemia mutant BCR-ABLT315I and exhibits a narrow resistance profile. Cancer Res 2011; 71(9): 3189-95.

[289] Lü S, Wang J. Homoharringtonine and omacetaxine for myeloid hematological malignancies. J Hematol Oncol 2014; 7: 2.

[290] Mahon FX, Réa D, Guilhot J, *et al.* Discontinuation of imatinib in patients with chronic myeloid leukaemia who have maintained complete molecular remission for at least 2 years: the prospective, multicentre Stop Imatinib (STIM) trial. Lancet Oncol 2010; 11(11): 1029-35.

[291] Rousselot P, Charbonnier A, Cony-Makhoul P, *et al.* Loss of Major Molecular Response As a Trigger for Restarting Tyrosine Kinase Inhibitor Therapy in Patients With Chronic-Phase Chronic Myelogenous Leukemia Who Have Stopped Imatinib After Durable Undetectable Disease J Clin Oncol 2014; 32(5): 424-30.

[292] Ross DM, Branford S, Seymour JF, *et al.* Safety and efficacy of imatinib cessation for CML patients with stable undetectable minimal residual disease: results from the TWISTER study. Blood 2013; 122(4): 515-22.

[293] Takahashi N, Kyo T, Maeda Y, *et al.* Discontinuation of imatinib in Japanese patients with chronic myeloid leukemia. Haematologica 2012; 97(6): 903-6.

[294] Yhim HY, Lee NR, Song EK, *et al.* Imatinib mesylate discontinuation in patients with chronic myeloid leukemia who have received front-line imatinib mesylate therapy and achieved complete molecular response. Leuk Res 2012; 36(6): 689-93.

[295] Raanani P, Granot G, Ben-Bassat I. Is cure of chronic myeloid leukemia in the third millennium a down to earth target (ed) or a castle in the air? Cancer Lett 2014; 3835(14): 00029-9.

[296] Yeung DT, Osborn MP, White DL, *et al.* TIDEL-II: first-line use of imatinib in CML with early switch to nilotinib for failure to achieve time-dependent molecular targets. Blood 2015; 125(6): 915- 23.

[297] Kantarjian HM, Hochhaus A, Saglio G, *et al.* Nilotinib *versus* imatinib for the treatment of patients with newly diagnosed chronic phase, Philadelphia chromosome-positive, chronic myeloid leukaemia: 24-month minimum follow-up of the phase 3 randomised ENESTnd trial. Lancet Oncol 2011; 12(9): 841-51.

[298] Breccia M, Alimena G. The role of comorbidities in chronic myeloid leukemia. Leuk Res 2013; 37(7): 729-30.

[299] Marin D, Ibrahim AR, Lucas C, *et al.* Assessment of BCR-ABL1 transcript levels at 3 months is the only requirement for predicting outcome for patients with chronic myeloid leukemia treated with tyrosine kinase inhibitors. J Clin Oncol 2012; 30(3): 232-8.

[300]   Breccia M, Alimena G. "To switch or not to switch: that is the question"-more than 10% of ratio @ 3 months: how to treat chronic myeloid leukemia patients with this response? Leuk Res 2013; 37(9): 995-7.

[301]   Hanfstein B, Müller MC, Hehlmann R, *et al.* Early molecular and cytogenetic response is predictive for long-term progression-free and overall survival in chronic myeloid leukemia (CML). Leukemia 2012; 26(9): 2096-102.

[302]   Jain P, Kantarjian H, Nazha A, *et al.* Early responses predict better outcomes in patients with newly diagnosed chronic myeloid leukemia: results with four tyrosine kinase inhibitor modalities. Blood 2013; 121(24): 4867-74.

[303]   Craig JM, Ed. Epigenetics: A Reference Manual. Caister Academic Press 2011.

[304]   Asimakopoulos FA, Shteper PJ, Krichevsky S, *et al.* ABL1 methylation is a distinct molecular event associated with clonal evolution of chronic myeloid leukemia. Blood 1999; 94(7): 2452-60.

[305]   Issa JP, Kantarjian H, Mohan A, *et al.* Methylation of the ABL1 promoter in chronic myelogenous leukemia: lack of prognostic significance. Blood 1999; 93(6): 2075-80.

[306]   Ben-Yehuda D, Krichevsky S, Rachmilewitz EA, *et al.* Molecular follow-up of disease progression and interferon therapy in chronic myelocytic leukemia. Blood 1997; 90(12): 4918-23.

[307]   Nguyen TT, Mohrbacher AF, Tsai YC, *et al.* Quantitative measure of c-abl and p15 methylation in chronic myelogenous leukemia: biological implications. Blood 2000; 95(9): 2990-2.

[308]   Jelinek J, Gharibyan V, Estecio MR, *et al.* Aberrant DNA methylation is associated with disease progression, resistance to imatinib and shortened survival in chronic myelogenous leukemia. PLoS One 2011; 6(7): e22110.

[309]   Sun B, Jiang G, Zaydan MA, La Russa VF, Safah H, Ehrlich M. ABL1 promoter methylation can exist independently of BCR-ABL transcription in chronic myeloid leukemia hematopoietic progenitors. Cancer Res 2001; 61(18): 6931-7.

[310]   Uchida T, Kinoshita T, Hotta T, Murate T. High-risk myelodysplastic syndromes and hyper-methylation of the p15Ink4B gene. Leuk Lymphoma 1998; 32(1-2): 9-18.

[311]   Tien HF, Tang JH, Tsay W, *et al.* Methylation of the p15(INK4B) gene in myelodysplastic syndrome: it can be detected early at diagnosis or during disease progression and is highly associated with leukaemic transformation. Br J Haematol 2001; 112(1): 148-54.

[312]   Herman JG, Civin CI, Issa JP, Collector MI, Sharkis SJ, Baylin SB. Distinct patterns of inactivation of p15INK4B and p16INK4A characterize the major types of hematological malignancies. Cancer Res 1997; 57(5): 837-41.

[313]   Oki Y, Kantarjian HM, Gharibyan V, *et al.* Phase II study of low-dose decitabine in combination with imatinib mesylate in patients with accelerated or myeloid blastic phase of chronic myelogenous leukemia. Cancer 2007; 109(5): 899-906.

[314]   Uehara E, Takeuchi S, Yang Y, *et al.* Aberrant methylation in promoter-associated CpG islands of multiple genes in chronic myelogenous leukemia blast crisis. Oncol Lett 2012; 3(1): 190-2.

[315]   Avramouli A, Tsochas S, Mandala E, *et al.* Methylation status of RASSF1A in patients with chronic myeloid leukemia. Leuk Res 2009; 33(8): 1130-2.

[316]   Dunwell T, Hesson L, Rauch TA, *et al.* A genome-wide screen identifies frequently methylated genes in haematological and epithelial cancers. Mol Cancer 2010; 9: 44.

[317]   Cadwell K, Liu JY, Brown SL, *et al.* A key role for autophagy and the autophagy gene Atg16l1 in mouse and human intestinal Paneth cells. Nature 2008; 456(7219): 259-63.

[318]   Gozuacik D, Kimchi A. DAPk protein family and cancer. Autophagy 2006; 2(2): 74-9.

[319]   Qian J, Wang YL, Lin J, Yao DM, Xu WR, Wu CY. Aberrant methylation of the death-associated protein kinase 1 (DAPK1) CpG island in chronic myeloid leukemia. Eur J Haematol 2009; 82(2): 119- 23.

[320]   Nelkin BD, Przepiorka D, Burke PJ, Thomas ED, Baylin SB. Abnormal methylation of the calcitonin gene marks progression of chronic myelogenous leukemia. Blood 1991; 77(11): 2431-4.

[321]   Issa JP, Zehnbauer BA, Kaufmann SH, Biel MA, Baylin SB. HIC1 hypermethylation is a late event in hematopoietic neoplasms. Cancer Res 1997; 57(9): 1678-81.

[322]   Issa JP, Zehnbauer BA, Civin CI, *et al.* The estrogen receptor CpG island is methylated in most hematopoietic neoplasms. Cancer Res 1996; 56(5): 973-7.

[323]   Strathdee G, Holyoake TL, Sim A, *et al.* Inactivation of HOXA genes by hypermethylation in myeloid and lymphoid malignancy is frequent and associated with poor prognosis. Clin Cancer Res 2007; 13(17): 5048-55.

[324]   Wang YL, Qian J, Lin J, *et al.* Methylation status of DDIT3 gene in chronic myeloid leukemia. J Exp Clin Cancer Res 2010; 29: 54.

[325]   Machova Polakova K, Koblihova J, Stopka T. Role of epigenetics in chronic myeloid leukemia. Curr Hematol Malig Rep 2013; 8(1): 28-36.

[326]   Cihák A. Biological effects of 5-azacytidine in eukaryotes. Oncology 1974; 30(5): 405-22.

[327]   Kaminskas E, Farrell AT, Wang Y-C, Sridhara R, Pazdur R. FDA drug approval summary: azacitidine (5-azacytidine, Vidaza) for injectable suspension. Oncologist 2005; 10(3): 176-82.

[328]   Kantarjian H, Issa JP, Rosenfeld CS, *et al.* Decitabine improves patient outcomes in myelodysplastic syndromes: results of a phase III randomized study. Cancer 2006; 106(8): 1794-803.

[329]   Kantarjian HM, O'Brien S, Cortes J, *et al.* Results of decitabine (5-aza-2'deoxycytidine) therapy in   130 patients with chronic myelogenous leukemia. Cancer 2003; 98(3): 522-8.

[330]   Candelaria M, Herrera A, Labardini J, *et al.* Hydralazine and magnesium valproate as epigenetic treatment for myelodysplastic syndrome. Ann Hematol 2011; 90(4): 379-87.

[331]   Cervera E, Candelaria M, López-Navarro O, *et al.* Epigenetic therapy with hydralazine and magnesium valproate reverses imatinib resistance in patients with chronic myeloid leukemia. Clin   Lymphoma Myeloma Leuk 2012; 12(3): 207-12.

[332]   Linggi BE, Brandt SJ, Sun ZW, Hiebert SW. Translating the histone code into leukemia. J Cell   Biochem 2005; 96(5): 938-50.

[333]   Choi Y, Elagib KE, Goldfarb AN. AML-1-ETO-Mediated erythroid inhibition: new paradigms for   differentiation blockade by a leukemic fusion protein. Crit Rev Eukaryot Gene Expr 2005; 15(3): 207- 16.

[334]   Mithraprabhu S, Grigoriadis G, Khong T, Spencer A. Deactylase inhibition in myeloproliferative   neoplasms. Invest New Drugs 2010; 28 (Suppl. 1): S50-7.

[335]   Rahmani M, Reese E, Dai Y, *et al.* Cotreatment with suberanoylanilide hydroxamic acid and 17- allylamino 17-demethoxygeldanamycin synergistically induces apoptosis in Bcr-Abl+ Cells sensitive   and resistant to STI571 (imatinib mesylate) in association with down-regulation of Bcr-Abl,   abrogation of signal transducer and activator of transcription 5 activity, and Bax conformational   change. Mol Pharmacol 2005; 67(4): 1166-76.

[336]   Rasheed WK, Johnstone RW, Prince HM. Histone deacetylase inhibitors in cancer therapy. Expert   Opin Investig Drugs 2007; 16(5): 659-78.

[337]   Mann BS, Johnson JR, Cohen MH, Justice R, Pazdur R. FDA approval summary: vorinostat for   treatment of advanced primary cutaneous T-cell lymphoma. Oncologist 2007; 12(10): 1247-52.

[338]   Nimmanapalli R, Fuino L, Stobaugh C, Richon V, Bhalla K. Cotreatment with the histone deacetylase inhibitor suberoylanilide hydroxamic acid (SAHA) enhances imatinib-induced apoptosis of Bcr-Ab- -positive human acute leukemia cells. Blood 2003; 101(8): 3236-9.

[339]   Fiskus W, Pranpat M, Balasis M, *et al.* Cotreatment with vorinostat (suberoylanilide hydroxamic acid) enhances activity of dasatinib (BMS-354825) against imatinib mesylate-sensitive or imatinib mesylate-resistant chronic myelogenous leukemia cells. Clin Cancer Res 2006; 12(19): 5869-78.

[340]   Bali P, Pranpat M, Bradner J, *et al.* Inhibition of histone deacetylase 6 acetylates and disrupts the   chaperone function of heat shock protein 90: a novel basis for antileukemia activity of histone   deacetylase inhibitors. J Biol Chem 2005; 280(29): 26729-34.

[341]   Lee RC, Feinbaum RL, Ambros V. The *C. elegans* heterochronic gene *lin-4* encodes small RNAs with antisense complementarity to lin-14. Cell 1993; 75(5): 843-54.

[342]   Wightman B, Bürglin TR, Gatto J, Arasu P, Ruvkun G. Negative regulatory sequences in the lin-14 3'- untranslated region are necessary to generate a temporal switch during Caenorhabditis elegans development. Genes Dev 1991; 5(10): 1813-24.

[343]   Reinhart BJ, Slack FJ, Basson M, *et al.* The 21-nucleotide *let-7* RNA regulates developmental timing in *Caenorhabditis elegans.* Nature 2000; 403(6772): 901-6.

[344]   Pasquinelli AE, Reinhart BJ, Slack F, *et al.* Conservation across animal phylogeny of the sequence and temporal regulation of the 21 nucleotide let-7 heterochronic regulatory RNA. Nature 2000; 408(6808): 86-9.

[345]   Basyuk E, Suavet F, Doglio A, Bordonné R, Bertrand E. Human let-7 stem-loop precursors harbor features of RNase III cleavage products. Nucleic Acids Res 2003; 31(22): 6593-7.
[346]   Vandenboom Ii TG, Li Y, Philip PA, Sarkar FH. MicroRNA and Cancer: Tiny Molecules with Major Implications. Curr Genomics 2008; 9(2): 97-109.
[347]   Bartel DP. MicroRNAs: genomics, biogenesis, mechanism, and function. Cell 2004; 116(2): 281-97.
[348]   Croce CM. Causes and consequences of microRNA dysregulation in cancer. Nat Rev Genet 2009; 10(10): 704-14.
[349]   Calin GA, Dumitru CD, Shimizu M, *et al.* Frequent deletions and down-regulation of micro-RNA genes miR15 and miR16 at 13q14 in chronic lymphocytic leukemia. Proc Natl Acad Sci USA 2002; 99(24): 15524-9.
[350]   Cimmino A, Calin GA, Fabbri M, *et al.* miR-15 and miR-16 induce apoptosis by targeting BCL2. Proc Natl Acad Sci USA 2005; 102(39): 13944-9.
[351]   Mirnezami AH, Pickard K, Zhang L, Primrose JN, Packham G. MicroRNAs: key players in carcinogenesis and novel therapeutic targets. Eur J Surg Oncol 2009; 35(4): 339-47.
[352]   Calin GA, Sevignani C, Dumitru CD, *et al.* Human microRNA genes are frequently located at fragile sites and genomic regions involved in cancers. Proc Natl Acad Sci USA 2004; 101(9): 2999-3004.
[353]   Agirre X, Jiménez-Velasco A, San José-Enériz E, *et al.* Down-regulation of hsa-miR-10a in chronic myeloid leukemia CD34+ cells increases USF2-mediated cell growth. Mol Cancer Res 2008; 6(12): 1830-40.
[354]   Venturini L, Battmer K, Castoldi M, *et al.* Expression of the miR-17-92 polycistron in chronic myeloid leukemia (CML) CD34+ cells. Blood 2007; 109(10): 4399-405.
[355]   Eiring AM, Harb JG, Neviani P, *et al.* miR-328 functions as an RNA decoy to modulate hnRNP E2 regulation of mRNA translation in leukemic blasts. Cell 2010; 140(5): 652-65.
[356]   Chan JA, Krichevsky AM, Kosik KS. MicroRNA-21 is an antiapoptotic factor in human glioblastoma cells. Cancer Res 2005; 65(14): 6029-33.
[357]   Chang JS, Santhanam R, Trotta R, *et al.* High levels of the BCR/ABL oncoprotein are required for the MAPK-hnRNP-E2 dependent suppression of C/EBPalpha-driven myeloid differentiation. Blood 2007; 110(3): 994-1003.
[358]   Perrotti D, Cesi V, Trotta R, *et al.* BCR-ABL suppresses C/EBPalpha expression through inhibitory action of hnRNP E2. Nat Genet 2002; 30(1): 48-58.
[359]   Perrotti D, Neviani P. From mRNA metabolism to cancer therapy: chronic myelogenous leukemia shows the way. Clin Cancer Res 2007; 13(6): 1638-42.
[360]   Ferreira AF, Moura LG, Tojal I, *et al.* ApoptomiRs expression modulated by BCR-ABL is linked to CML progression and imatinib resistance. Blood Cells Mol Dis 2014; 53(1-2): 47-55.
[361]   Machová Poláková K, Lopotová T, Klamová H, *et al.* Expression patterns of microRNAs associated with CML phases and their disease related targets. Mol Cancer 2011; 10: 41.
[362]   Flamant S, Ritchie W, Guilhot J, *et al.* Micro-RNA response to imatinib mesylate in patients with chronic myeloid leukemia. Haematologica 2010; 95(8): 1325-33.
[363]   Scholl V, Hassan R, Zalcberg IR. miRNA-451: A putative predictor marker of Imatinib therapy response in chronic myeloid leukemia. Leuk Res 2012; 36(1): 119-21.
[364]   Rokah OH, Granot G, Ovcharenko A, *et al.* Downregulation of miR-31, miR-155, and miR-564 in chronic myeloid leukemia cells. PLoS One 2012; 7(4): e35501.
[365]   Suresh S, McCallum L, Lu W, Lazar N, Perbal B, Irvine AE. MicroRNAs 130a/b are regulated by BCR-ABL and downregulate expression of CCN3 in CML. J Cell Commun Signal 2011; 5(3): 183-91.
[366]   San José-Enériz E, Román-Gómez J, Jiménez-Velasco A, *et al.* MicroRNA expression profiling in Imatinib-resistant Chronic Myeloid Leukemia patients without clinically significant ABL1-mutations. Mol Cancer 2009; 8: 69-72.
[367]   Lopotová T, Záčková M, Klamová H, Moravcová J. MicroRNA-451 in chronic myeloid leukemia: miR-451-BCR-ABL regulatory loop? Leuk Res 2011; 35(7): 974-7.
[368]   Khalaj M, Tavakkoli M, Stranahan AW, Park CY. Pathogenic microRNA's in myeloid malignancies. Front Genet 2014; 5: 361.
[369]   Liu L, Wang S, Chen R, *et al.* Myc induced miR-144/451 contributes to the acquired imatinib resistance in chronic myelogenous leukemia cell K562. Biochem Biophys Res Commun 2012; 425(2): 368-73.

[370]   Ohyashiki JH, Ohtsuki K, Mizoguchi I, *et al.* Downregulated microRNA-148b in circulating PBMCs in chronic myeloid leukemia patients with undetectable minimal residual disease: a possible biomarker to discontinue imatinib safely. Drug Des Devel Ther 2014; 8: 1151-9.

[371]   Passegué E, Jamieson CH, Ailles LE, Weissman IL. Normal and leukemic hematopoiesis: are leukemias a stem cell disorder or a reacquisition of stem cell characteristics? Proc Natl Acad Sci USA 2003; 100 (Suppl. 1): 11842-9.

[372]   Al-Hajj M, Wicha MS, Benito-Hernandez A, Morrison SJ, Clarke MF. Prospective identification of tumorigenic breast cancer cells. Proc Natl Acad Sci USA 2003; 100(7): 3983-8.

[373]   Park CY, Tseng D, Weissman IL. Cancer stem cell-directed therapies: recent data from the laboratory and clinic. Mol Ther 2009; 17(2): 219-30.

[374]   Müller-Tidow C, Steffen B, Cauvet T, *et al.* Translocation products in acute myeloid leukemia activate the Wnt signaling pathway in hematopoietic cells. Mol Cell Biol 2004; 24(7): 2890-904.

[375]   Chiotaki R, Polioudaki H, Theodoropoulos PA. Cancer stem cells in solid and liquid tissues of breast cancer patients: characterization and therapeutic perspectives. Curr Cancer Drug Targets 2015; 15(3): 256-69.

[376]   Pérez-Caro M, Cobaleda C, González-Herrero I, *et al.* Cancer induction by restriction of oncogene expression to the stem cell compartment. EMBO J 2009; 28(1): 8-20.

[377]   Hu Y, Swerdlow S, Duffy TM, Weinmann R, Lee FY, Li S. Targeting multiple kinase pathways in leukemic progenitors and stem cells is essential for improved treatment of Ph+ leukemia in mice. Proc Natl Acad Sci USA 2006; 103(45): 16870-5.

[378]   Jamieson CH, Ailles LE, Dylla SJ, *et al.* Granulocyte-macrophage progenitors as candidate leukemic stem cells in blast-crisis CML. N Engl J Med 2004; 351(7): 657-67.

[379]   Minami Y, Stuart SA, Ikawa T, *et al.* BCR-ABL-transformed GMP as myeloid leukemic stem cells. Proc Natl Acad Sci USA 2008; 105(46): 17967-72.

[380]   Bhatia R, Holtz M, Niu N, *et al.* Persistence of malignant hematopoietic progenitors in chronic myelogenous leukemia patients in complete cytogenetic remission following imatinib mesylate treatment. Blood 2003; 101(12): 4701-7.

[381]   Copland M, Hamilton A, Elrick LJ, *et al.* Dasatinib (BMS-354825) targets an earlier progenitor population than imatinib in primary CML but does not eliminate the quiescent fraction. Blood 2006; 107(11): 4532-9.

[382]   Hu Y, Chen Y, Douglas L, Li S. beta-Catenin is essential for survival of leukemic stem cells insensitive to kinase inhibition in mice with BCR-ABL-induced chronic myeloid leukemia. Leukemia 2009; 23(1): 109-16.

[383]   Reya T, Duncan AW, Ailles L, *et al.* A role for Wnt signalling in self-renewal of haematopoietic stem cells. Nature 2003; 423(6938): 409-14.

[384]   Nüsslein-Volhard C, Wieschaus E. Mutations affecting segment number and polarity in Drosophila. Nature 1980; 287(5785): 795-801.

[385]   Echelard Y, Epstein DJ, St-Jacques B, *et al.* Sonic hedgehog, a member of a family of putative signaling molecules, is implicated in the regulation of CNS polarity. Cell 1993; 75(7): 1417-30.

[386]   Dierks C, Beigi R, Guo GR, *et al.* Expansion of Bcr-Abl-positive leukemic stem cells is dependent on Hedgehog pathway activation. Cancer Cell 2008; 14(3): 238-49.

[387]   Zhao C, Chen A, Jamieson CH, *et al.* Hedgehog signalling is essential for maintenance of cancer stem cells in myeloid leukaemia. Nature 2009; 458(7239): 776-9.

[388]   Cea M, Cagnetta A, Cirmena G, *et al.* Tracking molecular relapse of chronic myeloid leukemia by measuring Hedgehog signaling status. Leuk Lymphoma 2013; 54(2): 342-52.

[389]   Rådmark O, Werz O, Steinhilber D, Samuelsson B. 5-Lipoxygenase: regulation of expression and enzyme activity. Trends Biochem Sci 2007; 32(7): 332-41.

[390]   Wymann MP, Schneiter R. Lipid signalling in disease. Nat Rev Mol Cell Biol 2008; 9(2): 162-76.

[391]   Anderson KM, Seed T, Jajeh A, *et al.* An *in vivo* inhibitor of 5-lipoxygenase, MK886, at micromolar concentration induces apoptosis in U937 and CML cells. Anticancer Res 1996; 16(5A): 2589-99.

[392]   Anderson KM, Seed T, Plate JM, Jajeh A, Meng J, Harris JE. Selective inhibitors of 5-lipoxygenase reduce CML blast cell proliferation and induce limited differentiation and apoptosis. Leuk Res 1995; 19(11): 789-801.

[393] Chen Y, Hu Y, Zhang H, Peng C, Li S. Loss of the Alox5 gene impairs leukemia stem cells and prevents chronic myeloid leukemia. Nat Genet 2009; 41(7): 783-92.

[394] Bruns I, Czibere A, Fischer JC, *et al.* The hematopoietic stem cell in chronic phase CML is characterized by a transcriptional profile resembling normal myeloid progenitor cells and reflecting loss of quiescence. Leukemia 2009; 23(5): 892-9.

[395] Peng C, Chen Y, Yang Z, *et al.* PTEN is a tumor suppressor in CML stem cells and BCR-AB- -induced leukemias in mice. Blood 2010; 115(3): 626-35.

[396] Tothova Z, Kollipara R, Huntly BJ, *et al.* FoxOs are critical mediators of hematopoietic stem cell resistance to physiologic oxidative stress. Cell 2007; 128(2): 325-39.

[397] Miyamoto K, Araki KY, Naka K, *et al.* FoxO3a is essential for maintenance of the hematopoietic stem cell pool. Cell Stem Cell 2007; 1(1): 101-12.

[398] Naka K, Hoshii T, Muraguchi T, *et al.* TGF-beta-FOXO signalling maintains leukaemia-initiating cells in chronic myeloid leukaemia. Nature 2010; 463(7281): 676-80.

[399] Pellicano F, Scott MT, Helgason GV, *et al.* The antiproliferative activity of kinase inhibitors in chronic myeloid leukemia cells is mediated by FOXO transcription factors. Stem Cells 2014; 32(9): 2324-37.

[400] Platt N, Gordon S. Is the class A macrophage scavenger receptor (SR-A) multifunctional? - The mouse's tale. J Clin Invest 2001; 108(5): 649-54.

[401] Nikolic DM, Cholewa J, Gass C, Gong MC, Post SR. Class A scavenger receptor-mediated cell adhesion requires the sequential activation of Lyn and PI3-kinase. Am J Physiol Cell Physiol 2007; 292(4): C1450-8.

[402] Miki S, Tsukada S, Nakamura Y, *et al.* Functional and possible physical association of scavenger receptor with cytoplasmic tyrosine kinase Lyn in monocytic THP-1-derived macrophages. FEBS Lett 1996; 399(3): 241-4.

[403] Xu J, Zheng SL, Komiya A, *et al.* Germline mutations and sequence variants of the macrophage scavenger receptor 1 gene are associated with prostate cancer risk. Nat Genet 2002; 32(2): 321-5.

[404] Yuan H, Wang Z, Li L, *et al.* Activation of stress response gene SIRT1 by BCR-ABL promotes leukemogenesis. Blood 2012; 119(8): 1904-14.

[405] Li L, Wang L, Li L, *et al.* Activation of p53 by SIRT1 inhibition enhances elimination of CML leukemia stem cells in combination with imatinib. Cancer Cell 2012; 21(2): 266-81.

[406] Ito K, Bernardi R, Morotti A, *et al.* PML targeting eradicates quiescent leukaemia-initiating cells. Nature 2008; 453(7198): 1072-8.

[407] Nowicki MO, Falinski R, Koptyra M, *et al.* BCR/ABL oncogenic kinase promotes unfaithful repair of the reactive oxygen species-dependent DNA double-strand breaks. Blood 2004; 104(12): 3746-53.

[408] Koptyra M, Cramer K, Slupianek A, Richardson C, Skorski T. BCR/ABL promotes accumulation of chromosomal aberrations induced by oxidative and genotoxic stress. Leukemia 2008; 22(10): 1969-72.

[409] Nieborowska-Skorska M, Kopinski PK, Ray R, *et al.* Rac2-MRC-cIII-generated ROS cause genomic instability in chronic myeloid leukemia stem cells and primitive progenitors. Blood 2012; 119(18): 4253-63.

[410] Bolton-Gillespie E, Schemionek M, Klein HU, *et al.* Genomic instability may originate from imatinib- refractory chronic myeloid leukemia stem cells. Blood 2013; 121(20): 4175-83.

[411] Nieborowska-Skorska M, Flis S, Skorski T. Chronic Myeloid Leukemia Stem Cells (LSCs) and Leukemia Progenitor Cells (LPCs) Display Overlapping and Unique Mechanisms of Genomic Instability: The Role of PI3k-AKT and PI3k-Rac2-PAK pathway. Poster Abstracts n.1791 ASCO 2014 December;

[412] Jørgensen HG, Copland M, Allan EK, *et al.* Intermittent exposure of primitive quiescent chronic myeloid leukemia cells to granulocyte-colony stimulating factor *in vitro* promotes their elimination by imatinib mesylate. Clin Cancer Res 2006; 12(2): 626-33.

[413] Drummond MW, Heaney N, Kaeda J, *et al.* A pilot study of continuous imatinib *vs* pulsed imatinib with or without G-CSF in CML patients who have achieved a complete cytogenetic response. Leukemia 2009; 23(6): 1199-201.

[414] Hassane DC, Guzman ML, Corbett C, *et al.* Discovery of agents that eradicate leukemia stem cells using an in silico screen of public gene expression data. Blood 2008; 111(12): 5654-62.

[415]   Manne V, Lee FY, Bol DK, *et al.* Apoptotic and cytostatic farnesyltransferase inhibitors have distinct pharmacology and efficacy profiles in tumor models. Cancer Res 2004; 64(11): 3974-80.

[416]   Copland M, Pellicano F, Richmond L, *et al.* BMS-214662 potently induces apoptosis of chronic myeloid leukemia stem and progenitor cells and synergizes with tyrosine kinase inhibitors. Blood 2008; 111(5): 2843-53.

[417]   Pellicano F, Copland M, Jorgensen HG, Mountford J, Leber B, Holyoake TL. BMS-214662 induces mitochondrial apoptosis in chronic myeloid leukemia (CML) stem/progenitor cells, including CD34+38- cells, through activation of protein kinase Cbeta. Blood 2009; 114(19): 4186-96.

[418]   Neviani P, Santhanam R, Trotta R, *et al.* The tumor suppressor PP2A is functionally inactivated in blast crisis CML through the inhibitory activity of the BCR/ABL-regulated SET protein. Cancer Cell 2005; 8(5): 355-68.

[419]   Magill L, Lynas J, Morris TC, Walker B, Irvine AE. Proteasome proteolytic activity in hematopoietic cells from patients with chronic myeloid leukemia and multiple myeloma. Haematologica 2004; 89(12): 1428-33.

[420]   Cortes J, Thomas D, Koller C, *et al.* Phase I study of bortezomib in refractory or relapsed acute leukemias. Clin Cancer Res 2004; 10(10): 3371-6.

[421]   Takimoto CH, Calvo E. Principles of Oncologic Pharmacotherapy In: Pazdur R, Wagman LD, Camphausen KA, Hoskins WJ, Eds. Cancer Management: A Multidisciplinary Approach (11 ed.), 2008.

[422]   Gatto S, Scappini B, Pham L, *et al.* The proteasome inhibitor PS-341 inhibits growth and induces apoptosis in Bcr/Abl-positive cell lines sensitive and resistant to imatinib mesylate. Haematologica 2003; 88(8): 853-63.

[423]   Heaney NB, Pellicano F, Zhang B, *et al.* Bortezomib induces apoptosis in primitive chronic myeloid leukemia cells including LTC-IC and NOD/SCID repopulating cells. Blood 2010; 115(11): 2241-50.

[424]   Heisterkamp N, Jenster G, ten Hoeve J, Zovich D, Pattengale PK, Groffen J. Acute leukaemia in bcr/abl transgenic mice. Nature 1990; 344(6263): 251-3.

[425]   Krause DS, Van Etten RA. Bedside to bench: interfering with leukemic stem cells. Nat Med 2008; 14(5): 494-5.

[426]   Rousselot P, Huguet F, Rea D, *et al.* Imatinib mesylate discontinuation in patients with chronic myelogenous leukemia in complete molecular remission for more than 2 years. Blood 2007; 109(1): 58-60.

[427]   Moriuchi M, Moriuchi H, Turner W, Fauci AS. Cloning and analysis of the promoter region of CXCR4, a coreceptor for HIV-1 entry. J Immunol 1997; 159(9): 4322-9.

[428]   Caruz A, Samsom M, Alonso JM, *et al.* Genomic organization and promoter characterization of human CXCR4 gene. FEBS Lett 1998; 426(2): 271-8.

[429]   Tamamis P, Floudas CA. Elucidating a key component of cancer metastasis: CXCL12 (SDF-1α) binding to CXCR4. J Chem Inf Model 2014; 54(4): 1174-88.

[430]   Greenbaum A, Hsu YM, Day RB, *et al.* CXCL12 in early mesenchymal progenitors is required for haematopoietic stem-cell maintenance. Nature 2013; 495(7440): 227-30.

[431]   Simanovsky M, Berlinsky S, Sinai P, Leiba M, Nagler A, Galski H. Phenotypic and gene expression diversity of malignant cells in human blast crisis chronic myeloid leukemia. Differentiation 2008; 76(8): 908-22.

[432]   Zhang B, Li M, McDonald T, *et al.* Microenvironmental protection of CML stem and progenitor cells from tyrosine kinase inhibitors through N-cadherin and Wnt-beta-catenin signaling. Blood 2013; 121(10): 1824-38.

[433]   Puissant A, Dufies M, Fenouille N, *et al.* Imatinib triggers mesenchymal-like conversion of CML cells associated with increased aggressiveness. J Mol Cell Biol 2012; 4(4): 207-20.

[434]   Peled A, Hardan I, Trakhtenbrot L, *et al.* Immature leukemic CD34+CXCR4+ cells from CML patients have lower integrin-dependent migration and adhesion in response to the chemokine SDF-1. Stem Cells 2002; 20(3): 259-66.

[435]   Ptasznik A, Urbanowska E, Chinta S, *et al.* Crosstalk between BCR/ABL oncoprotein and CXCR4 signaling through a Src family kinase in human leukemia cells. J Exp Med 2002; 196(5): 667-78.

[436]   Geay JF, Buet D, Zhang Y, *et al.* p210BCR-ABL inhibits SDF-1 chemotactic response *via* alteration of CXCR4 signaling and down-regulation of CXCR4 expression. Cancer Res 2005; 65(7): 2676-83.

[437]   Jin L, Tabe Y, Konoplev S, *et al.* CXCR4 up-regulation by imatinib induces chronic myelogenous leukemia (CML) cell migration to bone marrow stroma and promotes survival of quiescent CML cells. Mol Cancer Ther 2008; 7(1): 48-58.

[438]   Tabe Y, Jin L, Iwabuchi K, *et al.* Role of stromal microenvironment in nonpharmacological resistance of CML to imatinib through Lyn/CXCR4 interactions in lipid rafts. Leukemia 2012; 26(5): 883-92.

[439]   Chen YY, Malik M, Tomkowicz BE, Collman RG, Ptasznik A. BCR-ABL1 alters SDF-1alph- -mediated adhesive responses through the beta2 integrin LFA-1 in leukemia cells. Blood 2008; 111(10): 5182-6.

[440]   Vianello F, Villanova F, Tisato V, *et al.* Bone marrow mesenchymal stromal cells non-selectively protect chronic myeloid leukemia cells from imatinib-induced apoptosis *via* the CXCR4/CXCL12 axis. Haematologica 2010; 95(7): 1081-9.

[441]   Weisberg E, Azab AK, Manley PW, *et al.* Inhibition of CXCR4 in CML cells disrupts their interaction with the bone marrow microenvironment and sensitizes them to nilotinib. Leukemia 2012; 26(5): 985- 90.

[442]   Beider K, Darash-Yahana M, Blaier O, *et al.* Combination of imatinib with CXCR4 antagonist BKT140 overcomes the protective effect of stroma and targets CML *in vitro* and *in vivo.* Mol Cancer Ther 2014; 13(5): 1155-69.

[443]   Abraham M, Biyder K, Begin M, *et al.* Enhanced unique pattern of hematopoietic cell mobilization induced by the CXCR4 antagonist 4F-benzoyl-TN14003. Stem Cells 2007; 25(9): 2158-66.

[444]   Pellicano F, Mukherjee L, Holyoake TL. Concise review: cancer cells escape from oncogene addiction: understanding the mechanisms behind treatment failure for more effective targeting. Stem Cells 2014; 32(6): 1373-9.

[445]   Duy C, Hurtz C, Shojaee S, *et al.* BCL6 enables Ph+ acute lymphoblastic leukaemia cells to survive BCR-ABL1 kinase inhibition. Nature 2011; 473(7347): 384-8.

[446]   Mohyeldin A, Garzón-Muvdi T, Quiñones-Hinojosa A. Oxygen in stem cell biology: a critical component of the stem cell niche. Cell Stem Cell 2010; 7(2): 150-61.

[447]   Eliasson P, Jönsson JI. The hematopoietic stem cell niche: low in oxygen but a nice place to be. J Cell Physiol 2010; 222(1): 17-22.

[448]   Eliasson P, Rehn M, Hammar P, *et al.* Hypoxia mediates low cell-cycle activity and increases the   proportion of long-term-reconstituting hematopoietic stem cells during *in vitro* culture. Exp Hematol 2010; 38(4): 301-10. e2

[449]   Takubo K, Goda N, Yamada W, *et al.* Regulation of the HIF-1alpha level is essential for hematopoietic stem cells. Cell Stem Cell 2010; 7(3): 391-402.

[450]   Simon MC, Keith B. The role of oxygen availability in embryonic development and stem cell function. Nat Rev Mol Cell Biol 2008; 9(4): 285-96.

[451]   Semenza GL. Targeting HIF-1 for cancer therapy. Nat Rev Cancer 2003; 3(10): 721-32.

[452]   Keith B, Simon MC. Hypoxia-inducible factors, stem cells, and cancer. Cell 2007; 129(3): 465-72.

[453]   McBrien KD, Berry RL, Lowe SE, *et al.* Rakicidins, new cytotoxic lipopeptides from Micromonospora sp. fermentation, isolation and characterization. J Antibiot 1995; 48(12): 1446-52.

[454]   Takeuchi M, Ashihara E, Yamazaki Y, *et al.* Rakicidin A effectively induces apoptosis in hypoxia adapted Bcr-Abl positive leukemic cells. Cancer Sci 2011; 102(3): 591-6.

[455]   White E. Deconvoluting the context-dependent role for autophagy in cancer. Nat Rev Cancer 2012; 12(6): 401-10.

[456]   Cuervo AM. Autophagy: in sickness and in health. Trends Cell Biol 2004; 14(2): 70-7.

[457]   Guo JY, Chen HY, Mathew R, *et al.* Activated Ras requires autophagy to maintain oxidative metabolism and tumorigenesis. Genes Dev 2011; 25(5): 460-70.

[458]   Lock R, Roy S, Kenific CM, *et al.* Autophagy facilitates glycolysis during Ras-mediated oncogenic transformation. Mol Biol Cell 2011; 22(2): 165-78.

[459]   Yang S, Wang X, Contino G, *et al.* Pancreatic cancers require autophagy for tumor growth. Genes Dev 2011; 25(7): 717-29.

[460]   Rubinsztein DC, Gestwicki JE, Murphy LO, Klionsky DJ. Potential therapeutic applications of autophagy. Nat Rev Drug Discov 2007; 6(4): 304-12.

[461]   Nencioni A, Cea M, Montecucco F, *et al.* Autophagy in blood cancers: biological role and therapeutic implications. Haematologica 2013; 98(9): 1335-43.

[462]   Helgason GV, Karvela M, Holyoake TL. Kill one bird with two stones: potential efficacy of BCR- ABL and autophagy inhibition in CML. Blood 2011; 118(8): 2035-43.

[463]   Ertmer A, Huber V, Gilch S, *et al.* The anticancer drug imatinib induces cellular autophagy. Leukemia 2007; 21(5): 936-42.

[464]   Altman BJ, Jacobs SR, Mason EF, *et al.* Autophagy is essential to suppress cell stress and to allow BCR-Abl-mediated leukemogenesis. Oncogene 2011; 30(16): 1855-67.

[465]   Apperley JF. Chronic myeloid leukaemia. Lancet 2015 Apr 11; 385(9976): 1447-59. Epub 2014 Dec 5.

[466]   Sobrinho-Simões M, David M, Ross DM, Junia V, Melo VJ. Chronic myeloid leukemia: the possibility of a cure. Advances in the Treatment of Chronic Myeloid Leukemia 2013 July; 114-28.

[467]   Kamitsuji Y, Kuroda J, Kimura S, *et al.* The Bcr-Abl kinase inhibitor INNO-406 induces autophagy and different modes of cell death execution in Bcr-Abl-positive leukemias. Cell Death Differ 2008; 15(11): 1712-22.

[468]   Mishima Y, Terui Y, Mishima Y, *et al.* Autophagy and autophagic cell death are next targets for elimination of the resistance to tyrosine kinase inhibitors. Cancer Sci 2008; 99(11): 2200-8.

[469]   Sheng Z, Ma L, Sun JE, Zhu LJ, Green MR. BCR-ABL suppresses autophagy through ATF5-mediated regulation of mTOR transcription. Blood 2011; 118(10): 2840-8.

[470]   Yu Y, Yang L, Zhao M, *et al.* Targeting microRNA-30a-mediated autophagy enhances imatinib activity against human chronic myeloid leukemia cells. Leukemia 2012; 26(8): 1752-60.

[471]   Carayol N, Vakana E, Sassano A, *et al.* Critical roles for mTORC2- and rapamycin-insensitive mTORC1-complexes in growth and survival of BCR-ABL-expressing leukemic cells. Proc Natl Acad Sci USA 2010; 107(28): 12469-74.

[472]   Crowley LC, Elzinga BM, O'Sullivan GC, McKenna SL. Autophagy induction by Bcr-Abl-expressing cells facilitates their recovery from a targeted or nontargeted treatment. Am J Hematol 2011; 86(1): 38-47.

[473]   Stoklosa T, Glodkowska-Mrowka E, Hoser G, Kielak M, Seferynska I, Wlodarski P. Diverse mechanisms of mTOR activation in chronic and blastic phase of chronic myelogenous leukemia. Exp Hematol 2013; 41(5): 462-9.

[474]   Carella AM, Beltrami G, Pica G, Carella A, Catania G. Clarithromycin potentiates tyrosine kinase inhibitor treatment in patients with resistant chronic myeloid leukemia. Leuk Lymphoma 2012; 53(7): 1409-11.

[475]   Hamad A, Sahli Z, El Sabban M, Mouteirik M, Nasr R. Emerging therapeutic strategies for targeting chronic myeloid leukemia stem cells Stem Cells Int 2013; 2013: 724360.

[476]   Puissant A, Robert G, Fenouille N, *et al.* Resveratrol promotes autophagic cell death in chronic myelogenous leukemia cells *via* JNK-mediated p62/SQSTM1 expression and AMPK activation. Cancer Res 2010; 70(3): 1042-52.

[477]   Goussetis DJ, Gounaris E, Platanias LC. BCR-ABL1-induced leukemogenesis and autophagic targeting by arsenic trioxide. Autophagy 2013; 9(1): 93-4.

[478]   Goussetis DJ, Gounaris E, Wu EJ, *et al.* Autophagic degradation of the BCR-ABL oncoprotein and generation of antileukemic responses by arsenic trioxide. Blood 2012; 120(17): 3555-62.

[479]   Ciarcia R, Damiano S, Montagnaro S, *et al.* Combined effects of PI3K and SRC kinase inhibitors with imatinib on intracellular calcium levels, autophagy, and apoptosis in CML-PBL cells. Cell Cycle 2013; 12(17): 2839-48.

[480]   Fan J, Dong X, Zhang W, *et al.* Tyrosine kinase inhibitor Thiotanib targets Bcr-Abl and induces apoptosis and autophagy in human chronic myeloid leukemia cells. Appl Microbiol Biotechnol 2014; 98(23): 9763-75.

[481]   Townsend KN, Hughson LR, Schlie K, Poon VI, Westerback A, Lum JJ. Autophagy inhibition in cancer therapy: metabolic considerations for antitumor immunity. Immunol Rev 2012; 249(1): 176-94.

[482]   Blommaart EF, Krause U, Schellens JP, Vreeling-Sindelárová H, Meijer AJ. The phosphatidylinositol 3-kinase inhibitors wortmannin and LY294002 inhibit autophagy in isolated rat hepatocytes. Eur J Biochem 1997; 243(1-2): 240-6.

[483] Caro LH, Plomp PJ, Wolvetang EJ, Kerkhof C, Meijer AJ. 3-Methyladenine, an inhibitor of autophagy, has multiple effects on metabolism. Eur J Biochem 1988; 175(2): 325-9.

[484] Köchl R, Hu XW, Chan EY, Tooze SA. Microtubules facilitate autophagosome formation and fusion of autophagosomes with endosomes. Traffic 2006; 7(2): 129-45.

[485] Tanida I, Minematsu-Ikeguchi N, Ueno T, Kominami E. Lysosomal turnover, but not a cellular level, of endogenous LC3 is a marker for autophagy. Autophagy 2005; 1(2): 84-91.

[486] Klionsky DJ, Abeliovich H, Agostinis P, *et al.* Guidelines for the use and interpretation of assays for monitoring autophagy in higher eukaryotes. Autophagy 2008; 4(2): 151-75.

[487] Bellodi C, Lidonnici MR, Hamilton A, *et al.* Targeting autophagy potentiates tyrosine kinase inhibitor-induced cell death in Philadelphia chromosome-positive cells, including primary CML stem cells. J Clin Invest 2009; 119(5): 1109-23.

CHAPTER 2

Bispecific and Multivalent Antibodies – New Swords to Combat Hematological Malignancies

Michael Stanglmaier* and **Juergen Hess**

Lerchenstraße 3, D-85368 Moosburg, Germany

Abstract: With about 40 approved monoclonal antibodies these biologicals are of high significance especially for cancer treatment. Nevertheless, since anti-tumor responses to antibody therapies are often limited and a lot of patients finally relapse, much effort is spent to increase the therapeutic efficacy, *e.g.,* by optimizing the Fc-part of antibodies, thereby developing second generation antibodies. Albeit, monoclonal antibodies, even second generation formats are limited inasmuch as immune effector cells could only be unspecifically redirected *via* Fc-receptor binding and even more important, T cells as the most potent immune effector cells cannot be addressed at all. Thus, the idea arose to create antibodies with two different specificities. This enables specifically redirecting selected immune effector cells to target cells like tumor cells. A lot of different bispecific and multispecific antibody formats and platforms were designed to circumvent these constraints, like diabodies, single chain diabodies, chemically crosslinked F(ab)s fragments, tandem single chain Fv fragments, bispecific T cell engagers (BiTEs), triplebodies or knob in a hole as well as the dock and lock platforms. Despite major efforts in producing and testing new bsAb formats, the first approved bispecific antibody was the quadroma produced trifunctional antibody catumaxomab (anti-EpCAM x anti-CD3), demonstrating the eminent therapeutic potency of this full length antibody format. The recent approval of the anti-CD3 x anti-CD19 BiTE blinatumomab by the FDA furthermore emphasized the eminent role of bispecific antibodies. In this chapter an overview of the current development of bi- and multispecific antibodies for treatment of B-cell malignancies will be presented.

Keywords: Bispecific antibodies, B cell malignancies, antibody formats, cancer therapy, T-cell binding.

INTRODUCTION: MONOCLONAL ANTIBODY DEVELOPMENT

Monoclonal antibodies (mAb) are widely used in tumor therapy. With about 40 antibodies or immunoglobulin derivates approved in the United States or the European Union this class of molecules is of high significance as novel anti-tumor pharmaceuticals [1, 2]. Thereof, 19 antibodies or derivates are approved for

---

*Corresponding author Michael Stanglmaier: Lerchenstraße 3, D-85368 Moosburg, Germany; Tel: ++49 8761 66577; Fax: ++49 8761 759251; E-mail: info@michael-stanglmaier.de

Atta-ur-Rahman (Ed)

treatment of a variety of tumors. Interestingly, albeit lymphoma and leukemia account for only about 10% of all tumors, 8 out of this 18 approved anti-tumor antibodies are directed against hematological malignancies. This fact strongly emphasizes the importance of leukemias and lymphomas in the field of antibody anti-tumor drug development.

Nevertheless, since anti-tumor responses to antibody therapies are often limited and a lot of patients finally relapse [3], many efforts are spent to improve the therapeutic efficacy, *e.g.,* by optimizing the Fc-part of antibodies [4]. Already shortly after the engineering of mAbs the idea arose to create antibodies with two specificities. This concept allows to specifically re-direct selected immune effector cells to target cells like tumor cells. Selected effector cell populations bound by these antibodies could be natural killer (NK) cells (via CD16), FcγRI$^+$ cells like macrophages, dendritic cells (via CD64) or T cells (via CD3) [5]. Already eight years after the invention of the hybridoma technology for producing mAbs, Milstein and Cuella in 1983 established a technique that allowed the production of bispecific antibodies (bsAb). Fusing hybridoma cells that secrete anti-somatostatin antibodies with another hybridoma cell secreting anti-peroxidase antibodies results in quadroma cells that produce bispecific IgG-shaped antibodies [6, 7]. Shortly thereafter, Staerz and colleagues created a bispecific antibody by sodium dodecyl sulfate (SDS) mediated coupling of two antibodies directed against the T-cell receptor and the Thy-1 antigen, respectively [8]. This success then paved the way towards targeted use of T cells as the most potent anti-tumor effector cell of the immune system. Antibody-mediated re-directing of T cells to the tumor would allow a major histocompatibility complex (MHC) class I and T-cell-receptor independent tumor cell destruction. However, despite promising *in vitro* results, bsAbs in early clinical trials showed only limited success. One obstacle was the difficulty to produce these molecules in sufficiently high amounts and purity. On the other hand, tumor cell destruction, even in *in vitro* assays, relies on pre- and/or co-stimulation, *e.g., via* targeting CD28 or CD40 or addition of exogenous IL-2 [9, 10]. Therefore, a lot of different bispecific antibody formats were designed to circumvent these constraints, like diabodies, single chain diabodies, chemically crosslinked F(ab)s fragments or tandem single chain Fv-fragments [11].

Bispecific T-cell Engagers (BiTEs)

One of these new bispecific formats are bispecific T-cell engaging (BiTE) molecules consisting of a tandem array of two single chain variable fragments (scFv) domains connected by a non-immunogenic Glycin-Serin (Gly$_4$Ser)$_n$-linker, that allows rotational flexibility, targeting tumor associated antigens (TAA) and CD3 on T cells. BiTEs have the following domain order (Fig. **1A**): V$_L$TAA-V$_H$TAA-V$_H$CD3-V$_L$CD3 and display a molecular mass of around 55 kDa [12].

A

D

B

C

E

F

G

H

I

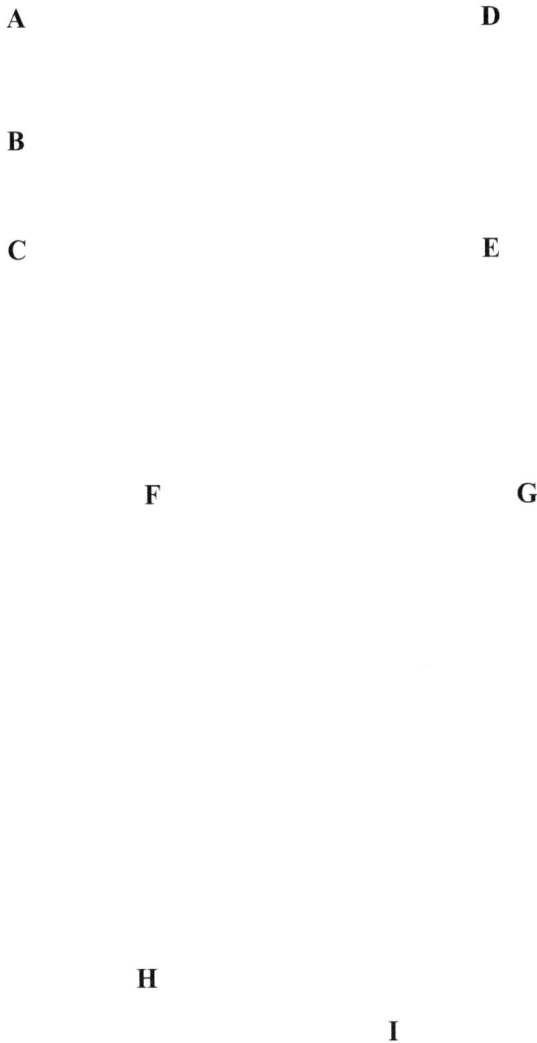

Figure 1: Schematic representation of bispecific antibody formats. **A**: BiTE; N-terminal domains target the TAA, C-terminal domains target CD3. **B**: single chain triplebodies (sctb); **C**: TandAb; **D**: bispecific immunotoxin DT2219ARL, **E**: DART; **F**: trifunctional antibody (trAb); **G**: hexavalent dock and lock antibody; **H**: bispecific Xencor XmAbs; **I**: Tribody, upper chain represents the heavy chain construct, lower chain the light chain construct. V_{L1} denotes a variable light chain domain of an antibody 1, V_{H2} a variable heavy chain domain of an antibody 2.

The most advanced BiTE antibody is the anti-CD19 x anti-CD3 BiTE blinatumomab (MT103), derived from a B-lineage monoclonal mouse antibody, for the treatment of B-cell lymphomas and leukemias. CD19 is a well characterized B-lineage-restricted antigen and represents the earliest expressed antigen on the surface of B-lymphocytes but not on hematopoietic stem cells and plasma cells, making it an appropriate target for bsAb-mediated redirection of T cells to tumor B cells [13, 14].

B-cell binding affinity is in the range of $10^{-8} - 10^{-9}$ M, while affinity to T cells is one to two magnitudes of order lower with a range of $10^{-7} - 10^{-8}$ M [15]. Blinatumomab showed a high cytotoxicity against tumor B cells using tumor B-cell lines and primary CLL patient samples in *in vitro* systems without any pre- or co-stimulation [12]. This antibody exerts its cytotoxic activity already in the pico- and femtomolar concentration range [16]. Cytotoxicity is mediated by CD4 as well as CD8 T cells, albeit the fastest target cell killing is mediated by CD8 T cells. Target B-cell lysis was shown to be dependent on the perforin-granzyme B pathway, but is not dependent on death ligands like FasL or TRAIL or TNFα [17]. Blinatumomab mediates its cytotoxicity by retargeting T cells to the tumor and building an immunological synapse between T cell and tumor target cell [18]. Of note, blinatumomab-mediated cytotoxicity is independent of major histocompatibility complex (MHC) class I/peptide/T-cell receptor complex and thus allowing redirection of CD4 and CD8 T cells independent of antigen presentation.

T-cell mediated cytotoxicity is accompanied by activation, as measured by up-regulation of activation markers CD25 and CD69 and proliferation of T cells as well as secretion of pro-inflammatory cytokines as IL-2, IL-6, IFNγ and TNFα and inhibitory cytokines like IL-4 and IL-10 [12, 19]. Video-assisted microscopy revealed that blinatumomab alters the motility and activity of T cells and mediates serial killing of several target cells by a single cytotoxic T cell [20], explaining how BiTEs can induce complete target cell lysis without pre- or co-stimulation even at very low effector: target ratios (E:T) of 50:1 and low antibody concentrations of 5 ng/ml treating primary chronic lymphocytic leukemia (CLL) patient cells for 6 days [21]. In the presence of IL-2, activity of the anti-CD19 x anti-CD3 BiTE was further enhanced.

Blinatumomab was additionally evaluated in a NOD/SCID mouse model with subcutaneously (S.C.) or intravenously (I.V.) inoculated tumor B-cell line and healthy donor peripheral blood mononuclear cells (PBMCs). Blinatumomab efficiently suppressed tumor cell growth in the S.C. xenograft model in a dose dependent fashion at low bsAb concentrations (5 x 0.1 μg). Furthermore, this BiTE prolonged survival of I.V. inoculated mice in a dose dependent manner [22]. This BiTE molecule has been evaluated in clinical studies with non-Hodgkin's lymphoma (NHL) and acute lymphoplastic leukemia (ALL) patients displaying impressive response rates in both malignancies.

Clinical development of blinatumomab started with three dose escalating phase I studies including 22 patients with refractory or relapsed NHL or CLL [21 x NHL, 1 x CLL] with conventional short term I.V. infusions from 0.75 to 13 $\mu g/m^2$ three times a week [23]. Most common adverse events (AEs) observed were pyrexia, rigor and fatigue and were mild to moderate. AEs mechanistically seemed to be related to polyclonal T-cell activation and were accompanied by an early and transient increase in inflammatory cytokines. Of note, several neurologic events like aphasia, ataxia, disorientation and seizure occurred that lead to drug discontinuation in 12 patients (55%). These events were reported to be fully reversible and might be due to transmigration across the blood-brain-barrier. Blinatumomab-mediated redirection of T cells to $CD19^+$ cells in the central nervous system (CNS) and thus activation of T cells might cause local cytokine secretion and inflammatory event in the CNS. Noteworthy, despite the impressive potency of blinatumomab in *in vitro* cytotoxicity assays and animal models no objective responses were observed in these studies, probably due to the short serum half-life of about 2 hours in humans. Because of the unfavorable risk/benefit ratio these studies were terminated early.

To circumvent the problems emerging in the first clinical evaluations blinatumomab was administered in the following studies by continuous infusions using a portable pump over several weeks to ensure a steady and prolonged exposure.

A phase I-dose escalation study in patients with refractory or relapsed NHL was conducted with seven dose escalation steps within a dose range of 0.5 up to 90 $\mu g/m^2$/day with continuous infusion for 4 to 8 weeks [24] (Table **1**, NCT00274742). This corresponds to a whole body dose of 14 $\mu g/m^2$ up to 5,040 $\mu g/m^2$. Taking into account the low molecular weight of BiTEs this normalized amount corresponds to 40 to 13,800 $\mu g/m^2$ for a normal IgG antibody. This dose equivalence is magnitudes of orders lower than the dose range of the anti-CD20 IgG-shaped but chimeric antibody rituximab (375,000 to 500,000 $\mu g/m^2$ per dose) demonstrating the high potency of this class of bsAbs compared to monovalent mAbs. As already stated due to the non-beneficial serum half-life BiTE antibodies

Table 1. Clinical studies performed with bsAbs in hematological malignancies currently ongoing or recently closed registered at clinicaltrials.gov. The corresponding homepage was screened on January, 8ᵗʰ 2015.

Drug	Phase	Patients	Pat. Number	Start	End	Status	Ident.
FBTA05	I/II	r/r NHL/CLL	30 (12)	2010	2014	recruiting	NCT01138579
MT103	I	rel NHL	76	2004	2012	completed	NCT00274742
MT103	II	B pre ALL with MRD	21	2007	2014	completed	NCT00560794
MT103	II	r/r B pre ALL	36	2010	2016	ongoing, not recruiting	NCT01209286
MT103	II	B pre ALL with MRD	130	2010	2016	ongoing, not recruiting	NCT01207388
MT103	II	r/r B pre ALL	225	2011	2017	ongoing, not recruiting	NCT01466179
MT103	II	r/r DLBCL	25	2012	2016	ongoing, not recruiting	NCT01741792
MT103	I/II	ped/adol r/r B pre ALL	84	2012	2016	ongoing, not recruiting	NCT01471782
MT103	-	ped/adol r/r B pre ALL	-	-	-	expanded access	NCT02187354
MT103	II	Ph⁺/BCR-ABL⁺ ALL	41	2013	2017	recruiting	NCT02000427
MT103	III	r/r B pre ALL	400	2013	2017	recruiting	NCT02013167
MT103	III	newly diagnosed BCR-Abl⁻ ALL	360	2913	2018	recruiting	NCT02003222
MT103	III	younger patients with rel. ALL,	598	2014	2018	recruiting	NCT02101853
MT103	II	older patients with newly diagnosed ALL	50	2014	2019	not yet open for recruitment	NCT02143414
CD20 Bi	I	hr NHL	24	2004	2012	unknown	NCT00244946
CD20 Bi	I	MM	12	2009	2013	completed	NCT00938626
CD20 Bi	I	hr NHL	0	2007	-	withdrawn	NCT00521261
AFM11	I	r/r NHL and B-pre ALL	40	2014	2016	recruiting	NCT02106091
AFM13	I	r/r HL	28	2010	2013	completed	NCT01221571
AFM13	II	r/r HL	39	2015	2017	not yet recruiting	NCT02321592
DT2219ARL	I	CD19⁺/CD22⁺ lymphoma/leukemia	30	2009	2014	completed	NCT00889408
MGD006	I	r/r AML	58	2014	2017	recruiting	NCT02152956

Abbreviations: adol: adolescent; hr: high risk; ped: pediatric; pre: precursor; r/r: relapsed/refractory; ALL: acute lymphoblastic leukemia; AML: acute myeloid leukemia; CLL: chronic lymphocytic leukemia; MM: Multiple Myeloma, NHL: Non Hodgkin's lymphoma; Ident: Identifier ClinicalTrials.gov.

like MT103 (blinatumomab) had to be administered in a continuous I.V. infusion over 4 to 8 weeks. A total of 76 patients were included in this clinical study between 2004 and 2012 [23] Table **1**. Initially only indolent and mantle cell lymphoma (MCL) were included. In addition, patients with the highly aggressive diffuse large B-cell lymphoma (DLBCL) were included later resulting in the following histological subtypes of lymphomas: follicular lymphoma (FL, 37%), mantle cell (32%), DLBCL (18%) and other indolent lymphomas (13%). All patients were heavily pretreated with a median of three prior regimens.

Blinatumomab induced response correlates with antibody dose. First responses consisting of stable diseases (SD) were observed already at dose levels of 1.5 $\mu g/m^2/day$ in four out of six patients and at 5 $\mu g/m^2/day$ in two of three patients [24]. Two short partial (PR) and one complete responses (CR) could be observed at a dose level of 15 $\mu g/m^2/day$ (n=15 patients). Twenty out of 28 patients exposed to 60 $\mu g/m^2/day$ showed an objective response (71%) [25] with four out of 12 patients with follicular lymphoma had a complete remission (33%) and six displayed a partial remission (50%). Patients suffering from aggressive MCL performed comparable well with an overall response rate of 71% (5/7 patients), three patients obtained a complete remission (43%) and two out of seven a partial remission (29%). Eight out of nine patients with relapsed follicular or mantle cell lymphoma showed an objective response (3 x CR, 5 x PR) at a dose of 60 $\mu g/m^2/day$ with a response duration of more than 2 years in 4 patients [26, 27].

A rapid and long-lasting eradication of B cells from peripheral blood could be observed during blinatumomab treatment. Decrease of tumor B-cell count and simultaneous increase of the apoptosis marker annexin-V showed that B-cell reduction was not caused by relocation but by lysis of tumor cells. Anti-tumor activity could be detected not only in the peripheral blood, but also in lymph nodes, liver and also in the bone marrow at doses of 15 $\mu g/m^2/day$ and higher. All patients responding to treatment nearly showed decrease of tumor burden within the first 4 weeks [24].

A short initial disappearance of peripheral T cells was observed. T cells returned within several days displaying a strong inter-patient difference. In several patients T-cell number exceeded baseline level several fold, accompanied by up-regulation of activation markers CD25, CC69 and HLA-DR. T-cell expansion was mainly due to $CD4^+$ and $CD8^+$ effector memory T cells with CD45RA and CCR7 negative phenotype.

The most common AEs observed were pyrexia (75%), headache (45%) and fatigue (37%). The most common laboratory abnormality AEs regardless of causality were lymphopenia (75%), leukopenia (57%), thrombocytopenia (39%),

C-reactive protein increase (53%) and fibrin-D dimer increase (37%). The medically most important AEs that resulted in permanent discontinuation of blinatumomab infusions were CNS events. Signs and symptoms observed included kinetic tremor, speech impairment, disorientation, apraxia and seizure. All CNS events were reported to be fully reversible during or shortly after discontinuation of treatment and without sequelae. Nine out of the 52 patients had to discontinue treatment permanently in the first cycle due to these CNS events. At a dose of 90 µg/m^2/day two DLTs consisting of CNS events were observed during the dose limiting toxicity (DLT) period of the first 2 weeks of treatment. Therefore, 60 µg/m^2/day was recommended as maximum tolerated dose (MTD) for further blinatumomab treatment. Adverse event of grade 3 and 4 occurred in 36 patients (95%) with the most frequent being lymphopenia (68%), increased C-reactive protein (34%), leucopenia (24%), neutropenia and thrombocytopenia (each 16%) [24]. No cytokine release syndrome was observed in 39 patients as reported by Bargou and colleagues. An initial transient elevation of the anti-inflammatory cytokine IL-10 was observed in 25 patients (64%), of whom 19 also showed an increase in IL-6 or IFNγ.

Due to the promising responses in NHL patients especially with follicular and mantle cell lymphoma this study was subsequently amended to include patients with relapsed or refractory DLBCL [23]. These patients had already received a median of 4 previous regimens, all had been treated with rituximab, and eight had undergone autologous stem cell transplantation. Thirteen patients were treated by 4 to 8 weeks continuous I.V. infusions with a stepwise dosing receiving 5 µg/m^2/day within the first week, 15 µg/m^2/day within the second week and 60 µg/m^2/day within the remaining treatment period. Patients were stratified in two cohorts with different dosing and schedule regarding corticosteroid treatment (prednisolone *versus* dexamethasone) to mitigate AEs. Two patients were not evaluable due to discontinued infusion because of reversible CNS events that were qualified as DLT [27]. Six of the 11 evaluable patients responded (ORR 55%) and two patients displayed stable disease. Thereof, four patients (37%) received a complete remission with a median duration of response of 7.8 months. The most common clinical adverse events were pyrexia (82%), fatigue (55%), constipation, headache, tremor and weight increase (each 36%).

Treatment of DLBCL patients with blinatumomab was further investigated in a phase II study (NCT01741792) with 25 patients refractory to treatment or relapsed with a median range of three prior treatments. Patients received 8 weeks continuous I.V. infusions, comparing stepwise with flat dosing [23] within 3 cohorts. Patients in cohort 1 (n=9) received 9, 28 and 122 µg/day within week 1, week 2 and remaining treatment, respectively. Patients in cohort 2 (n=2) were treated with 112 µg/day during the complete cycle. In a second stage of the study the stepwise dosing was chosen for cohort 3 (n=14). All patients received prophylactic dexamethasone [28]. Four out of the 25 patients were not evaluable due to early treatment discontinuation. Fourteen out of the 25 patients died (56%). Nine of the 21 evaluable patients displayed an objective response (ORR 43%) with four complete (19%) and five (24%) partial remissions with median response duration of 11.6 months.

Despite prophylactic corticosteroids all patients experienced at least one AE, serious AEs occurred in 23 patients (92%). The observed AEs were mainly tremor (52%), pyrexia (44%), diarrhea, fatigue, edema and pneumonia (each 24%). Twenty-four patients showed grade 3 (96%), and five patients (20%) grade 4 AEs. Of note, 7 patients (28%) displayed grade 3 neurologic AEs, including both patients treated with flat dosing.

Minimal residual disease (MRD) after induction chemotherapy is the most important adverse prognostic factor in ALL patients. Patients often relapse and their clinical outcome even after allogeneic stem cell transplantation (allo-SCT) remains poor. In a phase II study (Table **1**, NCT00560794) 21 MRD$^+$ patients were treated with four week continuous I.V. infusion of 15 µg/m^2/day of blinatumomab and a two weeks infusion free period defining one treatment cycle [29]. A median number of three treatment cycles were applied. ALL patients eligible for the study had to be in complete hematological response (CRh) and either molecular refractory (never achieved MRD, 15 patients) or in molecular relapse (MRD positivity after previous MRD negativity). A MRD response defined as MRD negativity could be induced in 16 patients (80%) already after the first treatment cycle. Even after a long-term follow-up with a median of 33 months the response rate is still as high as 61%. No increased incidence of AEs was observed in subsequent treatment cycles.

After a median observation of 33 months 12 patients were still in complete remission resulting in a relapse free survival (RFS) of 61% [30]. RFS was significantly different between MRD responders with 19.1 months and non-responders with 3.2 months. Nine of the 20 evaluable patients proceeded to allo-SCT with six of the nine patients presenting with a complete remission after a median follow-up of 33 months. Six of the 11 patients without allo-SCT displayed a hematological CR (CRh) with partial recovery of peripheral blood counts after a median follow-up of 31 months. Interestingly, no marked difference regarding long-term disease control was observed between patients who did or did not receive allo-SCT, albeit with a small patient number.

The most common adverse events independent of grade consisted of pyrexia, chills, decrease of blood immunoglobulin and hypokalemia. Most of the patients (81%) developed grade 3 and 4 AEs with lymphopenia representing the most common one with an incidence of 33%.

Blinatumomab induced a broad variety of B- and T-cell responses that where similar to those observed in NHL patients, *e.g.,* a rapid drop of T cells to a nadir within one day after onset of blinatumomab infusion that recovered to baseline after several days and at average doubled within the following weeks of drug infusion. Expanded T cells largely were effector memory T cells of the CD45⁻/CD197⁻ phenotype. No further expansion of T-cell counts could be observed in subsequent treatment cycles. Cytokine release was characterized by a short term increase at the initiation of the first treatment cycle and consisted mainly of IL-10, IL-6, IFNγ and to a lower extent IL-2 and TNFα.

However, the reactions did not differ in clinical responders and non-responders [31]. Despite the mouse origin of blinatumomab, no human-anti-mouse antibodies (HAMA) could be detected in any of the patients. To evaluate recovery of humoral immune response due to blinatumomab induced B-cell elimination, serum immunoglobulins (IgM, IgG, IgA, IgE) were measured in 6 patients not receiving allo-SCT after blinatumomab treatment with a median follow-up of 458 days [32]. The performed analysis suggests that naïve B cells tend to regenerate immediately after antibody administration as indicated by IgM-level recovery. Nevertheless memory B cells and plasma cells may need more time for recovery as mirrored by IgG- and IgA-levels. Thus, blinatumomab might improve complete response duration and overall survival compared to standard treatment protocols, which is now tested in a confirmatory phase II study with an estimated enrollment of 130 ALL patients (NCT01207388).

Beyond that, adult patients with refractory or relapsed ALL have a very unfavorable prognosis with a median survival of only 2 to 8 months. Therefore, blinatumomab was tested in a phase II study in this group of patients (Table **1**, NCT01209286) [33].

Patients enrolled had to have >5% leukemic blast cells in the bone marrow and had to display primary refractory disease or relapse after induction or consolidation therapy or after allo-SCT. Blinatumomab was administered in a 4 week continuous I.V. infusion within three different treatment schedules during the dose-testing phase of the study: cohort 1 (n=7) with 15 µg/m²/day for 28 days, cohort 2a (n=5) with 5 µg/m²/day for 7 days and 15 µg/m²/day for 21 days and cohort 2b (n=6) starting with 5 µg/m²/day, then increasing to 15 µg/m²/day and a final step to 30 µg/m²/day in week 3. Due to the low number of AEs cohort 2a was selected for extending the study with a further enrollment of 18 patients. Patients received 2 cycles and responders were offered another 3 cycles as maintenance therapy.

Patients displayed a high response rate as 25 patients achieved a CR (n=15) or a CRh (n=10). The portion of responders was highest in patients among first salvage therapy (100%) and lowest in patients with relapse after previous allo-SCT (53%). Of note, 22 of the 25 responding patients also achieved MRD negativity. The median relapse free survival was 7.6 months with a median follow-up of 9.7 months with no significant difference between patients achieving CR compared to CRh. Albeit a market difference between these two groups of responders was observed regarding overall survival as patients with CR had an OS of 13.2 months compared to 8.3 months for patients with CRh. Non-responders displayed an OS of 6.5 months. Thirteen of the 25 responders proceeded to allo-SCT during remission, six of them died due to treatment related mortality (46%) and two relapsed. All in all 22 of the 36 patients (61%) died until study cut-off.

Irrespective of grade the most common AEs were pyrexia (81%), fatigue (50%), headache (47%) and tremor (36%) that were mostly transient and developed within the first days of treatment. AEs were largely connected to blinatumomab mode of action. Grade 3 and worse AEs were observed in 27 patients (75%) and primarily consisted of infections (37%) and nervous system and psychiatric disorders (22%). Blinatumomab treatment has to be interrupted because of nervous system or psychiatric disorders in six patients (17%), but all patients could be re-exposed to blinatumomab after clinical resolution of all events. Severe cytokine release syndromes (grade 4) occurred in two patients.

Based on the results from this study, a further single-arm, open-label phase II study was initiated (Table **1**, NCT01466179). So far, 189 patients with Philadelphia-chromosome negative primary refractory or relapsed ALL were enrolled, including relapse within 12 months of first remission, relapse after allo-SCT or no response to or relapse after first salvage therapy. Eligible patients were required to have at least 10% bone marrow (BM) blasts, contrary to only 5% in the previous phase II study. Patients were administered a continuous I.V. infusion with stepwise application of blinatumomab with 9 µg/day for the first seven days and 28 µd/day for the following 21 days every six week up to a maximum of five cycles. 98 patients received at least 2 cycles and 43 patients at least 3 cycles [34]. To avoid the occurrence of a severe cytokine release syndrome patients with bone marrow blasts equal to or exceeding 50% or peripheral blood blast above 15,000 cells/µl received prephase treatment with dexamethasone and dexamethasone premedication within one hour before treatment initiation of each cycle.

At the interim study data cut-off 115 patients (61%) were dead. 81 patients (43%) achieved a CR or a CRh, 64 of them within the first treatment cycle. No significant differences regarding response in different subgroups related to age, sex, previous salvage therapy or disease state could be observed. Of note, a significant difference was detected regarding bone marrow blast counts, as 73% of patients (n=43) with less than 50% of BM blasts displayed a CR or CRh, but only 29% of patients (n=38) with 50% or more BM blasts did.

After a median follow-up of 8.9 months 37 patients (45%) with CR or CRh were still alive and in remission. 32 patients (40%) displaying CR or CRh underwent subsequent allo-SCT with overall 100 day mortality after allo-SCT of 32 patients. Within the median follow-up the relapse free survival (RFS) of the responders was 5.9 months, with patients achieving a CR having a higher RFS (6.9 months) compared to patients with CRh (5.0 months). Within a median follow-up of 9.8 months for all patients the median OS was 6.1 months. As minimal residual disease is associated with an unfavorable outcome during first line therapy the MRD status was evaluated in 73 patients with CR or CRh demonstrating a clear difference regarding outcome between MRD responders and non-responders as RFS for MRD responders was 6.9 months compared to 2.3 month for non-responders. Furthermore, MRD responders displayed an OS of 11.5 months *versus* 6.7 months for non-responders.

Adverse events observed were consistent with those previously reported and were observed in all but one patient in this study. Grade 3 and 4 AEs arose in 71 (38%) and 56 (30%) patients, respectively and mostly consisted of neutropenia (30%), febrile neutropenia (25%), anaemia (14%) and pneumonia (9%). 23 patients (12%) displayed fatal AEs, mainly infections. 98 patients (52%) had neurological events mostly grade 1 and 2 (76% of patients) and mostly in treatment cycle one, which could be managed with dexamethasone treatment. 20 patients (11%) displayed grade 3 neurological events, but no fatal neurological events occurred. Only three patients showed a cytokine release syndrome that might be due to dexamethasone prephase treatment and stepwise dosing. In summary, blinatumomab monotherapy displayed a very favorable outcome in this group of ALL patients presenting with substantial disease severity and selected for negative prognostic factors that could not be achieved with standard of care.

Due to the very encouraging results obtained with blinatumomab in the treatment of ALL patients in July 2014 the U.S. Food and Drug Administration (FDA) has granted blinatumomab the "breakthrough therapy" designation based on phase II data from relapsed or refractory B-cell precursor ALL patients, thereby falling under the FDA accelerated approval program. Already in December 2014, blinatumomab (brand name Blincyto^{TM}) was approved by the FDA [35] for patients with Philadelphia chromosome-negative relapsed or refractory precursor B-cell ALL making blinatumomab the first recombinant bsAb and the first bsAb against hematological malignancies to be approved.

Several more BiTE antibodies are so far in preclinical or early clinical early testing, MT110 (anti-EpCAM x anti-CD3, NCT00635596) [36] in advanced solid tumors, MT111/MEDI 565 (anti-CEA x anti-CD3, NCT01284231) against gastrointestinal cancer [37] and MT112/BAY2010112 (anti-PSCA x anti-CD3) for prostate cancer [38]. Further tumor antigens are already tested as MCSP (melanoma-associated chondroitin sulfate proteoglycan), EphA2 (ephrin type A receptor tyrosin kinase) for solid tumors, IGF-1R (insulin like growth factor 1 receptor) for solid tumors or HER2/neu [39].

Another BiTE molecule targeting hematological malignancies is AMD330 (former MT114), which besides CD3 targets the sialic acid-binding lectin CD33 frequently expressed on AML blast cells and leukemic stem cells, the latter showing higher CD33 expression in comparison with healthy stem cells [40]. AMG330 binds with comparable affinity to human and macaque CD3 and CD33 molecules, with dissociation constants in the range of 5 to 8 nM [41, 42]. AMG330 mediated lysis of tumor target cells using AML cell lines and human PBMCs depleted from CD33$^+$ cells is strictly T-cell dependent, as no lysis with AMG 330 occurred in the absence of T cells or using an inactive analog compound instead of AMG330. Lysis is accompanied by activation of T cells, as measured by CD25 and CD69 up-regulation, and cytokine release, mainly IFNγ, TNFα, IL-6, IL-2 and IL-10. Concordantly to blinatumomab, AMG330 mediated EC$_{50}$ values of cell lysis lay in the low picomolar and subpicomolar range. This BiTE-molecule furthermore induced depletion of CD33$^+$ cells from cynomolgus monkey bone marrow aspirates and proliferation of autologous T cells.

Interestingly, AMG330 was further tested in a NOD/SCID mouse model utilizing MOLM-13 ALM cell line and resting human T cells. Already with the lowest dose used (2 µg/kg/day) a significantly extended survival of mice compared to the vehicle control was observed. Higher doses administered (20 and 200 µg/kg/day) even more improved survival of the animals as 30% or 50% of mice survived after the end of observation period of 111 days.

Cytotoxicity assays with feeder cell-based long-term cultures to circumvent the short *in vitro* survival of primary AML patient blasts, revealed that AMG330 efficiently recruits and expands CD45RA⁻/CCR7⁺ memory T cells. Even at low effector:target ratios of 1:3.4 with autologous T cells AMG330-induced efficient blast cell lysis [43].

Using lentivirus-transduced cell lines to enable CD33 expression at various surface densities, AMG330 activity was shown to be dose and effector:target ratio-dependent, as a direct relationship between AMG333-induced cytotoxicity and CD33 expression level was detected [44]. Nevertheless even at low CD33 surface levels AMG330 was able to lyse target cells. Of note, cytotoxicity was not affected by CD33 polymorphism. In contrast to bivalent anti-CD33 antibodies [45, 46] binding of AMG330 to CD33 did not reduce CD33 surface expression suggesting that no or only minor limitation or reduction of AMG330s anti-AML activity due to reduced target-binding sites has to be expected in clinical use of this BiTE-molecule. Therefore, BiTE-molecules like AMG330 might be a powerful new tool to treat AML in which overall survival of patients is still relatively poor and new therapeutic options are strongly needed.

Trivalent and Trispecific Formats

Although the bispecific single chain BiTEs display impressive anti-tumor activity in *in vitro* assays and in the clinic, there might be still room for improvement. The low molecular weight of about 55 kDa leads to an extremely short serum half-life, probably due to renal clearance. Secondly, BiTE molecules might penetrate the blood brain barrier leading to AEs regarding the CNS, like aphasia, ataxia, disorientation and seizure, which lead to termination of treatment in 12 of 21 patients in early blinatumomab studies [23]. In an attempt to overcome the limitations of scFv-formats like BiTEs, a new class of antibody format, trispecific or trivalent single chain antibodies called single chain triplebodies (sctb) (Table **2**) was developed [47].

Table 2. Different single chain triplebody (sctb) construct currently under investigation. Triplebodies bind to target antigens like CD19 and CD16 as for example indicated by 19 x 16 x 19.

Triplebody	Malignancy	Mechanism	Lit.
19 x 16 x 19	B-cell	NK-cell redirection	[48]
HLA-DR x 16 x 19	B-cell	Dual targeting	[51, 52]
33 x 16 x 19	B-cell	Dual targeting	[53]
19 x 3 x 19	B-cell	T-cell redirection	[49]
33 x 16 x 33	AML	NK-cell redirection	[54]
123 x 16 x 123	AML	NK-cell redirection	[50]
123 x 16 x 33	AML	Dual targeting	[50]

By adding a third scFv-fragment in the same polypeptide chain a triplebody with a molecular mass up to 90 kDa was created (Fig. **1B**). This is above the molecular weight threshold for renal clearance, resulting in a prolonged serum half-life of this new format. These artificial triplebodies consist of two distal specific binding site for antigens and one central binding site to recruit effector cells, separated by a flexible linker mostly 20 amino acids in length built by multiple (Gly4Ser)-entities.

The first triplebody produced was a 19 x 16 x 19 construct addressing Fcγ receptor III (CD16), triggering NK cells as effector cells [48]. *In vitro* cytotoxicity assays with this sctb revealed efficient tumor cells lysis using leukemia cell lines and primary tumor cells from leukemia and lymphoma patients with an EC_{50} in the low picomolar range. Equal lysis mediated by the sctb 19 x 16 x 19 and the scFv 19 x 16 was induced with 10- to 40-fold lower concentrations of the sctb molecule. One reason might be that the triple body displayed a 3-fold greater avidity to CD19 compared with the bispecific 19 x 3 scFv, containing only one CD19 targeting arm, while both molecules display equal affinity to CD16. Of note, serum half-life in mice for 19 x 16 x 19 was 4 hours compared to 2 hours with the monovalent 19 x 3 antibody indicating that addition of a third scFv indeed improves pharmacokinetic characteristics by extending serum half-life.

Several triplebodies were produced so far and tested *in vitro* (Table **2**). Most of these antibodies utilize CD16 as target antigen to address effector cells. The first triplebody utilizing T cells as effector cells was a recently described 19 x 3 x 19 antibody [49]. Performing *in vitro* analysis a 19 x 3 scFv was used to mimic the anti-CD19 x anti-CD3 binding BiTE-formate blinatumomab.

19 x 3 x 19 sctb induced efficient tumor cell lysis with B-cell lines and primary patient cells of CLL and NHL patients with an EC_{50} in the low picomolar range. No significant difference between the triplebody and the BiTE-like molecule could be detected in this setting, albeit the triplebody showed a 2-fold greater binding strength compared to the monovalent scFv. The 19 x 3 x 19 sctb induced activation of T cells without a pre- or costimulation but only in the presence of B cells. No unspecific T-cell activation occurred. In agreement with blinatumumab 19 x 3 x 19 induced serial lysis of tumor cells. Furthermore, incubation of a patient sample with the triple body for seven days leads to expansion of CD45RO-positive memory T cells.

The 19 x 3 x 19 displays a similar T-cell response and cytotoxic profile as BiTE, nevertheless one might speculate that the triplebody displays a more favorable response due to increased serum half-life in animal models and in the clinic.

The value of triplebodies targeting CD3 might be increased by the capability of this antibody format of dual targeting. The hypothesis of "dual targeting" is to further increase tumor selectivity and drug specificity. Targeting two different epitopes on one tumor cell should allow the binding of tumor cells in a more prominent way and at the same time should reduce unwanted engagement of healthy cells. On one hand this fact might lead to improved clinical outcome in terms of enhanced response rates and long-lasting remissions and on the other hand might increase drug safety and decrease number and severity of AEs.

A challenging task in the fight against cancer still remains the eradication of cancer stem cells (SCs) to prevent relapse and achieve long-lasting remissions or even cure of cancer. As cancer stem cells and in particular leukemic stem cells have characteristics similarly to healthy tissue or hematopoietic SCs, dual targeting anti-cancer drugs might be capable to discriminate between hematopoietic and leukemic SCs and leave sufficient SCs for hematopoietic reconstitution avoiding the need of transplantation. Therefore, suitable pairs of antigens have to be found. For some types of leukemia like subtypes of AML the first antigens *e.g.,* CD123, CD33, CD96 or CD47 have been identified that might meet these requirements.

Several dual targeting triplebodies have so far been produced, like 19 x 16 x 33, HLA-DR x 16 x 19 or 123 x 16 x 33 (Table **2**) [50-53].

It could be already demonstrated for the dual targeting triplebodies 33 x 16 x 19 and 123 x 16 x 3 that sctbs simultaneously bind to one copy of each antigen on double positive cells [50]. The sctb 19 x 16 x 33 displayed a 2-fold higher avidity for double positive cells compared to the monovalent controls 19 x 16 and 33 x 16. Furthermore, dual binding, albeit not in all cell lines tested, contributed to target cell lysis that was significantly more effective for the sctb compared to 19 x 16 and 33 x 16 with EC_{50} values that were 25-fold and 1.5-fold lower for sctb than for 33 x 19 and 33 x16, respectively [53].

These data could be confirmed with another triplebody construct HLA-DR x 16 x 19. An equilibrium binding constant (K_D) of 37 nM compared to 71 nM for the scFv HLA-DR and 61 nM (19 x 16) could be determined. A preferential binding to double positive cells *versus* single positive cells with comparable antigen densities, even in the case of a 20-fold excess of the single positive cells, could be demonstrated. In line with these results, the sctb efficiently destroyed double positive Raji cells with an EC_{50} which was 136-fold lower (8 *vs.* 1,102 pM) for scFv 19 x 16 and 3-fold lower for the scFv HLA-DR x 16 construct. Furthermore, using stably transfected cells to obtain comparable cell surface densities for target antigens HLA-DR and CD19, preferential lysis of the double-positive cells compared to singe positive cells could be shown [52].

The triplebodies 123 x 16 x 33 and 123 x 16 x 123 were designed to eliminate AML cells [50]. Performing antibody dependent cellular cytotoxicity (ADCC) assays both constructs induced potent NK-cell dependent cell lysis using double-positive AML cell lines with the dual targeting triblebody showing a 2.5-fold lower EC_{50} (21 pM *vs.* 50 pM) and an increased maximum lysis (46% *vs.* 37%) with MOLM-13 cells and comparable results with a second cell line (THP-1). Competition assays revealed that binding to both antigens contributed to target cell lysis. The higher potency of the dual targeting antibody could be confirmed using primary samples from AML patients. In this setting, the bivalent 33 x 16 x 33 construct was also tested and displayed no statistically significant lower potency as compared to the dual targeting triplebody [54].

It will be interesting to see if the promising *in vitro* results could be verified in appropriate animal models and clinical studies. Furthermore, this antibody format might also allow generating recombinant trispecific antibodies targeting tumor cells, T cells and accessory cells combining the corresponding single chain fragments for CD19, CD3 and a suitable Fcγ-receptor.

Tribodies

Another smart approach creating multifunctional recombinant antibody derivates takes advantage of the natural *in vivo* heterodimerization of the heavy and light chain of a Fab fragment.

Each chain of a Fab-fragment can be extended at the C-terminus by additional binders like single chain variable fragments forming bibodies (one additional binder) or tribodies (two additional binders). The V_H and V_L modules of each of the scFv- as well as scFvs and the Fab-fragment are connected by flexible linkers made of non-immunogenic amino acid sequences like glycin-serine repeats (*e.g.,* G_4S) up to 20 amino acids in length (($G_4S)_4$). Using scFvs as additional binders (Fab-scFv$_2$) tribodies with an expected mass of around 110 kDa are synthesized (Fig. **1I**). This disulphide stabilized bsAbs can be easily produced in mammalian cells with a high yield [55, 56].

In addition to the specific heterodimerization of the Fab fragment another advantage of this kind of multifunctional antibody format is represented by the higher affinity and greater stability of a Fab-fragment compared to a scFv-derivate [57].

A tribody targeting BCL1 on murine myeloma cells *via* the Fab and one scFv-fragment and murine CD3 *via* the second scFv-fragment was synthesized and compared to the corresponding bibody (BCL1 x CD3) and a bispecific tandem scFv (bsscFv) as an analog to the BiTE format of bsAbs [58]. Binding studies revealed a significantly stronger binding of the tribody (B$_{50}$ binding value = 3 nM) in contrast to the bibody (B$_{50}$ = 30 nM). Of note, the BiTE analogue displayed an even weaker binding. Blood clearance and blood elimination half-life were determined in healthy Balb/c mice. The bispecific scFv displayed the shortest half-life (T$_{1/2}$ = 1.5 h) whereas the bibody (T$_{1/2}$ = 2.9 h) and especially the tribody (T$_{1/2}$ = 5.7 h) remained significantly longer in circulation. Using radiolabeled proteins to measure biodistribution revealed that the bispecific scFv is rapidly accumulated in the kidneys contrary to the bibody and tribody. Kidney accumulation is even lower for a full length anti BCL1 IgG1 antibody control. In line with these results, analysis of a limb-injected tumor nodule demonstrated a low accumulation of the bispecific scFv and an increased one of the IgG1control as well as for the tribody format.

To evaluate the therapeutic efficacy of tribodies Balb/c mice were inoculated intraperitoneal (i.p.) with BCL1 cells and treated with different doses of the bispecific antibody formats. Control animals without any antibody treatment developed terminal illness with huge increase of tumor load in the spleen. In contrast, antibody-treated animals showed prolonged dose-related survival accompanied with a significant protection against splenic tumor formation. At the highest antibody dose tested (200 pmol), 60% of mice treated with the bibody or the bispecific scFv survived, whereas 100% of mice treated with the tribody survived.

The latter animal did not show any reappearance of tumors until the end of the observation period. Administering lower doses (100, 50 pmol) 85% and 70% of mice treated with the tribody survived, but only 45% and 20%, respectively, treated with the Fab-scFv. Of note, no animals in both groups treated with the BiTE analogue bispecific scFv survived. Anti-tumor activity in this mouse model correlated well with the serum half-lives of the various antibody formats evaluated, but increased activity of the tribody might not only be explained by increased serum half-life but also by a greater anti-tumor potency *per se*, at least in part mediated *via* a stronger tumor cell binding. Taken together, first results, albeit limited, show impressive potency of this novel antibody format that seems to be superior to tandem scFv bispecific antibodies at least under these conditions.

As in some clinical settings like treatment of ALL patients after stem cell transplantation, NK cells are still prominent effectors in comparison with T cells, tribodies addressing CD16 expressed on NK cells were developed [59].

Disulphide stabilized scFvs targeting CD16 and CD19, respectively, were fused to a Fab targeting CD19 [60]. This [(19)$_2$ x 16] tribody format displayed a 3-fold greater CD19 binding avidity compared to a corresponding bibody (19 x 16) and a 5-fold greater one compared to a tandem bispecific scFv (8 nM *vs*. 24 nM *vs*. 42.4 nM). Binding affinity for CD16 was identical for all three antibody constructs. The higher binding avidity went along with prolonged surface retention time on CD19-positive cells. Moreover, EC$_{50}$-values in ADCC assays showed the same pattern with a 6.3-fold greater anti-tumor cell activity of the tribody compared to the bibody (55.5 pM *vs*. 348.2 pM) and 12.2-fold greater one compared to bispecific scFv. In 3 out of 6 primary NHL patient samples (5 x CLL, 1 x DLBCL) tested tumor cell destruction was significantly higher with the tribody than with the bibody.

A slightly different architecture was used for a [(CD20)$_2$ x CD16] tribody with both scFv binders targeting CD20 and a Fab targeting CD16 [61]. Binding to CD16a isoforms was identical, irrespective of the V/F polymorphism at amino acid 158, which resulted in significantly diverse clinical responses regarding rituximab treatment [62]. Cell separation and blocking experiments identified NK cells as crucial effector cell population to mediate cytotoxicity. The tribody could not induce complement dependent cytotoxicity (CDC). Compared to rituximab the tribody showed 9-fold greater activity in ADCC assays and in addition was more pronounced with primary CLL tumor cells. Even in whole blood assays with active complement the tribody had greater anti-tumor cell activity. By transplanting highly purified CD34-positive human cord blood cells in

NOD/SCID mice a humanized *in vivo* model was established to analyze B-cell depletion. Administration of the CD20 tribody, but not of a [(Her)$_2$ x CD16] control tribody induced significant and selective B-cell depletion.

The data available so far identify tribodies and triplebodies as promising tools to fight B-cell malignancies due to favorable serum half-lives and higher avidity compared to tandem scFvs like the BiTE antibodies. A further advantage might be a better tissue penetration potential compared to monospecific or bispecific full-length IgG antibodies due to the smaller molecular mass of about 110 kDa in the case of tribodies or 90 kDa in the case of triple bodies. Therefore, more data, especially *in vivo* data obtained with appropriate animal models and clinical settings will be necessary to fully elucidate the clinical potential of these new platform technologies.

Dock and Lock

With the Dock and Lock (DNL) method another bispecific platform technology was established to create bispecific multivalent antibodies (Fig. **1G**). The DNL-method is based on the specific protein/protein interaction between the regulatory subunits (R) of cAMP-dependent protein kinase A and the anchoring domain (AD) of an interactive A-kinase anchoring protein, that only binds to dimeric regulatory subunits [63]. The dimerization and AD binding domain are located within the same 44 amino acid sequence by the way building up the dimerization as well as the docking domain (DDD). The DDD and AD form a non-covalent aggregate that could be covalently linked by additional cysteine residues in the DDD and AD to improve *in vivo* stability without interfering with the bioactivity of the immunological molecules [64]. The DDD and AD were fused to the corresponding immunoglobulin chain by 12 to 15 amino acid peptide linkers composed of glycin and serine. The Dock and Lock allows the manifold site-specific self-assembly of a multitude of immunoglobulins from Fab-, F(ab)$_2$-fragments to complete immunoglobulin G antibodies, generating tri- up to hexavalent constructs.

As a combination of the anti-CD20 Ab rituximab and the anti-CD22 Ab epratuzumab in NHL patients showed enhanced anti-tumor activity without correspondent increase in toxicity [65], different anti-CD20 x anti-CD22 bispecific, hexavalent antibodies [IgG – (Fab)$_4$] with a molecular mass of 365 kDa were generated by the DNL-method [66].

Anti-CD20 antibody veltuzumab and anti-CD22 epratuzumab were linked at their C-terminal ends to the AD and fused to a pair of epratuzumab or veltuzumab Fab

DDD dimers, creating bispecific antibodies with bivalent binding to CD20 or CD22 and tetravalent binding *via* the Fab fragments to CD22 and CD20, respectively, denoted 20-22 (20-22-22) or 22-20 (22-20-20). As control a hexavalent CD20 construct 20-20 (20-20-20) was generated. The number of CD20 binding arms correlated with the dissociation rate from bound Raji-cells, as the off-rate of 20-20 and 22-20 was about 2-fold lower compared to veltuzumab. Performing competition assays the bispecific binding to CD20- and CD22-positive cells could be demonstrated.

Compared to parental antibodies 20-22 and 22-20 more significantly inhibited proliferation of tumor cell lines, with EC_{50}-values 78-fold (22-20) and 12-fold (20-22) lower than for the combination of veltuzumab and epratuzumab. 20-22 and 22-20 induced 2- to 3-fold greater apoptosis in tumor cell lines (*e.g.,* Burkitt lymphoma lines Raji, Ramos) compared to the parental antibodies alone or in combination. Contrary to rituximab and veltuzumab 20-22 and 22-20 displayed preferential killing of tumor B cells compared to healthy B cells in blood samples from healthy doors supplemented with tumor B-cell lines. Of note, in this setting rituximab and veltuzumab showed a significantly higher anti-tumor potency.

Analyzing apoptotic and survival signaling pathways a marked difference between 20-22 and 22-20 on one side and rituximab or veltuzumab on the other side could be detected, *i.e.* a significant increase in the levels of phosphorylated p38 or PTEN [67].

Although, veltuzumab exhibited potent CDC-activity, neither 22-20 nor 20-22 showed this feature. Remarkably, given the greater molecular mass of the hexavalent antibodies compared to normal IgG immunoglobulins (365 kDa *vs.* 150 kDa), both 20-22 and 22-20 have an about 2-fold shorter serum half-life in mice. *In vivo* efficacy was analyzed in a Burkitt lymphoma xenograft model with SCID mice administered with Daudi cells. Application of a single dose of 22-20 or 20-22 antibody displayed a significantly increased median survival time compared to control but not compared to veltuzumab or a hexavalent 20-14 antibody. In a second set of experiments with three applications within one week efficacy of 22-20 was compared with epratuzumab and a panel of hexavalent control antibodies (734-20, 22-14). All treatments improved survival compared to saline with 22-20 being much more effective than epratuzumab (median OS 66 days *vs.* 32 days), 22-14 (32 days), 734-20 (42 days) but not compared to a mixture of 734-20 and 22-14 (68.5 days). The later combination used the double amount of antibody and provided the equal number of CD20 and CD22 binding groups. A strong reduction of overall survival was induced by NK-cell depletion in mice, indicating that *in vivo* cytotoxicity was mainly due to ADCC [66].

Comparable results could be obtained with hexavalent antibodies targeting CD20 and CD74 that were constructed due to the observation that a combination of anti-CD20 Ab rituximab and anti-CD74 Ab milatuzumab showed improved anti-tumor activity compared with either antibody alone with MCL cell lines and patient samples [68, 69]. A dose-dependent increase of survival was seen in SCID mice inoculated with MCL JeKo-1 cells with a median survival of 53 days for the 20-74 format and 43.5 days in case of the 74-20 construct compared to 34 days for saline control. As controls with the parental antibodies veltuzumab and milatuzumab were not performed no conclusions could be drawn from augmented *in vivo* activities of the hexavalent antibodies in comparison with the parental formats.

Immunotoxin-Coupled Bispecific Antibodies

A further strategy improving tumor cell elimination by dual targeting is represented by the concept of bispecific ligand-directed toxins coupling a toxin to a bispecific antibody directed against two different antigens displayed on one cell with the aim of an increasing targeting capacity. The immunotoxin DT2219ARL was generated by fusing a truncated form of the diphtheria toxin (DT) that contains the A fragment of native DT which catalyzes ADP ribosylation of elongation factor 2 (EF-2) leading to complete blockade of protein synthesis followed by cell death, to the N-terminus of a bispecific anti-CD19 x anti-CD22 antibody (Fig. **1D**). This antibody was made by fusing two scFvs directed against CD19 and CD22, respectively [70]. The original immunotoxin DT2219 (DT2219EA) was genetically improved for superior anti-tumor activity. For this purpose the original arrangement of the V_H and V_L domains was changed from CD22-V_H-V_L-CD19-V_H-V_L to CD22-V_L-V_H-CD19-V_L-V_H and the linker between the corresponding V_H and V_L domains was replaced by an ARL-motif that contained charged amino acids enabling increased protein yields due to improved folding [71].

DT2219ARL was tested in a recently completed phase I study (Table **1**, NCT00889408) with 22 patients suffering from chemorefractory CD19 and CD22 B-cell leukemia and lymphoma with 4 daily infusions with escalating doses from 0.5 µg/kg/dose up to 80 µg/kg/dose [72].

Within a dose range of 0.5 µg/kg/dose up to 20 µg/kg/dose (9 patients) no responses or dose limiting toxicities were observed. 13 patients were treated with 40 (n=5), 60 (n=5) and 80 µg/kg/dose (n=3). Two partial remissions could be achieved, one at a dose of 40 μg/kg/dose and one at a dose of 60 μg/kg/dose. The latter was accompanied by a DLT due to a grade 3 capillary leaking syndrome. This patient received a second cycle of DT2219ARL with 40 μg/kg/dose, this time resulting in a complete remission. Both patients were in remission after 6 and 4 months, respectively. Adverse events occurred in all patients treated with doses ≥40 μg/kg/dose consisting mainly of grade 1-2 capillary leak syndrome and hematological toxicities, elevated liver function and fatigue. Two DLTs were observed, one at 40 μg/kg/dose (grade 3 lower extremity weakness) and one at 60 μg/kg/dose (grade 3 capillary leak syndrome and grade 4 neutropenia). All adverse reactions resolved completely within one week. With a determined biological active dose of 40 to 80 μg/kg/dose, DT2219ARL is intended to be tested in a phase II study.

Dual Affinity Retargeting Antibodies (DART)

Another strategy to produce recombinant small bsAb harnesses the heterodimerization of the immunoglobulin V_L and V_H domain. Linking the V_H domain of an antibody Ab1 to the V_L domain of Ab2 and on the complementary chain the V_H domain of Ab2 to the V_L domain of Ab1 (V_H1-V_L2; V_H2-V_L1) generates bispecifc diabodies with two different antigen binding sites. Several diabodies targeting B cells *via* CD19 or CD20, T cells *via* CD3 or NK cells *via* CD16 were investigated in preclinical studies [9, 10, 73-75]. Nevertheless, despite proof-of-concept in *in vitro* and animal studies diabodies have not played any role in the clinical development of anti-tumor therapies towards hematological malignancies so far. Beyond that scFv-based bispecific compounds might be impaired by the tendency to form aggregates or by the linker sequences connecting the V-regions impairing antigen recognition or potency.

To overcome limitations of diabodies or scFv constructs an alternative platform technology was developed by Macrogenics, called dual affinity retargeting (DART) [76]. DART antibodies are based on the diabody format. A short cysteine-containing peptide was added to the C-terminus of each of the two chains (Fig. **1E**). The resulting covalent linkage limits the freedom to undergo domain exchange resulting in a greater stability independent of the V_H-V_L interaction. In addition, due to the absence of an intervening linker this construct resembles more the natural IgG molecules.

It was recently demonstrated that DART antibodies could successfully target B cells [77, 78]. As a model system a DART construct targeting human CD16 on NK cells and human CD32B on B cells was generated. This dart molecule displayed potent and dose dependent cytotoxicity against lymphoma B-cell lines using human PBMCs of healthy donors as effector cells. In addition, efficient B-cell depletion was induced in a mouse model, depending on the transgenic expression of human CD16 and human CD32B.

To compare the DART format with the successfully applied BiTE format, an anti-CD19 x anti-CD3 DART was generated using the identical variable antibody regions as those used for blinatumomab [79]. A BiTE-like molecule was constructed as control. The DART antibody displayed an increased affinity for CD19 and CD3 due to an increased association rate for CD3 and a decreased dissociation rate for CD19. Furthermore, a significant difference in cytotoxicity between the two formats was established with EC_{50}-values of the DART molecule being 11-fold lower in dependence on the tumor cell line used. Concordant with this, the DART-molecule induced a stronger T-cell activation as determined by CD69 up-regulation on CD4 and CD8 T cells and IFNγ-release and a 3-fold stronger B-cell/T-cell association. The more favorable features of the DART molecule may be caused by its more rigid configuration with limited flexibility between the two antigen binding domains [80]

As the $TCR_{\alpha\beta}$ complex displays a more limited expression than $CD3_{\varepsilon}$ and in some studies showed lower side effects when used as therapeutic target, a $TCR_{\alpha\beta}$ targeting DART was tested [81]. Both CD19 x CD3 and CD19 x TCR displayed identical cytotoxicity *in vitro*. Furthermore, in a NOD/SCID mouse model with subcutaneously implanted Raji tumor cells and human PBMCs as effector cells an efficient block of tumor growth could be induced by I.V. application of the TCR-targeting DART.

To achieve longer serum half-lives a modified CD19 x CD3 DART was engineered by attaching of a modified Fc-domain on the C-terminal end of one chain that is inactive for Fcγ-receptor binding but does bind to the FcRn neonatal receptor of the immunoglobulin salvage pathway [82]. Furthermore, this new molecule with a mass around 90 kDa lies significantly above the kidney threshold enabling convenient application schemes with once a week or longer infusion intervals. As expected this DART antibody exhibited prolonged pharmacokinetic properties in cynomolgus monkeys. This modified DART molecule named MGD011 mediates potent *in vitro* cytotoxicity against B-cell lymphoma cell lines and in an autologous setting that outperform blinatumomab-induced cell killing. A favorable response could be evaluated in two mouse models. Firstly, MGD011 efficiently inhibited growth of newly implanted tumor xenografts, like HBL-2

(MCL cell line) and Raji. High complete response rates with no relapses over the study duration were detected in human PBMC-reconstituted mice bearing pre-established intradermal HBL-2 xenografts. MGD011 is intended to enter clinical development in 2015.

The first DART antibody entering the clinical stage addressing hematological malignancies is the CD3 x CD123 targeting MGD006 (Table **1**, NCT02152956) that is currently tested in a phase I study with relapsed or refractory AML patients. In preclinical studies this antibody showed efficient activation and proliferation of T cells and killing of CD123 positive K562 cells and primary AML blasts. In a mouse xenograft model with CD123 expressing K562 cells tumor growth was completely inhibited by MGD006 but not by control DARTs that only addressed CD3 or CD123 [83]. A side-by-side comparison *in vivo* would be necessary to uncover if higher *in vitro* potency of the DART antibody is reflected in a clinical setting.

Ex vivo **Arming of T cells with Bispecific Antibodies**

Common application of bsAbs should induce retargeting of T cells or other effector cells addressed by the antibody to the tumor cell. An enlargement of this concept is the combination of antibody infusion and application of lymphocytes or T cells of an allogeneic donor (donor lymphocyte infusion DLI) to arm theses T cells with the shortly before injected bsAb *in vivo* [84]. A further enlargement depicts the isolation of T cells from autologous or allogeneic donors, expanding and arming them *ex vivo* with a bsAb.

The first clinical trial was already published in 1990 using autologous lymphokine activated killer cells (LAK) *ex vivo* armed with an anti-CD3 x anti-glioma F(ab)$_2$ bsAb and locally administered in the brain tumor demonstrating proof-of-concept as patients treated with armed LAKs showed an improved overall and progression-free-survival than patients treated with unarmed LAKs [85]. In the following years a variety of clinical phase I and II studies with patients suffering from different solid tumors (*i.e.* glioma, ovarian, head and neck) with different bsAbs, even with the 2009 approved anti-EpCAM x anti-CD3 catumaxomab, were performed and showed promising responses accompanied by mild side effects [86-90].

In recent years the treatment concept was intensely investigated at the Barbara Ann Karmanos Cancer Institut [91]. For that purpose bsAbs were produced by chemical crosslinking of full length IgG antibodies. To address CD3 the antibody OKT-3 was used and as tumor antigens CD20 and HER2/neu were chosen so far, using herceptin to generate HER2Bi (anti-CD3 x anti-HER2) and rituximab to generate the anti-CD3 x anti-CD20 bsAb CD20Bi. Therefore, anti-CD3 antibody OKT3 was coupled to Trauts reagent and an anti-TAA antibody like rituximab was coupled to Sulpho-SMCC, both at the C-terminus of the corresponding antibodies, which allows specific heteroconjugation to form a stable bispecific antibody dimer [92].

Autologous PBMCs were isolated, activated with soluble anti-CD3 antibody and expanded adding exogenous IL-2 *ex vivo* up to two weeks to generate activated T-cells (ATCs) [93]. ATCs were armed with in average with $50ng/10^6$ cells. To remove unbound bsAb ATCs were washed before application. This strategy aims to create an artificial tumor- associated antigen specific T-cell receptor without genetic modification of T cells to overcome tumor escape or resistance mechanisms and to avoid cytokine storm due to binding to and activation of T cells and accessory cells *in vivo* induced by infusion of unbound bsAb.

Preclinical studies with HER2Bi revealed that *ex vivo* armed ATCs could bind to and kill tumor cells (*i.e.* breast-, prostate-, pancreatic cancer-cell lines) *in vitro* [93-95] and prevented tumor development in SCID/Beige mice and induced remissions in established prostate xenografts [96]. Phase I studies with multiple infusions of armed ATCs revealed a favorable toxicity profile [97, 98].

To target CD20-positive B-cell malignancies the bsAb CD20Bi was developed. Arming ATCs established from PBMCs of healthy donors and cancer patients induced efficient lysis of B-cell lines like Raji or ARH-77 but not of breast cancer cell line SK-BR-3 [99]. This demonstrates that patient-specific T cells from an immunosuppressive cancer environment *per se* retain their functional activity and even more could regain their cytotoxic capability in a stimulating environment. Of note, blocking experiments with an excess of rituximab interfered with tumor cell binding but to a much lesser extent with cytotoxicity. A significant decrease in cytotoxicity was only seen with rituximab concentrations that were equivalent to an 8,000 to 64,000 excess. Furthermore, T-cell stimulation with anti-CD3 Abs in the presence of CTLA-4 targeting ipilimumab enhanced T-cell proliferation and significantly increased tumor cell-directed cytotoxicity [100] accompanied by an increase in Th1 cytokines (IFNγ, IL-12) and decrease of immunosuppressive IL-10 with increasing concentrations of ipilimumab. This fact might be due to inhibition of immunosuppressive pre-existing regulatory T cells.

Two resent reports described the clinical use of CD20Bi armed ATCs in high risk or refractory NHL-patients after autologous SCT [92, 101] which investigated safety, affection of immune recovery and anti-lymphoma effect.

In the first trial 12 patients, 11 with DLBCL, 1 with FL, were treated with 4 weekly infusions of armed ATCs in 4 dose escalation groups with 5, 10, 15 or 20 $x10^9$ ATCs. PBMCs were harvested by leukapheresis, expanded for 2 weeks in the presence of OKT3 and IL-2, washed and cryopreserved. ATC infusions were well tolerated with main side effects consisting of fever, chills, hypotension and fatigue. Engraftment of stem cells was not impaired. No dose limiting toxicities were observed and the maximum tolerated dose was not reached. With a median follow up time of 24 months, five out of the 12 patients died with a median survival of 112 days, while 7 patients (58%), all presenting with a complete remission survived with a median survival of 914 days. Interestingly, there was a marked difference regarding survival between two subgroups, one with remission at time of transplantation, and one with refractory or persistence disease, with the first one performing much better.

PBMCs taken two to twelve weeks after SCT mediated a significantly higher cytotoxicity against B-cell lines compared to PBMCs before SCT. This was in line with results obtained with IFNγ-Elispot assays, demonstrating a higher number of IFNγ-secreting T cells in PBMCs after SCT than before SCT. No increase in anti-lymphoma activity was seen in patients receiving only SCT without armed ATCs, indicating that treatment with bsAb armed ATCs induced endogenous anti-lymphoma activity. Application of armed ATCs was accompanied by a short term increase of Th1 cytokines like IFNγ and IL-12 and long-term increase of chemokines like CXCL-9 and CXCL-10.

A pilot phase I trial was conducted in three high-risk and refractory DLBCL patients with 15 infusions of 5 x 10^9 armed ATCs (3 infusions/per week for 3 weeks and 6 weekly infusions) 4 days after autologous transplantation in combination with IL-2 [101]. No DLTs could be observed with this increased number of applications, albeit with a limited number of patients. One patient died after 612 days despite a complete remission of 9 month. The two other patients with ongoing complete remissions were still alive at time of publication with OS of 73 and 77 months. Infused ATCs persisted up to 7 days. Again, application of CD20Bi armed T cells induced specific cytotoxicity against lymphoma targets.

In a phase Ib study (NCT00938626) with 12 multiple myeloma (MM) patients the practicability of CD20Bi-armed ATCs could be confirmed [102]. In contrast to the previous studies armed ATCs were administered before SCT with two infusions of 10^{10} ATCs one week apart and within 4 weeks prior to autologous transplantation. Four patients received an additional booster of 10^{10} armed ATCs

6 to 12 month after SCT. Again, an increase in IFNγ-specific Elispots and cytotoxicity against B-cell lines could be detected after armed ATC infusion and SCT. Interestingly, a significant increase in concentration of antibodies against the MM-antigen SOX-2 in responders compared to patients who relapsed could be detected, which suggests that SOX-2 levels might be suitable as biomarker for clinical outcome in multiple myeloma.

Interestingly, *ex vivo* expansion of T cells, isolated from untreated CLL patients, could also be successfully performed using the BiTE format blinatumomab [103]. Starting from only 10 ml of patient blood, 5×10^8 T-cells could be expanded within 18 to 25 days in the presence of IL-2 consisting mostly of effector and central memory T cells. Co-culture with blinatumomab led to a rapid decrease and in most cases to a complete depletion of CLL tumor cells. The blinatumomab expanded T cells (BETs) displayed a normalization of the synapse inhibitors CD272 and CD279 compared to the starting T-cell population [104]. BET cells in the presence of blinatumomab were highly effective against B-cell lines and primary CLL cells and showed therapeutic efficacy in a systemic DLBCL model in NOD/SCID mice. In this setting, BET cells or blinatumomab alone had no beneficial effect. Albeit, contrary to the experiments performed by Lum and colleagues [92, 101], BET cells were not armed *ex vivo* with blinatumomab that was administered by I.V. infusions together with BET inoculation and the following 4 days. It would be interesting to see how blinatumomab-armed BETs would perform in comparison with separate BET cells and blinatumomab infusions.

Tandabs

To overcome problems of bispecific diabodies regarding quantitative heterodimer formation due to utilizing two different gene products, tandem diabodys (TandAb®) were developed by Affimed Therapeutics. Therefore the light and heavy chain variable domains representing two different specificities (*e.g.,* anti-CD3 and anti-TAA) were linked on one chain in a sequence preventing intramolecular Fv-formation but favoring their intermolecular dimerization by using suitable linkers between the variable fragments (Fig. **1C**). The thereby created molecules display molecular masses of 105 to 110 kDa, which is above the threshold for first pass renal clearance [105-107]. An anti-CD3 x anti-CD19 tandem diabody was generated with domain order V_H3-V_L19-V_H19-V_L3 forming the tetravalent bispecific molecule anti-CD3 x anti-CD19 x anti-CD19 x anti-CD3 [108].

Using the RECRUIT-TandAb platform the humanized anti CD3 x anti-CD19 TandAb AFM11 was produced, that showed improved pharmacokinetic properties in comparison with smaller bsAb formats [109]. AFM11 exhibits a 4-fold higher affinity to CD19 B cells and a 27-fold higher affinity to CD3 T cells compared to a bivalent bispecific scFv-dimer. Cytotoxicity assays displayed that AFM11 mediates target cell lysis by $CD4^+$ and $CD8^+$ T cells with EC_{50} values in the low to sub-picomolar range that is independent of the target cell CD19 density. AFM11 showed T cell activation only in the presence of $CD19^+$ target cells resulting in up-regulation of activation markers CD25 and CD69, and specifically mediates lysis of these cells without affecting antigen-negative bystander cells. T-cell activation is accompanied by cytokine releases, *i.e.* mainly of Th1-cytokines IFNγ, TNFα, IL-2, IL-6. Of note, AFM11 displayed a 70-fold lower EC_{50} value relative to CD19 x CD3 tandem scFv, used as control [110, 111].

In a xenograft NOD/SCID mouse model with implanted Raji cells, low doses of AFM11 (0.1–5 µg/kg) induced a 60% delay of tumor growth while higher doses (10 µg and 5 mg/kg) induced a complete protection. AFM11 is currently investigated in a phase I study (NCT02106091) in relapsed or refractory NHL and B-precursor ALL patients.

A second RECRUIT-TandAb is currently under clinical investigation, AFM13 targeting CD30 on Reed Sternberg cells in Hodkin lymphoma and CD16A, lacking CD16B binding on granulocytes. Therefore AFM13 is protected from removal from the circulation by binding on CD16B granulocytes. The anti-CD30 x anti-CD16A TandAb retained longer on NK cells than a corresponding diabody resulting in improved potency and efficacy of this tetravalent molecule [111]. Cytotoxicity of this TandAb was independent of the CD16A allotyp. Concordant with the anti-CD3 x anti-CD19 TandAb activation of NK cells is strictly dependent on the presence of target cells.

AFM13 was tested in a phase I study with relapsed or refractory Hodgkin lymphoma (HL) patients (NCT01221571). 28 HL patients received stepwise escalated doses of intravenous AFM13 (0.01 to 7.0 mg/kg BW) weekly (n=24) or 4.5 mg/kg twice weekly (n=4) over 4 weeks. AFM13 was well tolerated and a DLT was not reached. AFM13 showed activity mainly at doses ≥1.5 mg/kg (n=13), with three partial responses and disease control in further ten patients. Importantly, AFM13 was also active in Hodgkin lymphoma patients refractory to brentuximab vedostin as last treatment. The number of activated NK cells in the peripheral blood increased 3-fold, with the kinetics of NK-cell activation corresponding to AFM13 serum levels [112]. Based on this phase I data a phase II study will be conducted in 2015 (NCT02321592).

Interestingly, as anti-CD3 x anti-CD19 TandAb AFM11 is stated to be more active in *in vitro* cytotoxicity assays as a corresponding anti-CD3 x anti-CD19 tandem scFv, Mølhøj *et al.* also performed a side by side analysis of blinatumomab and a tandem anti-CD3 x anti-CD19 diabody generated according to published literature [113, 114]. In control assays this tandem diabody displayed comparable activity as the literature described construct. Albeit, in the cytotoxicity assays performed by Mølhøj and colleagues [113], blinatumomab showed a cytotoxic activity magnitudes of order higher than observed with the tandem diabody. Using un-stimulated T cells as effector cells EC_{50}-values for the BiTE molecule were 736 – 2,606-fold lower than for the tandem diabody. Using stimulated T cells, blinatumomab was even more potent by a factor between 2,218 up to 8,062-fold, dependent on the target B-cell line used.

The observed discrepancy may in part be explained by the fact, that the tandem diabody used in this analysis is not identical with AFM11, which for example is a humanized molecule. Furthermore, the linkers between the Fv-domains used for AFM11 and the newly constructed tandem might not be identical. AFM11 displayed a 4-fold higher affinity to CD19 compared to the tandem scFv used as control by Reusch [109], the affinity of blinatumomab and the control tandem diabody reported by Mølhøj *et al.* was nearly identical.

Triomabs

Despite the major effort in developing, producing and testing new bsAb formats, the first and until recently only approved bispecific antibody is the quadroma produced trifunctional antibody (trAb) catumaxomab (anti-EpCAM x anti-CD3), demonstrating the eminent therapeutic potency of this full length antibody format [115]. A major breakthrough in the quadroma technology was the observation that the isotype combination of a rat (r) IgG2b and a mouse (m) IgG2a antibody enables a preferential species-specific heavy and light chain pairing (Fig. **1F**) [116]. This significantly reduced mismatch (antibodies formed by undesired chain assembly) content down to 4 - 10% and facilitated a production process based on a two-step purification sequence performing protein A and ion exchange chromatography. Thus, the rIgG2b x mIgG2a isotype combination enabled a GMP-compliant production platform.

Trifunctional antibodies could efficiently re-direct T cells to tumor cells leading to major histocompatibility (MHC) complex independent activation of CD4 and CD8 T-cells. These new full length bispecific antibodies display efficient binding to human Fcγ-RI, Fcγ-RIIa and Fcγ-RIII receptors, but not to inhibitory Fcγ-RIIb.

Monocytes and macrophages contribute to trAb induced tumor cell killing [117]. Direct phagocytosis by $CD14^+$ monocytes was observed after trAb addition. Monocyte activation could be demonstrated by IL-8 secretion and up-regulation of CD25 and CD40 [118]. In addition, T cells receive a second, accessory cell mediated co-stimulatory signal leading to a profound activation and proliferation. This concerted action of different immune effector cells mediates a highly efficient destruction of tumor cells at a very low antibody concentration in the pico- and femto-molar range without the need of any further co-stimulation [119].

Catumaxomab was tested in a pivotal phase II/III study in 258 patients with epithelial ovarian and non-ovarian cancers with the manifestation of malignant ascites (MA) [120, 121]. Besides its significant morbidity MA is associated with impairment of quality of life. Catumaxomab was administered intraperitoneal (i.p.) in four escalating doses from 10µg to 150µg, resulting in a cumulative total dose of only 230 µg [122]. Catumaxomab treatment induced a prolongation in puncture-free-survival of 46 days compared to 11 days in the control group and a puncture free time of 77 days *versus* 13 days. Catumaxomab was therefore approved by the EMA (European Medicines Agency) in 2009. Remarkably, a statistical significant prolongation of overall survival could be demonstrated in the gastric cancer subgroup. Anti-tumor activity could also be demonstrated by almost complete eradication of tumor cells in the highly immunosuppressed microenvironment of ascites fluids. Moreover, putative $CD133^+/EpCAM^+$ cancer stem cells were completely eliminated in peritoneal fluids.

Beside direct tumor cell eradication the induction of a long-term systemic tumor immunity is one of the major goals in cancer therapy. Numerous vaccination strategies, like DNA-based vaccines [123] or dendritic cell based vaccines [124] are under intensive investigation. However, despite the potential clinical efficacy, a lot of these immunotherapeutics are likely to require a particularly laborious and cost-intensive production process in order to comply with GMP-regulations. Therefore, biologicals like trAbs that could be produced with a proven process platform and administered with standard dosing procedures would be of great advantage. Trifunctional antibodies not only re-direct T cells to the tumor cells, resulting in efficient T-cell lysis, but also bind to activating Fcγ-receptors on immune effector cells like macrophages or dendritic cells, which furthermore leads to enhanced T-cell activity due to costimulation and phagocytosis of apoptotic or necrotic tumor cells. This might result in antigen processing and subsequent presentation of immunogenic peptides to T cells independent of further trAb recruitment and thus inducing a long lasting anti-tumor immunity.

This hypothesis was evaluated in different mouse models using surrogate trAbs that are capable of binding of the human tumor target and to mouse CD3 [125]. Syngenic mice were inoculated with EpCAM-transfested A16 melanoma or A20 lymphoma cells and treated with the trAb BiLU (anti-human EpCAM x anti-mouse CD3). Application of BiLu into mice bearing a lethal dose of tumor cells resulted in 100% survivors. If these mice were re-challenged with a second lethal dose of EpCAM$^+$ tumor cells, this time in the absence of BiLu, tumor cells were successfully rejected. Analysis of mice sera before the re-challenge revealed the existence of a strong humoral anti-tumor response, mainly of IgG2a subclass composition. Mice also survived if EpCAM-negative A20 tumor cells were used as second challenge. Furthermore, an immune response against the lymphoma cell idiotype, a well-known tumor associated antigen (TAA), could be detected in the sera of this mice. This clearly indicates that tumor antigen targeted by a trAb does not necessarily function as TAA for a T-cell response. The induction of a humoral immune response was furthermore demonstrated by infusion of sera of the surviving mice together with tumor cells and without trAb into naive mice which resulted in a better survival compared to control mice receiving sera from naive animals. The immune response was concluded to depend on the T cells, as depletion of CD4 or CD8 T cells reduced survival. In addition, the functional Fc-part of the trAb was a prerequisite for occurrence of an anti-tumor immunity. If the corresponding F(ab)2 fragment was used instead of BiLu no anti-idiotype response could be detected and F(ab)2 treated mice did not survive a second tumor cell challenge. Remarkably, the parental antibodies also failed in inducing an immune response, albeit in *in vitro* cytotoxicity assays, the F(ab)2 fragment and the mixture of the parental antibodies displayed comparable activity as BiLu. Taken together, this indicates that the use of trAb formats compared to the use of monospecific or of bsAbs without functional Fc-part might be beneficial regarding anti-tumor vaccination.

A long-lasting anti-tumor immunity could also be induced in mice inoculated with EpCAM- expressing melanoma cells and treated with donor splenocytes and BiLu. The majority of the long-term surviving mice were resistant to a second challenge with a lethal dose of tumor cells 201 days after the first inoculation [126]. Of note, a catumaxomab triggered anti- tumor response could not only be detected in animal models, but also in patients with malignant ascites recently treated in the phase IIIb CASIMAS study [127]. The majority of patients displayed an enhancement of a pre-existing anti-EpCAM and a *de novo* anti-HER2/neu response. Remarkably, the anti HER2/neu response not only correlated with an improvement of progression free survival but also with a prolongation of the overall survival. Comparable results could also be obtained in a mouse melanoma model using a trifunctional antibody directed against the disialogangliosid antigen GD2 [128-130]. Likewise, CD4$^+$ as well as CD8$^+$ T cells were essential for therapeutic effectiveness.

The induction of a T-cell response was recently also shown with the anti-CD20 antibody rituximab in follicular lymphoma patients [131] and a mouse model [132]. Contrary to trAbs TAA-independent induced immunity to a tumor challenge rituximab induced immunity in the mouse model was dependent on CD20 expression on the tumor cells used for restimulation indicating a monovalent immune response, therefore rendering rituximab induced immunity susceptible to tumor escape mechanisms like CD20 downregulation. This furthermore indicates that the use of trifunctional antibody formats compared to monospecific or Fc-depleted bsAb is likely to be beneficial for anti-tumor vaccination.

Due to the encouraging results obtained with the trAb catumaxomab the anti-CD20 x anti-CD3 targeting trAb FBTA05 was generated. This trAb consists of the identical rat IgG2b CD3 binding arm as catumaxomab and has an anti-CD20 mouse IgG2a arm that targets the same epitope as rituximab. FBTA05 displayed very potent cytotoxicity as analyzed with tumor B-cell lines and CLL patient cells and PBMCs of healthy donors with tumor cell elimination already at concentrations in the low picomolar range that was magnitudes of orders lower than for rituximab used as a control [118]. Of importance, FBTA05 induced tumor cell elimination, concordant to results obtained with blinatumomab, is independent from MHC restriction and of any pre- or coactivation. Using autologous CLL patient samples efficient tumor cell killing was induced in 5 of 8 cases, while rituximab did not show any activity.

In addition, FBTA05 displayed efficient tumor cell killing even at very low surface expression of tumor antigen, demonstrated *in vitro* with low expressing CD20$^+$ CLL patient cells that was also seen with the HER2/neu-targeting trAb ertumaxomab using HER2/neu-positive cell lines [133], whereas rituximab or trastuzumab were ineffective in the same setting. Therefore, TAAs that only display a low expression might be suitable targets for trAb mediated tumor therapy, thereby allowing the use of TAAs not suitable for monospecific antibody therapy. Tumor cell killing is accompanied by activation and proliferation of CD4 and CD8 T cells, albeit with preferential proliferation of the CD8 T-cell subset. The observed monocyte activation was dependent on concurrent T-cell binding but independent of concomitant tumor-cell binding. FBTA05 induced a Th1-like cytokine profile with high secretion of IFNγ, IL-6 and IL-2 and no IL-4. Based on that promising preclinical data a proof-of-concept clinical investigation in relapsed or refractory NHL patients after allogeneic SCT was initiated [84].

Allogeneic SCT currently represents the only therapeutic measure for indolent lymphomas, especially CLL. Moreover, donor lymphocyte infusions (DLI) display high treatment efficacy in certain leukemias like CML [134]. In spite of major advancements regarding conditioning regimens, alloSCTs and also DLIs are often accompanied by life threatening graft-*versus*-host-disease (GvHD). Of note, during GvHD graft-*versus*-leukemia (GvL) responses driven by donor T cells are induced that correlate with improved therapeutic outcome. Thus, the most important therapeutic challenge is represented by the dissection of GvHD and GvL. Importantly, by therapeutic intervention with bsAb this therapeutic aim could be achieved as donor T cells could be specifically redirected to the tumor, thereby at the same time GvHD effects are reduced and GvL are strengthened. This dissection could be demonstrated with BiLu in the A16 melanoma mouse model [126, 135]. The administration of BiLu prevented GvHD caused by application of haplo-identical lymphocytes.

Six patients all pretreated with chemotherapy and rituximab or alemtuzumab, three CLL patients with mutated p53, two with DLBCL and on with Burkitt lymphoma, were treated with one course of escalating doses of FBTA05 from 10 to 2,000 µg in combination with DLI or allogeneic stem cells in two cases after FBTA05 course in compassionate use setting. One patient was administered two courses and three patients were treated with additional single applications of FBTA05 in combination with DLI. The antibody was administered as a continuous 6-12 hour I.V. infusion. Number of FBTA05 infusions ranged from 4 to 24 with a total dose from 820 µg up to 13,610 µg. Number of DLIs ranged from one (n=3) up to 5. Cell counts of the DLI administration were in a range from 1×10^6/kg to 1×10^8/kg. Side effects were tolerable and mainly consisted of fever and chills, which appeared at dose levels between 40 and 160 µg in case of CLL patients and at doses ≥200 µg for the other patients. In one DLBCL patient a prolonged period of leucopenia (grade 4) and transient granulocytopenias in all three CLL patients were observed.

All CLL patients showed only transient clinical and hematological responses. In one DLBCL patient a stable disease, lasting for four months was observed. The cytokine release pattern showed temporary increase of IL-6, IL-8 and IL-10 that resolved to base line within 24 hours after FBTA05 application. IL 8 secretion indicates an activation of monocytes *in vivo*, which is in agreement with the *in vitro* data. Contrary to the observed *in vitro* cytokine profile, no IL-2 secretion could be observed *in vivo*. Despite repeated applications of FBTA05 no human-anti-mouse-antibodies could be observed, indicating a favorable immunogenicity of FBTA05 that would allow repeated treatment courses and maintenance therapies. Furthermore, despite repeated and high doses of allogeneic T cells or CD34-positive SCs infused no GvHDs were observed, indicating that redirection of T cells by the bispecific antibody to CD20 positive target cells might indeed suppress GvHD and favor GvL effect. According to the low patient number these findings have to be confirmed in a clinical study with a higher patient number.

This therapeutic regime is currently evaluated in a dose escalation phase I/II study (NCT01138579) for patients with CLL, low and high grade NHL with relapse or refractory disease after allo-SCT [136]. After a first safety cohort with patients receiving 10, 20 and 50 µg doses within one week, patients are medicated with 8 weekly infusions from 50 µg in the first cohort up to 300 µg in the fourth cohort. Donor lymphocyte infusions are administered within one day after the third, seventh and eleventh trAb infusion. FBTA05 showed promising clinical response rates in these heavily pretreated patients already within the first dose escalation cohort with 50 µg per single dose, resulting in an overall dose of 480 µg after completion of the complete treatment cycle [115]. Nine patients, comprising four patients with CLL, two with MCL, two with DLBCL and one with follicular lymphoma, were treated so far within the first study cohort.

Four patients had to be taken off the study before completing treatment, two patients (CLL and FL/DLBCL) due to massive tumor progression, one patient due to the detection of HAMA after the fourth infusion and one patient because of a severe rash. One patient suffered from DLBCL previously transformed from FL and was removed from the study due to tumor progression. He was subsequently treated with higher doses of FBTA05 in compassionate use whereby a temporary SD could be achieved. Interestingly, one CLL patient, who completed the complete treatment cycle without application of DLI, displayed an ongoing SD. So far no DLTs were reported. Adverse events were restricted to fever, chills, nausea and hypotension. In one case, a short episode of bradycardia was reported. According to preliminary data, three patients developed GvHD which could be controlled by subsequent immunosuppressive therapy. So far three patients died, two due to fulminate tumor progression with a median OS in this patient group of 166 days.

Four out of the five patients who completed the entire treatment schedule displayed a complete remission. Of note, two of them (MCL, CLL) are still in remission without any further therapy with an overall survival of 1,270 days and 1,096 days, respectively. Two patients displayed stable disease. Taken together, on overall response rate in this group of heavily pretreated and high risk patients of 67% was achieved so far, albeit with a limited number of patients.

Thus, FBTA05 might not only be a potent anti-tumor drug in stand-alone application but may pave the way for new treatment options like SCT and/or DLI for otherwise therapy resistant CLL- and NHL-patients by significantly reducing risks of severe AEs like GvHD.

Xencor Platform

Besides the triomab platform another technology platform for bispecific antibodies possessing an Fc-part and targeting T cells was developed by the US biotech company Xencor. These antibodies possess a full Fc-domain that was genetically engineered to exchange several amino acids, to abolish binding to Fcγ-receptors in order to reduce the risk for non-selective T-cell activation (Fig. **1H**). On the other hand, to maintain the long serum half-life of full length antibodies, binding to the neonatal Fcn-receptor was preserved. The tumor antigen binding arm of these antibodies consists of a complete Fab-fragment, while the CD3 binding arm was made of a single chain variable fragment. So far, several of these bispecific antibodies were generated, targeting CD20 for treatment of B-cell lymphoma and leukemia [137], CD123 for treatment of acute myelogenous leukemia [138] and CD38 for treatment of multiple myeloma [139] and characterized in a preclinical setting in *in vitro* binding and cytotoxicity assays and in mice and cynomolgus monkeys.

Two anti-CD20 bispecific antibodies were engineered, one with high affinity for CD20 (XmAb 13677) and one with low affinity (XmAb 13676). Both have the identical binding arm and binding affinity for CD3. Both antibodies mediated *in vitro* B-cell binding and killing using B-cell lines and human PBMCs or purified T cells with the high affinity XmAb13677 displaying a 26-fold greater potency compared to XmAb13676. As expected both anti-CD20 antibodies displayed a prolonged serum half-life in mice of 6.7 and 6.6 days, respectively. As both molecules were designed to bind to cynomolgus monkey antigens efficacy was evaluated in these animals with increasing doses from 0.03 to 3 mg/kg body weight.

After bsAb application CD4 as well as CD8 T cells exhibited short term (up to 2 days) activation as depicted by up-regulation of activation markers CD25 and CD69. T cells were rapidly removed from the circulation and returned to baseline within 2 to 7 days indicating that T cells were not lysed but re-directed. A dose-dependent depletion of the circulating healthy B cells up to 97% was observed. Furthermore, B cells in lymph nodes and the bone marrow were depleted up to

90% at all doses tested and at higher dose levels did not recover during the study duration of 29 days. Similar results were obtained with the corresponding anti-CD3 x anti-CD123 and anti-CD3 x anti-CD38 antibodies.

Regeneron Platform

A further technology platform to provide full length humanized bispecific IgG antibodies was invented by Regeneron Pharmaceuticals, based on modified Fc-regions that differ by at least one amino acid leading to different affinities with respect to protein A as principle for efficient purification [140]. The first representative of this class of antibodies (REGN1979) is an analog of FBTA05 and directed against CD20 and CD3 molecules and was evaluated in tumor mouse models and cynomolgus monkeys [141].

As expected this full length IgG bsAb displayed a long serum half-life in cynomolgus monkeys of more than 14 days. Growth of Raji tumor cells in NOD/SCID mice implanted with human PBMCs was completely inhibited if low dose of REGN1979 were administered together with implanted Raji cells. Absence of T cells prevented tumor growth inhibition depicting the expected requirement of T cells. Furthermore, this bispecific antibody induced tumor regression in large advanced Raji tumors associated with long-lasting tumor control. Expression of inhibitory receptors Tim-3 and PD-1 expressed on tumor infiltrating T-lymphocytes was reduced by REGN1979 treatment and lead to tumor regression. REGN1979 was further compared to rituximab and a blinatumomab like bispecific antibody and performed significantly better suppressing established tumors than rituximab albeit the later was administered in a 50-fold greater dosing and was as effective as the BiTE-like molecule. In addition, REGN1979 was also more effective in depleting B-cells in mesenteric lymph nodes of cynomolgus monkeys than rituximab.

It will be interesting to further elucidate the potential of this two classes of antibodies in suitable tumor animal models and in a clinical setting and compare it with smaller bsAb formats like BiTEs, DARTs or tribodies and on the other side with bAbs like triomabs possessing a fully functional Fc-part the enables binding and activating of accessory effector cells like monocytes and macrophages.

Pediatric Patients

Pediatric patients suffering from B cell NHL (B-NHL) and B cell acute lymphoblastic leukemia (B-ALL) show a prominent therapeutic outcome after first-line chemotherapy, with cure rates of approximately 75% and above 90% for high-risk disease and for others, respectively [142-146]. Nevertheless, patients with relapse or refractory disease display a very bad prognosis, especially for subtypes of Burkitt-lymphoma (BL) and B-ALL, as survival rates are below 30% [147-150]. In children with relapse after transplantation re-transplantation offers potential of cure, but relapse rate is still high and represents the major cause of death. The development of targeted mAbs has opened up new therapeutic options by reducing toxicity of current first-line treatment regimens. Thereby, the anti-CD20 antibody rituximab showed promising treatment benefits in both newly diagnosed and refractory B-NHL, when added to a variety of pediatric treatment regimens [151, 152]. Nevertheless, children suffering from refractory or relapsed B-cell malignancies need alternative treatment regimes. In this line, bsAbs represent promising therapeutic measures.

Therefore, blinatumomab was tested in pediatric patients presenting with ALL post-transplant relapse on compassionate use basis [153, 154]. Nine patients with a median age of 10.4 years (range 4.3 up to 18.5 years), all presenting with post-transplant relapses were included. Concordant with adult patients, blinatumomab in pediatric patients was administered as continuous I.V. infusion for 28 days with a dosing schedule from 5 $\mu g/m^2$/day to 15 $\mu g/m^2$/day and 7-14 days without blinatumomab infusions defining one treatment cycle. Patients received total of 18 cycles (1 cycle up to 4 cycles). For second and following cycles the initial dose was increased to 15 $\mu g/m^2$/day. Patients with a molecular remission were intended for haplo-identical SCT.

As minimal residual disease defines an important parameter for long-term event-free- survival, MRD status was monitored before, during and after therapy by multicolor flow cytometry, real time or reverse transcriptase polymerase chain reaction. Four out of 9 patients displayed remission within the first cycle of blinatumomab. Albeit, one of these patients achieving a hematological CR died from gram-negative sepsis during the application of blinatumomab due to a pre-existing long-lasting multidrug resistance *Pseudomonas* infection.

Remarkably, in two further patients resistant to the first cycle of blinatumomab, chemotherapy was administered to stop tumor progression. In one patient, chemotherapy was followed by a SC-boost, before both patients received a second cycle of blinatumomab, finally achieving remission. In 5 of these 6 responders a MRD conversion from positive to negative could be detected. In one of the responders remission was consolidated by two further cycles of blinatumomab and treatment with the tyrosine kinase nilotinib, with an ongoing remission at time of publication of 1,490 days. The other four responders in sustained molecular remission went on to haplo-identical SCT. The two patients responding to the first cycle of blinatumomab were still in molecular remission at time of publication with an OS of 675 and 1,851 days, respectively.

Three patients did not show any response to blinatumomab, albeit one patient received a second cycle with up to 30 $\mu g/m^2/day$ for one day. In summary, six of the 9 patients died with a median overall survival of 166 days, two due to sepsis and the other four children due to tumor progression or relapse. Of note, there was a significant difference between the non-responding and the responding group according to pharmacodynamic responses. Non-responders displayed no change in the composition of mononuclear cells. Regarding responders there was an increase in T-cell number, which peaked within the first week and stayed stable during further cycles and in parallel a decrease in $CD19^+$ blast and non-leukemic cells that were efficiently suppressed in further cycles.

Albeit there was no remarkable difference between the responding and non-responding cycles regarding absolute T-cell count in the peripheral blood, percentage and absolute blast count prior to onset of blinatumomab infusion, there was a significant difference regarding T cell to blast ratio between both groups. A remarkable shift in the T cell to blast ratio in the two second cycle responders could be observed from below 0.2 before the first cycle to over 10 and 100 before the second cycle, respectively. Furthermore the median ratio of all responding cycles (n=12) compared to all non-responding cycles (n=6) was significantly lower. Therefore, the T cell to blast ratio might serve as a biomarker to predict to probability of response to blinatumomab. Thus, in order to achieve a favorable ratio it might be useful in case of refractory patients to reduce the number of leukemic blast before onset of blinatumomab treatment to increase the response rate in this group of patients.

Blinatumomab administration was generally well tolerated in this class of high risk pediatric patients, with grade 3 and 4 toxicities (number of affected patients) consisting of neutropenia (3), anemia (8), low platelet count (6), elevation of bilirubin (3), seizure (2) and cytokine release syndrome (2). Seven of nine patients displayed fever at the beginning of the first cycle. However, blinatumumab administration has not to be terminated in any patient due to treatment-induced AEs.

As blinatumomab showed high response rates in these, albeit restricted number of patients with poor outcome and might facilitate long term survival by inducing complete remissions for subsequent allo-SCT, an ongoing multicenter phase I/II study was initiated (Table **1**, NCT01471782). 41 patients suffering from B-cell precursor ALL presenting with second or later bone marrow relapse (17%), marrow relapse after allo-SCT (63%) or refractoriness to induction or re-induction therapy (20%) were included in the phase I part and administered a total of 73 blinatumomab courses [155]. Dose escalation from 5 to 15 and 30 $\mu g/m^2/day$ and stepwise dosing from 5-15 and 15-30 $\mu g/m^2/day$ were evaluated. Due to grade 4 cytokine release syndromes the MTD was established at 15 $\mu g/m^2/day$ with a recommendation for a stepwise dosing of 5-15 $\mu g/m^2/day$ (5 $\mu g/m^2/day$ for 7 days, 15 $\mu g/m^2/day$ for 21 days) for the phase II part. This dosage form was further evaluated in 18 patients with only one patient developing a grade 3 cytokine release syndrome.

Across all dosing levels 13 patients (32%) achieved a CR, with 10 of these patients becoming MRD negative. Nine of these responders underwent stem cell transplantation. Currently, median OS in the responder group is at 5.7 months and relapse free survival is at 8.3 months with a median follow up of 12.4 months. The most common AEs were pyrexia (78%), headache (37%), hypertension (32%), nausea (29%), abdominal pain (27%), pain in the extremity (27%), and anemia (27%). In addition to the 18 patients already tested in the phase I part with the recommended dosing further 21 patients were so far included in the phase II part of the study [156].

During the first two treatment cycles, 12 patients achieved complete remissions (31%), mainly during the first cycle, with five of them turning to MRD negativity. Six of the 12 responders proceeded to SCT. So far median OS reached 4.3 months with a study follow up time of 6 months. Profile of AEs was comparable to the phase I part, the most frequent grade ≥ 3 AEs included anemia (26%), pyrexia (21%), increased transaminases (18%) and febrile neutropenia (15%). Two patients (5%) acquired a grade 3 cytokine-release syndrome. Besides blinatumomab the trifunctional bispecific antibody FBTA05 is clinically evaluated in CD20-positive pediatric leukemia and lymphoma patients, albeit contrary to blinatumomab focusing on lymphoma patients [157].

Front-line chemotherapy in childhood mature B-NHL results in a beneficial treatment outcome of approximately 90% cure rates [146, 158-159]. However, in case of recurrence or refractoriness the current survival rates for DLBCL are 40-60% and for BL and B-ALL even below 30%, respectively. Ten refractory or relapsed patients, intensively pretreated including chemotherapy (n=10), rituximab (n=5), radiation (n=3) and SCT (n=5), were included in this compassionate use based clinical investigation. Patients were stratified into two cohorts.

Six patients suffering from Burkitt-Lymphoma (n=3), Burkitt leukemia (n=2) and pre B-acute lymphoblastic leukemia (n=1) were treated with one course of daily escalated doses of FBTA05 (dose range: 10-1,000 µg) for 5 to 7 days followed by intensive chemotherapy to eradicate residual disease before allo-SCT, except in one patient (pre-B-ALL) where FBTA05 was followed by repeated applications of dose-adjusted mercaptopurine (50 mg/m^2). One patient received 4 cycles of daily application.

In another four patients, diseased with diffuse large B-cell lymphoma (n = 3) and Burkitt leukemia (n = 1), as well as one patient of the "daily cohort" suffering from Burkitt leukemia, FBTA05 treatment (10-300 µg) was initiated weekly after preceding SCT. After a drug introduction part (safety part) of three FBTA05 administrations within one week (10, 20 and 50 µg; absolute doses), FBTA05 was applied weekly (dose range: 100-300 µg). Total FBTA05 applications varied from 80 µg (3 infusions) up to 6520 µg (27 infusions). In addition, two of these patients received post SCT donor lymphocyte infusions in escalating doses (2 x 10^4/kg – 5 x 10^6/kg CD3$^+$ T cells) every fourth week. Interestingly, in one patient DLI and FBTA05 applications were further combined with the immunomodulatory drug lenalidomide (5-15 mg/kg/day) within a second treatment course.

In five of 10 pediatric patients (50%) who obtained FBTA05 amongst their diverse multimodal treatment regimes complete remissions were induced or sustained. One patient with Burkitt leukemia achieved a complete hematological and molecular response after stand-alone treatment with FBTA05, as assessed by the complete molecular remission of the c-myc marker in blast cells of the bone marrow. However, the patient died due to a CNS relapse. Of note, after the seventh FBTA05 application (1,000 µg) the antibody could not be detected in the cerebrospinal fluid, indicating a prevention of penetration of the blood brain barrier. Interestingly, impaired CNS antibody infiltration was also observed for rituximab [159-161].

In two patients a partial remission and in one patient a long lasting stable disease was observed. 4 out of the 5 patients with CR were still alive at time of publication [157] with a median overall survival of 1482 days (1205 -1717 days) without any further anti-tumor treatments. Four patients died to tumor progression, one due to pulmonary embolism and one due to pneumonia. FBTAA05 applications were well tolerated with main adverse reactions restricted to fever (90%) and chills (30%). Patients displayed a similar cytokine release pattern as adult patients treated with FBTA05, characterized by transient IL-6, IL-8 and IL-10 secretion [84], which resolved to initial value within 12 to 24 hours with no cytokine release syndrome ≥ grade 3. Thereby IL-8 elevation indicates activation of monocytes and involvement of neutrophil granulocytes which may be induced by Fcγ-receptor engagement *via* the intact Fc-part of FBTA05. Interestingly, phagocytosis of tumor B cells opsonized with FBTA05 may also contribute to tumor cell destruction in pediatric patients. An increase of transaminases (grade 3) after FBTA05 application could be measured in three patients (30%). A grade 3 granulocytopenia was observed in one patient and a decline of platelets (grade 4) in two patients. In one patient FBTA05 treatment had to be stopped after 3 infusions due to caput medusa. Remarkably, this patient achieved an ongoing complete remission at time of publication without any further treatment since then. One patient receiving FBTA05 in combination with DLI displayed a severe GvHD (grade 3-4) after the last FBTA05 infusion and DLI application that could be resolved by combined immunosuppressive treatment with prednisolone, extracorporal photopheresis (ECP) and everolimus. This GvHD reaction might be caused by low tumor burden at the end of the FBTA05 cycle which would be in line with observations by Morecki and colleagues [126, 135], that low tumor burden increases the risk of GvHD due to a reduced number of targets for trAb mediated redirection of allogeneic T cells.

In line with results obtained with adult patients HAMA development only plays a minor role in pediatric patients treated with FBTA05 despite the mouse/rat nature of this antibody. Only one of the 10 patients developed a significant HAMA reaction after the fifth infusion that peaked at 18,200ng/ml and resolved to baseline within four month. Remarkably, this patient was administered two further doses of FBTA05 without any adverse reactions. In one other patient a weak HAMA reaction marginally above the detection limit was observed after the 35th infusion and resolved within weeks in spite of ongoing FBTA05 therapy. Due to the low immunogenicity of FBTA05 in adult and pediatric patients repeated applications of this trAb could be administered, allowing not only initial anti-tumor therapies but also long-lasting maintenance treatments without the need of antibody humanization.

Interestingly, combination of FBTA05 with lenalidomide was well tolerated, especially after adjusting lenalidomide doses from 15 to 5 mg daily. The combination of this immunomodulatory drug that among other mechanisms improves formation of the immunological synapse between T cell and tumor cell and a T-cell redirecting trifunctional antibody could be a promising future strategy to combat tumors even in pediatric patients.

Taken together, the currently available data, albeit with limited number of patients, regarding treatment of refractory and relapsed pediatric B-cell lymphoma and leukemia patients with the bispecific antibodies blinatumomab and FBTA05 show impressive response rates and duration in this group of high risk patients. Therefore, bispecific antibodies alone or in combination with for instance stem cell transplantation and/or immunomodulatory drugs might offer new treatment options to induce long lasting remissions or even cure in these patients.

CONCLUDING REMARKS

In the last fifteen years bispecific antibodies have proven their efficacy. A multitude of encouraging results, not only *in vitro* and in animal models but also in patients has demonstrated the potency of this antibody format initiating a lot of effort to develop new platforms, characterize new targets, and design new treatment strategies not only for treatment of cancer but also of other diseases like inflammatory or autoimmune diseases. Given the potency and the capabilities of bsAbs, this kind of molecules raised enormous interest and therefore, a lot of new platforms were created to generate new bispecific molecules. New approaches like DVD-Ig (dual-variable domain immunoglobulin), SEED (strand-exchanged engineered domain), the Azymetric scaffold, Kappa-lambda bodies, mAB^{2TM} (modular bsAbs), TBTI (tetravalent bispecific tandem Ig) or knob into hole (KIH) and CrossMab were designed and are currently under intensive investigation [162-165].

Furthermore, a plethora of new other drugs like genetically engineered T-cells expressing chimeric antigen receptors (CARs) [166], immunomodulatory drugs like lenalidomide or pomalidomide [167, 168], specific kinase inhibitors like PI-3 kinase inhibitor Idelalisib (Zydelig®) [169] or Bruton's tyrosin kinase inhibitor Ibrutinib (Imbruvica®) [170] or inhibitors of immunological checkpoints like programmed death 1 (PD-1) pathway [171] are under intense clinical evaluation or already approved [172]. Albeit, due to the multiplicity of bi- and multispecific platform technologies and the proven activity in clinical evaluations, this class of antibodies will be an indispensable tool to fight hematological malignancies as monotherapeutic or as combination therapy with cell therapies and small molecule drugs.

Furthermore, combination of bsAbs with these new drugs could create new synergies and might thus allow more efficient and targeted tumor cell destruction and enable personalized medicine to improve response rates and long term tumor free survival of patients.

ACKNOWLEDGEMENTS

Declared None.

CONFLICT OF INTEREST

The author(s) confirm that this chapter contents have no conflict of interest.

REFERENCES

[1]   Reichert J [Homepage on the Internet]. Waban Antibody Society c2007-13 [cited: 26th December 2014]. Available from: http://www.antibodysociety.org/news/approved_mabs.php
[2]   U S Food and Drug Administration [Homepage on the Internet]. Drug approvals and databasis; Hematology/Oncology (Cancer) Approvals & Safety Notifications. [cited: 26th December 2014] Available from: http://www.fda.gov/Drugs/InformationOnDrugs/ApprovedDrugs/ucm279174.htm
[3]   Smith MR. Rituximab (monoclonal anti-CD20 antibody): mechanisms of action and resistance. Oncogene 2003; 22(47); 7359-68.
[4]   Dalle S, Reslan L, Besseyre de Horts T, *et al*. Preclinical studies on the mechanism of action and the anti-lymphoma activity of the novel anti-CD20 antibody GA101. Mol Cancer Ther 2011; 10(1); 178-85.
[5]   Peipp M and Valerius T. Bispecific antibodies targeting cancer cells. Biochem Soc Trans 2002; 30(4); 507-11.
[6]   Milstein C and Cuello AC. Hybrid hybridomas and their use in immunohistochemistry. Nature 1983; 305(5934); 537-40.
[7]   Milstein C, Cuello AC. Hybrid hybridomas and the production of bi-specific monoclonal antibodies. Immunol Today 1984; 5(10); 299-304.
[8]   Staerz UD, Kanagawa O and Bevan MJ. Hybrid antibodies can target sites for attack by T cells. Nature 1985; 314(6012); 628-31.
[9]   Kipriyanov SM, Moldenhauer G, Strauss G and Little M. Bispecific CD3 x CD19 diabody for T cell-mediated lysis of malignant human B cells. Int J Cancer 1998; 77(5); 763-72.
[10]   Xiong D, Xu Y, Liu H, *et al*. Efficient inhibition of human B-cell lymphoma xenografts with an anti-CD20 x anti-CD3 bispecific diabody. Cancer Lett 2002; 177(1); 29-39.
[11]   Chames P and Baty D. Bispecific antibodies for cancer therapy. Curr Opin Drug Discov Devel 2009; 12(2); 276-83.
[12]   Löffler A, Kufer P, Lutterbüse R, *et al*. A recombinant bispecific single-chain antibody, CD19 x CD3, induces rapid and high lymphoma-directed cytotoxicity by unstimulated T lymphocytes. Blood 2000; 95(6); 2098-103.
[13]   Nagorsen D, Bargou R, Ruttinger D, Kufer P, Baeuerle PA, Zugmaier G. Immunotherapy of lymphoma and leukemia with T-cell engaging BiTE antibody blinatumomab. Leuk Lymphoma 2009; 50(6); 886-91.
[14]   d'Argouges S, Wissing S, Brandl C, *et al*. Combination of rituximab with blinatumomab (MT103/MEDI-538), a T cell-engaging CD19-/CD3-bispecific antibody, for highly efficient lysis of human B lymphoma cells. Leuk Res 2009; 33(3); 465-73.
[15]   Wolf E, Hofmeister R, Kufer P, Schlereth B, Baeuerle PA. BiTEs: bispecific antibody constructs with unique anti-tumor activity. Drug Discov Today 2005; 10(18); 1237-44.
[16]   Dreier T, Lorenczewski G, Brandl C, *et al*. Extremely potent, rapid and costimulation-independent cytotoxic T-cell response against lymphoma cells catalyzed by a single-chain bispecific antibody. Int J Cancer 2002; 100(6); 690-7.
[17]   Gruen M, Bommert K, Bargou RC. T-cell-mediated lysis of B cells induced by a CD19xCD3 bispecific single-chain antibody is perforin dependent and death receptor independent. Cancer Immunol Immunother 2004; 53(7); 625-32.
[18]   Baeuerle PA, Reinhardt C. Bispecific T-cell engaging antibodies for cancer therapy. Cancer Res 2009; 69(12); 4941-4.
[19]   Brandl C, Haas C, d'Argouges S, *et al*. The effect of dexamethasone on polyclonal T cell activation and redirected target cell lysis as induced by a CD19/CD3-bispecific single-chain antibody construct. Cancer Immunol Immunother 2007; 56(10); 1551-63.

[20]   Hoffmann P, Hofmeister R, Brischwein K, *et al*. Serial killing of tumor cells by cytotoxic T cells redirected with a CD19-/CD3-bispecific single-chain antibody construct. Int J Cancer 2005; 115(1); 98-104.

[21]   Löffler A, Gruen M, Wuchter C, *et al*. Efficient elimination of chronic lymphocytic leukaemia B cells by autologous T cells with a bispecific anti-CD19/anti-CD3 single-chain antibody construct. Leukemia 2003; 17(5); 900-9.

[22]   Dreier T, Baeuerle PA, Fichtner I, *et al*. T cell costimulus-independent and very efficacious inhibition of tumor growth in mice bearing subcutaneous or leukemic human B cell lymphoma xenografts by a CD19-/CD3- bispecific single-chain antibody construct. J Immunol 2003; 170(8); 4397-402.

[23]   Zimmerman Z, Maniar T, Nagorsen D. Unleashing the clinical power of T cells: CD19/CD3 bi-specific T cell engager (BiTE®) antibody construct blinatumomab as a potential therapy. Int Immunol 2015; 27(1); 31-37.

[24]   Bargou R, Leo E, Zugmaier G, *et al*. Tumor regression in cancer patients by very low doses of a T cell-engaging antibody. Science 2008; 321(5891); 974-7.

[25]   Viardot A, Goebeler M, Scheele JS, *et al*. Treatment of Patients with Non-Hodgkin Lymphoma (NHL) with CD19/CD3 Bispecific Antibody Blinatumomab (MT103): Double-Step Dose Increase to Continuous Infusion of 60 μg/m²/d Is Tolerable and Highly Effective. ASH annual meeting 2010; abstract 2880.

[26]   Göbeler M, Viardot A, Noppeney R, *et al*. Blinatumomab (CD3/CD19 Antibody) results in a high response rate in patients with relapsed Non Hodgkin Lymphoma (NHL) including MCL and DLBCL. Ann Oncol 2011; 22(suppl 4); iv105.

[27]   Viardot A, Göbeler M, Noppeney R, *et al*. Blinatumomab Monotherapy Shows Efficacy in Patients with Relapsed Diffuse Large B Cell Lymphoma. ASH annual meeting 2011; abstract 1637.

[28]   Viardot A, Goebeler M, Hess G, *et al*. Treatment of Relapsed/Refractory Diffuse Large B-Cell Lymphoma with the Bispecific T-Cell Engager (BiTE®) Antibody Construct Blinatumomab: Primary Analysis Results from an Open-Label, Phase 2 Study. ASH annual meeting 2014; abstract 4460.

[29]   Topp MS, Kufer P, Gökbuget N, *et al*. Targeted therapy with the T-cell-engaging antibody blinatumomab of chemotherapy-refractory minimal residual disease in B-lineage acute lymphoblastic leukemia patients results in high response rate and prolonged leukemia-free survival. J Clin Oncol 2011; 29(18); 2493-8.

[30]   Topp MS, Gökbuget N, Zugmaier G, *et al*. Long-term follow-up of hematologic relapse-free survival in a phase 2 study of blinatumomab in patients with MRD in B-lineage ALL. Blood 2012; 120(26); 5185-7.

[31]   Klinger M, Brandl C, Zugmaier G, *et al*. Immunopharmacologic response of patients with B-lineage acute lymphoblastic leukemia to continuous infusion of T cell-engaging CD19/CD3-bispecific BiTE antibody blinatumomab. Blood 2012; 119(26); 6226-33.

[32]   Zugmaier G, Topp MS, Alekar S, *et al*. Long-term follow-up of serum immunoglobulin levels in blinatumomab-treated patients with minimal residual disease-positive B-precursor acute lymphoblastic leukemia. Blood Cancer J 2014; 4; 244.

[33]   Topp MS, Gökbuget N, Zugmaier G, *et al*. Phase II Trial of the Anti-CD19 Bispecific T Cell-Engager Blinatumomab Shows Hematologic and Molecular Remissions in Patients With Relapsed or Refractory B-Precursor Acute Lymphoblastic Leukemia. J Clin Oncol 2014; 32(36); 4134-40.

[34]   Topp MS, Gökbuget N, Stein AS, *et al*. Safety and activity of blinatumomab for adult patients with relapsed or refractory B-precursor acute lymphoblastic leukaemia: a multicentre, single-arm, phase 2 study. Lancet Oncol 2015; 16(1); 57-66.

[35]   US Food and Drug Administration [Homepage on the Internet]. Drug approvals and databasis; Blinatumomab. [cited: 3rd December 2014] Available from: http://www.fda.gov/Drugs/InformationOnDrugs/ApprovedDrugs/ucm425597.htm

[36]   Herrmann I, Baeuerle PA, Friedrich M, *et al*. Highly efficient elimination of colorectal tumor-initiating cells by an EpCAM/CD3-bispecific antibody engaging human T cells. PLoS One 2010; 5(10); e13474.

[37]   Oberst MD, Fuhrmann S, Mulgrew K, *et al*. CEA/CD3 bispecific antibody MEDI-565/AMG 211 activation of T cells and subsequent killing of human tumors is independent of mutations commonly found in colorectal adenocarcinomas. MAbs 2014; 6(6); 1571-84.

[38]   Friedrich M, Raum T, Lutterbuese R, *et al*. Regression of human prostate cancer xenografts in mice by AMG 212/BAY2010112, a novel PSMA/CD3-Bispecific BiTE antibody cross-reactive with non-human primate antigens. Mol Cancer Ther 2012; 11(12); 2664-73.

[39]   Baeuerle P, Zugmaier G and Rüttinger D (2011) Bispecific T cell engager for cancer therapy. In: Kontermann RE, Ed. Bispecific Antibodies. Heidelberg: Springer 2011; pp 273-288.

[40] Walter RB. Biting back: BiTE antibodies as a promising therapy for acute myeloid leukemia. Expert Rev Hematol 2014; 7(3); 317-9.

[41] Aigner M, Feulner J, Schaffer S, *et al*. T lymphocytes can be effectively recruited for *ex vivo* and *in vivo* lysis of AML blasts by a novel CD33/CD3-bispecific BiTE antibody construct. Leukemia 2013; 27(5); 1107-15.

[42] Friedrich M, Henn A, Raum T, *et al*. Preclinical characterization of AMG 330, a CD3/CD33-bispecific T-cell-engaging antibody with potential for treatment of acute myelogenous leukemia. Mol Cancer Ther 2014; 13(6); 1549-57.

[43] Krupka C, Kufer P, Kischel R, *et al*.C D33 target validation and sustained depletion of AML blasts in long-term cultures by the bispecific T-cell-engaging antibody AMG 330. Blood 2014; 123(3); 356-65.

[44] Laszlo GS, Gudgeon CJ, Harrington KH, *et al*. Cellular determinants for preclinical activity of a novel CD33/CD3 bispecific T-cell engager (BiTE) antibody, AMG 330, against human AML. Blood 2014; 123(4); 554-61.

[45] Scheinberg DA, Lovett D, Divgi CR, *et al*. A phase I trial of monoclonal antibody M195 in acute myelogenous leukemia: specific bone marrow targeting and internalization of radionuclide. J Clin Oncol 1991; 9(3); 478-90.

[46] van Der Velden VH, te Marvelde JG, Hoogeveen PG, *et al*. Targeting of the CD33-calicheamicin immunoconjugate Mylotarg (CMA-676) in acute myeloid leukemia: *in vivo* and *in vitro* saturation and internalization by leukemic and normal myeloid cells. Blood 2001; 97(10); 3197-204.

[47] Stein C, Schubert I and Fey GH. Trivalent and trispecific antibody derivates for cancer therapy. In: Kontermann RE, Ed. Bispecific Antibodies. Heidelberg: Springer 2011; pp 65-82.

[48] Kellner C, Bruenke J, Stieglmaier J, *et al*. A novel CD19-directed recombinant bispecific antibody derivative with enhanced immune effector functions for human leukemic cells. J Immunother 2008; 31(9); 871-84.

[49] Roskopf CC, Schiller CB, Braciak TA, *et al*. T cell-recruiting triplebody 19-3-19 mediates serial lysis of malignant B-lymphoid cells by a single T cell. Oncotarget 2014; 5(15); 6466-83.

[50] Kügler M, Stein C, Kellner C, *et al*. A recombinant trispecific single-chain Fv derivative directed against CD123 and CD33 mediates effective elimination of acute myeloid leukaemia cells by dual targeting. Br J Haematol 2010; 150(5); 574-86.

[51] Schubert I, Saul D, Nowecki S, *et al*. A dual-targeting triplebody mediates preferential redirected lysis of antigen double-positive over single-positive leukemic cells. MAbs 2014; 6(1); 286-96.

[52] Schubert I, Kellner C, Stein C, *et al*. A recombinant triplebody with specificity for CD19 and HLA-DR mediates preferential binding to antigen double-positive cells by dual-targeting. MAbs 2012; 4(1); 45-56.

[53] Schubert I, Kellner C, Stein C, *et al*. A single-chain triplebody with specificity for CD19 and CD33 mediates effective lysis of mixed lineage leukemia cells by dual targeting. MAbs. 2011; 3(1); 21-30.

[54] Singer H, Kellner C, Lanig H, *et al*. Effective elimination of acute myeloid leukemic cells by recombinant bispecific antibody derivatives directed against CD33 and CD16. J Immunother 2010; 33(6); 599-608.

[55] R. Schoonjans, A. Willems, S. Schoonooghe, *et al*. Fab chains as an efficient heterodimerization scaffold for the production of recombinant bispecific and trispecific antibody derivatives, J Immunol 2000; 165(12); 7050–7.

[56] R. Schoonjans, A. Willems, S. Schoonooghe, *et al*. A new model for intermediate molecular weight recombinant bispecific and trispecific antibodies by efficient heterodimerization of single chain variable domains through fusion to a Fab-chain. Biomol Eng 2001; 17(6); 193–202.

[57] V. Quintero-Hernandez, V.R. Juarez-Gonzalez, M. Ortiz-Leon, *et al*. The change of the scFv into the Fab format improves the stability and *in vivo* toxin neutralization capacity of recombinant antibodies, Mol. Immunol 2007; 44(6); 1307–15.

[58] Mertens N. Tribodies: Fab-scFv fusion proteins as a platform to create multifunctional pharmaceuticals. In: Kontermann R, Ed. Bispecific Antibodies. Heidelberg: Springer 2011; pp. 135-49.

[59] Eyrich M, Lang P, Lal S, *et al*. A prospective analysis of the pattern of immune reconstitution in a paediatric cohort following transplantation of positively selected human leucocyte antigen disparate haematopoietic stem cells from parental donors. Br. J. Haematol 2001; 114(2); 422–32.

[60] Kellner C, Bruenke J, Horner H, *et al*. Heterodimeric bispecific antibody-derivatives against CD19 and CD16 induce effective antibody-dependent cellular cytotoxicity against B-lymphoid tumor cells. Cancer Lett 2011; 303(2);128-39.

[61] Glorius P, Baerenwaldt A, Kellner C, *et al*. The novel tribody [(CD20)2xCD16] efficiently triggers effector cell-mediated lysis of malignant B cells. Leukemia 2013; 27(1); 190-201,

[62] Cartron G, Dacheux L, Salles G, *et al*, Therapeutic activity of humanized anti-CD20 monoclonal antibody and polymorphism in IgG Fc receptor FcgammaRIIIa gene. Blood 2002; 99(3); 754–8.

[63]   Chang CH, Rossi EA, Sharkey RM, *et al*. The Dock-and-Lock (DNL) approach to novel bispecific antibodies. In: Kontermann RE, Ed. Bispecific Antibodies. Heidelberg: Springer 2011; pp 199-216.

[64]   Chang CH, Rossi EA, Goldenberg DM. The dock and lock method: a novel platform technology for building multivalent, multifunctional structures of defined composition with retained bioactivity. Clin Cancer Res 2007; 13(18 pt 2); 5586s-91s.

[65]   Leonard JP, Coleman M, Ketas J, *et al*. Combination antibody therapy with epratuzumab and rituximab in relapsed or refractory non-Hodgkin's lymphoma. J Clin Oncol 2005; 23(22); 5044-51.

[66]   Rossi EA, Goldenberg DM, Cardillo TM, *et al*. Hexavalent bispecific antibodies represent a new class of anticancer therapeutics, 1: Properties of anti-CD20/CD22 antibodies in lymphoma. Blood 2009; 113(24); 6161-71.

[67]   Gupta P, Goldenberg DM, Rossi EA, *et al*. Multiple signaling pathways induced by hexavalent, monospecific, anti-CD20 and hexavalent, bispecific, anti-CD20/CD22 humanized antibodies correlate with enhanced toxicity to B-cell lymphomas and leukemias. Blood 2010; 116(17); 3258-67.

[68]   Alinari L, Yu B, Christian BA, *et al*. Combinationanti-CD74 (milatuzumab) and anti-CD20 (rituximab) monoclonal antibody therapy has *in vitro* and *in vivo* activity in mantle cell lymphoma. Blood 2011; 117(17); 4530-41.

[69]   Gupta P, Goldenberg DM, Rossi EA, *et al*. Dual-targeting immunotherapy of lymphoma: potent cytotoxicity of anti-CD20/CD74 bispecific antibodies in mantle cell and other lymphomas. Blood 2012; 119(16); 3767-78.

[70]   Vallera DA, Todhunter DA, Kuroki DW, *et al*. A bispecific recombinant immunotoxin, DT2219, targeting human CD19 and CD22 receptors in a mouse xenograft model of B-cell leukemia/lymphoma. Clin Cancer Res 2005; 11(10); 3879-88.

[71]   Vallera DA, Chen H, Sicheneder AR, *et al*. Genetic alteration of a bispecific ligand-directed toxin targeting human CD19 and CD22 receptors resulting in improved efficacy against systemic B cell malignancy. Leuk Res; 33(9); 1233-42

[72]   Bachanova V, Frankel A, Cao Q, *et al*. Remission Induction in a Phase 1 Study of a Bispecific Single Chain Immunotoxin Targeting CD22 and CD19 (DT2219) for Refractory B-Cell Malignancies. ASH annual meeting 2014; abtract 3098.

[73]   Cochlovius B, Kipriyanov SM, Stassar MJ, *et al*. Treatment of human B cell lymphoma xenografts with a CD3 x CD19 diabody and T cells. J Immunol 2000; 165(2); 888-95.

[74]   Kipriyanov SM, Cochlovius B, Schäfer HJ, *et al*. Synergistic anti-tumor effect of bispecific CD19 x CD3 and CD19 x CD16 diabodies in a preclinical model of non-Hodgkin's lymphoma. J Immunol 2002; 169(1); 137-44.

[75]   Schlenzka J, Moehler TM, Kipriyanov SM, *et al*. Combined effect of recombinant CD19 x CD16 diabody and thalidomide in a preclinical model of human B cell lymphoma. Anticancer Drugs 2004; 15(9); 915-9.

[76]   Müller D and Kontermann RE. Diabodies, single-chain diabodies, and their derivates. In: Kontermann RE, Ed. Bispecific Antibodies. Heidelberg: Springer 2011; pp 83-100.

[77]   Johnson S, Burke S, Huang L, *et al*. Effector cell recruitment with novel Fv-based dual-affinity re-targeting protein leads to potent tumor cytolysis and *in vivo* B-cell depletion. J Mol Biol. 2010;399(3); 436-49.

[78]   Veri MC, Burke S, Huang L *et al*. Therapeutic control of B cell activation *via* recruitment of Fcgamma receptor IIb (CD32B) inhibitory function with a novel bispecific antibody scaffold. Arthritis Rheum 2010; 62(7);1933-43.

[79]   Moore PA, Zhang W, Rainey GJ, *et al*. Application of dual affinity retargeting molecules to achieve optimal redirected T-cell killing of B-cell lymphoma. Blood. 2011; 117(17);4542-51.

[80]   Rader C. DARTs take aim at BiTE. Blood 2011; 117(17); 4403-04.

[81]   Zlabinger GJ, Stuhlmeier KM, Eher R, *et al*. Cytokine release and dynamics of leukocyte populations after CD3/TCR monoclonal antibody treatment. J Clin Immunol 1992;12(3); 170-7.

[82]   Liu L, Lam A, Alderson R, *et al*. MGD011, Humanized CD19 x CD3 DART® Protein with Enhanced Pharmacokinetic Properties, Demonstrates Potent T-Cell Mediated Anti-Tumor Activity in Preclinical Models and Durable B-Cell Depletion in Cynomolgus Monkeys Following Once-a-Week Dosing. ASH annual meeting 2014; abstract 1775.

[83]   Moore P, Chichili GR, Huang L, *et al*. Preclinical Activity and Safety of MGD006/S80880, a CD123 x CD3 Bispecific DART® Molecule for the Treatment of Hematological Malignancies. 26th EORTC Meeting 2014; abstract 138.

[84]   Buhmann R, Simoes B, Stanglmaier M, *et al.* Immunotherapy of recurrent B-cell malignancies after allo-SCT with Bi20 (FBTA05), a trifunctional anti-CD3 x anti-CD20 antibody and donor lymphocyte infusion. Bone Marrow Transplant 2009; 43(5); 383-97.

[85]   Nitta T, Sato K, Yagita H, *et al.* Preliminary trial of specific targeting therapy against malignant glioma. Lancet 1990; 335(8686); 368-71.

[86]   Jung G, Brandl M, Eisner W, *et al.* Local immunotherapy of glioma patients with a combination of 2 bispecific antibody fragments and resting autologous lymphocytes: evidence for *in situ* t-cell activation and therapeutic efficacy. Int J Cancer 2001; 91(2); 225-30.

[87]   Lamers CH, van de Griend RJ, Braakman E, *et al.* Optimization of culture conditions for activation and large-scale expansion of human T lymphocytes for bispecific antibody-directed cellular immunotherapy. Int J Cancer 1992; 51(6); 973-9.

[88]   Lamers CH, Bolhuis RL, Warnaar SO, Stoter G, Gratama JW. Local but no systemic immunomodulation by intraperitoneal treatment of advanced ovarian cancer with autologous T lymphocytes re-targeted by a bi-specific monoclonal antibody. Int J Cancer 1997; 73(2); 211-9.

[89]   Canevari S, Stoter G, Arienti F, *et al.* Regression of advanced ovarian carcinoma by intraperitoneal treatment with autologous T lymphocytes retargeted by a bispecific monoclonal antibody. J Natl Cancer Inst 1995; 87(19); 1463-9.

[90]   Riechelmann H, Wiesneth M, Schauwecker P, *et al.* Adoptive therapy of head and neck squamous cell carcinoma with antibody coated immune cells: a pilot clinical trial. Cancer Immunol Immunother 2007; 56(9); 1397-406.

[91]   Lum LG and Thakur A. Bispecific antibodies for arming activated T cells and other effector cells for tumor therapy. In: Kontermann RE, Ed. Bispecific Antibodies. Heidelberg: Springer 2011; pp 243-71.

[92]   Lum LG, Thakur A, Liu Q, *et al.* CD20-targeted T cells after stem cell transplantation for high risk and refractory non-Hodgkin's lymphoma. Biol Blood Marrow Transplant 2013; 19(6); 925-33.

[93]   Sen M, Wankowski DM, Garlie NK, *et al.* Use of anti-CD3 x anti-HER2/neu bispecific antibody for redirecting cytotoxicity of activated T cells toward HER2/neu+ tumors. J Hematother Stem Cell Res. 2001; 10(2); 247-60.

[94]   Lum HE, Miller M, Davol PA, *et al.* Preclinical studies comparing different bispecific antibodies for redirecting T cell cytotoxicity to extracellular antigens on prostate carcinomas. Anticancer Res. 2005; 25(1A); 43-52.

[95]   Grabert RC, Cousens LP, Smith JA, *et al.* Human T cells armed with Her2/neu bispecific antibodies divide, are cytotoxic, and secrete cytokines with repeated stimulation. Clin Cancer Res 2006; 12(2); 569-76.

[96]   Davol PA, Smith JA, Kouttab N, *et al.* Anti-CD3 x anti-HER2 bispecific antibody effectively redirects armed T cells to inhibit tumor development and growth in hormone-refractory prostate cancer-bearing severe combined immunodeficient beige mice. Clin Prostate Cancer 2004. 3(2); 112-21.

[97]   Lum LG and Thakur A. Targeting T cells with bispecific antibodies for cancer therapy. BioDrugs 2011; 25(6); 365-79.

[98]   Thakur A and Lum LG. Cancer therapy with bispecific antibodies: Clinical experience. Curr Opin Mol Ther 2010; 12(3); 340-9.

[99]   Gall JM, Davol PA, Grabert RC, *et al.* T cells armed with anti-CD3 x anti-CD20 bispecific antibody enhance killing of CD20+ malignant B cells and bypass complement-mediated rituximab resistance *in vitro*. Exp Hematol 2005; 33(4); 452-9.

[100]   Yano H, Thakur A, Tomaszewski EN, *et al.* Ipilimumab augments anti-tumor activity of bispecific antibody-armed T cells. J Transl Med 2014; 12; 191.

[101]   Lum LG, Thakur A, Pray C, *et al.* Multiple infusions of CD20-targeted T cells and low-dose IL-2 after SCT for high-risk non-Hodgkin's lymphoma: a pilot study. Bone Marrow Transplant 2014; 49(1); 73-9.

[102]   Lum LG, Thakur A, Al-Kadhimi Z, *et al.* Induction Of Anti-Myeloma Cellular and Humoral Immunity By Pre-Targeting Clonogenic Myeloma Cells Prior To Stem Cell Transplant With T Cells Armed With Anti-CD3 x Anti-CD20 Bispecific Antibody Leads To Transfer Of Cellular and Humoral Anti-Myeloma Immunity. ASH annual meeting 2013; abstract 139.

[103]   Golay J, D'Amico A, Borleri G, *et al.* A novel method using blinatumomab for efficient, clinical-grade expansion of polyclonal T cells for adoptive immunotherapy. J Immunol. 2014; 193(9); 4739 17.

[104]   Ramsay, A. G., A. J. Clear, R. Fatah, *et al.* Multiple inhibitory ligands induce impaired T-cell immunologic synapse function in chronic lymphocytic leukemia that can be blocked with lenalidomide: establishing a reversible immune evasion mechanism in human cancer. Blood 2012; 120 (7); 1412–21.

[105] Cochlovius B, Kipriyanov SM, Stassar MJ, *et al.* Cure of Burkitt's lymphoma in severe combined immunodeficiency mice by T cells, tetravalent CD3 x CD19 tandem diabody, and CD28 costimulation. Cancer Res 2000; 60(16); 4336-41.

[106] Kipriyanov SM, Moldenhauer G, Schuhmacher J, *et al.* Bispecific tandem diabody for tumor therapy with improved antigen binding and pharmacokinetics. J Mol Biol 1999; 293(1); 41-56.

[107] Le Gall F, Reusch U, Little M, Kipriyanov SM. Effect of linker sequences between the antibody variable domains on the formation, stability and biological activity of a bispecific tandem diabody. Protein Eng Des Sel 2004; 17(4); 357-66.

[108] Reusch U, Le Gall F, Hensel M, *et al.* Effect of tetravalent bispecific CD19xCD3 recombinant antibody construct and CD28 costimulation on lysis of malignant B cells from patients with chronic lymphocytic leukemia by autologous T cells. Int J Cancer 2004; 112(3); 509-18.

[109] Reusch U, Burkhardt C, Knackmuss, *et al.* High affinity CD3 RECRUIT-TandAbs for Tcell-mediated lysis of malignant CD19+ cells. AACR Annual meeting 2012; abstract 4624.

[110] Zhukovsky EA, Reusch U, Burkhardt C, *et al.* Bispecific TandAbs: a safe and potent platform for T cell-mediated killing of CD19+ cells. AACR Annual meeting 2013; abstract 1243.

[111] Reusch U, Burkhardt C, Fucek I, *et al.* A novel tetravalent bispecific TandAb (CD30/CD16A) efficiently recruits NK cells for the lysis of CD30+ tumor cells. MAbs 2014; 6(3); 728-39.

[112] Rothe A, Sasse S, Topp MS, *et al.* Specific NK Cell Activation to Treat Relapsed/Refractory (r/r) Hodgkin Lymphoma (HL) Patients – Final, Updated Data on Clinical Outcome, Pharmacokinetics and Pharmacodynamics of a Phase 1 Study Investigating AFM13, a Bispecific Anti-CD30/CD16A Tandab. ASH Annual meeting 2014; abstract 1753.

[113] Mølhøj M1, Crommer S, Brischwein K, *et al.* CD19-/CD3-bispecific antibody of the BiTE class is far superior to tandem diabody with respect to redirected tumor cell lysis. Mol Immunol 2007; 44(8); 1935-43.

[114] Kipriyanov, SM, Moldenhauer, G, Schuhmacher, J, *et al.* Bispecific tandem diabody for tumor therapy with improved antigen binding and pharmacokinetics. J Mol Biol 1999; 293(1); 41–56.

[115] Lindhofer H, Stanglmaier M, Buhmann R, *et al.* Trifunctional Antibodies: Combining Direct Tumor Cell Killing with Therapeutic Vaccination. In: Duebel S and Reichert JM, Eds. Handbook of Therapeutic Antibodies, 2nd ed. Weinheim: Wiley-VCH 2014.

[116] Lindhofer H, Mocikat R, Steipe B, Thierfelder S. Preferential species-restricted heavy/light chain pairing in rat/mouse quadromas. Implications for a single-step purification of bispecific antibodies. J Immunol 1995; 155(1); 219-25.

[117] Zeidler R, Mysliwietz J, Csánady M, *et al.* The Fc-region of a new class of intact bispecific antibody mediates activation of accessory cells and NK cells and induces direct phagocytosis of tumour cells. Br J Cancer 2000; 83(2); 261-6.

[118] Stanglmaier M, Faltin M, Ruf P, Bodenhausen A, Schröder P, Lindhofer H. Bi20 (fBTA05), a novel trifunctional bispecific antibody (anti-CD20 x anti-CD3), mediates efficient killing of B-cell lymphoma cells even with very low CD20 expression levels. Int J Cancer 2008; 123(5); 1181-9.

[119] Hess J, Ruf P, Lindhofer H. Cancer therapy with trifunctional antibodies: linking innate and adaptive immunity. Future Oncol 2012; 8(1); 73-85.

[120] Heiss MM, Murawa P, Koralewski P, *et al.* The trifunctional antibody catumaxomab for the treatment of malignant ascites due to epithelial cancer: Results of a prospective randomized phase II/III trial. Int J Cancer 2010; 127(9); 2209-21.

[121] Ott MG, Marmé F, Moldenhauer G, *et al.* Humoral response to catumaxomab correlates with clinical outcome: results of the pivotal phase II/III study in patients with malignant ascites. Int J Cancer 2012; 130(9); 2195-203.

[122] Seimetz D, Lindhofer H, Bokemeyer C. Development and approval of the trifunctional antibody catumaxomab (anti-EpCAM x anti-CD3) as a targeted cancer immunotherapy. Cancer Treat Rev 2010; 36(6); 458-67.

[123] Ahmad S, Sweeney P, Sullivan GC, Tangney M. DNA vaccination for prostate cancer, from preclinical to clinical trials - where we stand? Genet Vaccines Ther 2012; 10(1); 1-9.

[124] Bhargava A, Mishra D, Banerjee S, Mishra PK. Dendritic cell engineering for tumor immunotherapy: from biology to clinical translation. Immunotherapy 2012; 4(7); 703-18.

[125] Ruf P, Lindhofer H. Induction of a long-lasting anti-tumor immunity by a trifunctional bispecific antibody. Blood 2001; 98(8); 2526-34.

[126] Morecki S, Lindhofer H, Yacovlev E, Gelfand Y, Ruf P, Slavin S. Induction of long-lasting anti-tumor immunity by concomitant cell therapy with allogeneic lymphocytes and trifunctional bispecific antibody. Exp Hematol 2008; 36(8); 997-1003.

[127] Ruf P, Suckstorff I, Jäger M, Ernst C, Seimetz D, Lindhofer H. Association of humoral anti-tumor response with clinical benefit in catumaxomab-treated malignant ascites patients: Results from a phase IIIb study. J Clin Oncol 2012; 30: (suppl; a 2584).

[128]   Ruf P, Schäfer B, Eissler N, *et al*. Ganglioside GD2-specific trifunctional surrogate antibody Surek demonstrates therapeutic activity in a mouse melanoma model. J Transl Med 2012; 10; 219.

[129]   Eissler N, Mysliwietz J, Deppisch N, Ruf P, Lindhofer H, Mocikat R. Potential of the trifunctional bispecific antibody surek depends on dendritic cells: rationale for a new approach of tumor immunotherapy. Mol Med 2013; 19; 54-61.

[130]   Eissler N, Ruf P, Mysliwietz J, Lindhofer H, Mocikat R. Trifunctional bispecific antibodies induce tumor-specific T cells and elicit a vaccination effect. Cancer Res 2012; 72(16); 3958-66.

[131]   Hilchey SP, Hyrien O, Mosmann TR, *et al*. Rituximab immunotherapy results in the induction of a lymphoma idiotype-specific T-cell response in patients with follicular lymphoma: support for a "vaccinal effect" of rituximab. Blood 2009; 113(16); 3809-12.

[132]   Abès R, Gélizé E, Fridman WH, Teillaud JL. Long-lasting anti-tumor protection by anti-CD20 antibody through cellular immune response. Blood 2010; 116(6): 926-34.

[133]   Jäger M, Schoberth A, Ruf P, Hess J, Lindhofer H. The trifunctional antibody ertumaxomab destroys tumor cells that express low levels of human epidermal growth factor receptor 2. Cancer Res 2009; 69(10); 4270-6.

[134]   Kolb HJ. Graft-*versus*-leukemia effects of transplantation and donor lymphocytes. Blood 2008; 112(12); 4371-83.

[135]   Morecki S, Lindhofer H, Yacovlev E, Gelfand Y, Slavin S. Use of trifunctional bispecific antibodies to prevent graft *versus* host disease induced by allogeneic lymphocytes. Blood 2006; 107(4); 1564-9.

[136]   Buhmann R, Stanglmaier M, Hess J, Lindhofer H, Peschel C, Kolb HJ. Immunotherapy with FBTA05 (Bi20), a trifunctional bispecific anti-CD3 x anti-CD20 antibody and donor lymphocyte infusion (DLI) in relapsed or refractory B-cell lymphoma after allogeneic stem cell transplantation: study protocol of an investigator-driven, open-label, non-randomized, uncontrolled, dose-escalating Phase I/II-trial. J Transl Med 2013; 11; 160.

[137]   Chu SY, Lee SH, Rashid R, *et al*. Immunotherapy with Long-Lived Anti-CD20 × Anti-CD3 Bispecific Antibodies Stimulates Potent T Cell-Mediated Killing of Human B Cell Lines and of Circulating and Lymphoid B Cells in Monkeys: A Potential Therapy for B Cell Lymphomas and Leukemias. ASH annual meeting 2014, abstract 3111.

[138]   Chu SY, Miranda Y, Phung S, *et al*. Immunotherapy with Long-Lived Anti-CD38 × Anti-CD3 Bispecific Antibodies Stimulates Potent T Cell-Mediated Killing of Human Myeloma Cell Lines and CD38+ Cells in Monkeys: A Potential Therapy for Multiple Myeloma. ASH annual meeting 2014, abstract 4727.

[139]   Chu SY, Pong E, Chen H, *et al*. Immunotherapy with Long-Lived Anti-CD123 × Anti-CD3 Bispecific Antibodies Stimulates Potent T Cell-Mediated Killing of Human AML Cell Lines and of CD123+ Cells in Monkeys: A Potential Therapy for Acute Myelogenous Leukemia. ASH annual meeting 2014, abstract 2316.

[140]   US patent US 8568713 B2.

[141]   Varghese B, Menon J, Rodriguez L, *et al*. A Novel CD20xCD3 Bispecific Fully Human Antibody Induces Potent Anti-Tumor Effects Against B Cell Lymphoma in Mice. ASH annual meeting 2014; abstract 4501.

[142]   Woessmann W, Seidemann K, Mann G, *et al*. The impact of the methotrexate administration schedule and dose in the treatment of children and adolescents with B-cell neoplasms: a report of the BFM Group Study NHL-BFM95. Blood 2005; 105(3); 948-58.

[143]   Cairo MS, Gerrard M, Sposto R, *et al*. Results of a randomized international study of high-risk central nervous system B non-Hodgkin lymphoma and B acute lymphoblastic leukemia in children and adolescents. Blood 2007; 109(7); 2736-43.

[144]   Patte, C., Auperin, A., Gerrard, M., *et al*. (2007) Results of the randomized international FAB/LMB96 trial for intermediate risk B-cell non-Hodgkin lymphoma in children and adolescents: it is possible to reduce treatment for the early responding patients. Blood 109; 2773-80.

[145]   Anoop P, Sankpal S, Stiller C, *et al*. Outcome of childhood relapsed or refractory mature B-cell non-Hodgkin lymphoma and acute lymphoblastic leukemia. Leuk Lymphoma 2012; 53(10); 1882-8.

[146]   Ward E, Desantis C, Robbins A, Kohler B, Jemal A. (2014) Childhood and adolescent cancer statistics, 2014. CA Cancer J Clin 2014; 64(2); 83-103.

[147]   Atra A, Gerrard M, Hobson R, Imeson JD, Hann IM, Pinkerton CR. Outcome of relapsed or refractory childhood B-cell acute lymphoblastic leukaemia and B-cell non-Hodgkin's lymphoma treated with the UKCCSG 9003/9002 protocols. Br J Haematol 2001; 112(4), 965-8.

[148]   Kobrinsky NL, Sposto R, Shah NR, *et al*. Outcomes of treatment of children and adolescents with recurrent non-Hodgkin's lymphoma and Hodgkin's disease with dexamethasone, etoposide, cisplatin, cytarabine, and l-asparaginase, maintenance chemotherapy, and transplantation: Children's Cancer Group Study CCG-5912. J Clin Oncol 2001; 19(9); 2390-6.

[149]   Attarbaschi A, Dworzak M, Steiner M, *et al*. Outcome of children with primary resistant or relapsed non-Hodgkin lymphoma and mature B-cell leukemia after intensive first-line treatment: a population-based analysis of the Austrian Cooperative Study Group. Pediatr Blood Cancer 2005; 44(1); 70-6.

[150] Burkhardt B, Reiter A, Landmann E, *et al.* Poor outcome for children and adolescents with progressive disease or relapse of lymphoblastic lymphoma: a report from the berlin-frankfurt-muenster group. J Clin Oncol 2009; 27(20); 3363-9.

[151] Griffin TC, Weitzman S, Weinstein, *et al.* A study of rituximab and ifosfamide, carboplatin, and etoposide chemotherapy in children with recurrent/refractory B-cell (CD20+) non-Hodgkin lymphoma and mature B-cell acute lymphoblastic leukemia: a report from the Children's Oncology Group. Pediatr Blood Cancer 2009; 52(2); 177-81.

[152] Meinhardt A, Burkhardt B, Zimmermann M, *et al.* Phase II window study on rituximab in newly diagnosed pediatric mature B-cell non-Hodgkin's lymphoma and Burkitt leukemia. J Clin Oncol 2010; 28(19), 3115-21.

[153] Handgretinger R, Zugmaier G, Henze G, *et al.* Complete remission after blinatumomab-induced donor T-cell activation in three pediatric patients with post-transplant relapsed acute lymphoblastic leukemia. Leukemia 2011; 25(1); 181-4.

[154] Schlegel P, Lang P, Zugmaier G, *et al.* Pediatric posttransplant relapsed/refractory B-precursor acute lymphoblastic leukemia shows durable remission by therapy with the T-cell engaging bispecific antibody blinatumomab. Haematologica 2014; 99(7); 1212-9.

[155] Von Stackelberg A, Locatelli F, Zugmaier, G, *et al.* Phase 1/2 Study in Pediatric Patients with Relapsed/Refractory B-Cell Precursor Acute Lymphoblastic Leukemia (BCP-ALL) Receiving Blinatumomab Treatment. ASH annual meeting 2014; abstract 2292.

[156] Gore L, Locatelli F, Zugmaier G, *et al.* Initial Results from a Phase 2 Study of Blinatumomab in Pediatric Patients with Relapsed/Refractory B-Cell Precursor Acute Lymphoblastic Leukemia. ASH annual meeting 2014; abstract 3703.

[157] Schuster FR, Stanglmaier M, Woessmann W, *et al.* Immunotherapy with the trifunctional anti-CD20 x anti-CD3 antibody FBTA05 (Lymphomun) in pediatric high-risk patients with recurrent CD20 positive B cell malignancies. Brit J Haematol 2014; [Epub ahead of print].

[1598] Poirel HA, Cairo MS, Heerema NA, *et al.* Specific cytogenetic abnormalities are associated with a significantly inferior outcome in children and adolescents with mature B-cell non-Hodgkin's lymphoma: results of the FAB/LMB 96 international study. Leukemia 2009; 23(2); 323-31.

[159] Woessmann W. How to treat children and adolescents with relapsed non-Hodgkin lymphoma? Hematol Oncol 2013; 31 Suppl 1; 64-8.

[160] Harjunpaa A, Wiklund T, Collan J, *et al.* (2001) Complement activation in circulation and central nervous system after rituximab (anti-CD20) treatment of B-cell lymphoma. Leuk Lymphoma 2001; 42(4); 731-8.

[161] Rubenstein JL, Combs D, Rosenberg J, *et al.* Rituximab therapy for CNS lymphomas: targeting the leptomeningeal compartment. Blood 2003; 101(2); 466-8.

[162] Holmes D. Buy buy bispecific antibodies. Nature Rev Drug Discovery 2011; 10 (11); 798-800.

[163] Dhimolea E and Reichert JM. World Bispecific Antibody Summit 2011. MAbs 2012; 4(1); 4-13.

[164] Schaefer W, Regula JT, Bähner M, *et al.* Immunoglobulin domain crossover as a generic approach for the production of bispecific IgG antibodies. Proc Natl Acad Sci U S A 2011; 108(27); 11187-92.

[165] Zhao L, Tong Q, Qian W, *et al.* Eradication of non-Hodgkin lymphoma through the induction of tumor-specific T-cell immunity by CD20-Flex BiFP. Blood 2013; 122(26); 4230-6.

[166] Kochenderfer JN, Rosenberg SA. Treating B-cell cancer with T cells expressing anti-CD19 chimeric antigen receptors. Nat Rev Clin Oncol. 2013; 10(5); 267-76.

[167] Kater AP, Tonino SH, Egle A, Ramsay AG. How does lenalidomide target the chronic lymphocytic leukemia microenvironment? Blood 2014; 124(14); 2184-9.

[168] Mark TM, Coleman M, Niesvizky R. Preclinical and clinical results with pomalidomide in the treatment of relapsed/refractory multiple myeloma. Leuk Res 2014; 38(5); 517-24.

[169] Markham A. Idelalisib: first global approval. Drugs 2014; 74(14); 1701-7.

[170] Davids MS, Brown JR. Ibrutinib: a first in class covalent inhibitor of Bruton's tyrosine kinase. Future Oncol 2014; 10(6); 957-67.

[171] Ansell SM, Lesokhin AM, Borrello I, *et al.* PD-1 Blockade with Nivolumab in Relapsed or Refractory Hodgkin's Lymphoma. N Engl J Med 2015; 372(4); 311-9.

[172] U S Food and Drug Administration [Homepage on the Internet].FDA approved drug products [cited: 15th January 2015] Available from: http://www.accessdata.fda.gov/scripts/cder/drugsatfda/

CHAPTER 3

Natural Killer Cells as an Immune Cell Therapy Option for Bone Marrow Transplantation in Hematological Malignancies: Implications for New Therapies Based on Lipid Transfer and Cell-to-Cell Communication

Beatriz Martín-Antonio[1,2,*], Nuria Martínez-Cibrian[1,2], Alvaro Urbano-Ispizua[1,2] and Ciril Rozman[2,3]

[1]*Department of Hematology, Hospital Clinic, IDIBAPS, Barcelona, Spain;* [2]*Josep Carreras Leukaemia Research Institute, Barcelona, Spain and* [3]*University of Barcelona, Barcelona, Spain*

Abstract: Natural killer (NK) cells belong to the innate immune system. In recent years it has been suggested that their use could improve the outcome of stem cell transplantation for hematological malignancies, mainly due to the high anti-tumor effect they mediate and because they are not associated with an increase of the graft *versus* host disease, a complication which decreases the survival of patients submitted to this procedure. Once activated by several mechanisms, NK cells deliver cytolytic molecules triggering different cell death pathways, which can be caspase-3 dependent or caspase-3 independent, such as endoplasmic reticulum stress or lysosomal cell death. Apparently, NK cells take advantage of the particular features of different tumor cells to attack them in a more efficient way. Understanding the factors that contribute to cancer cell death is therefore critical for the development of novel therapies and to circumvent chemo-resistance.

Lipids are emerging as new targets for anti-inflammatory and cancer therapies as they interact extensively with organelles, are involved in immune responses, and can lead to initiation of different types of cell death. Interactions between lipids with membranes are crucial for the effects they mediate in cell-to-cell communication. Our group demonstrated that cord blood derived NK cells (CB-NK) can be efficiently expanded *in vitro* and exert anti-tumor activity both *in vitro* and *in vivo* in a multiple myeloma (MM) model thus, providing a clinically applicable strategy for the generation of highly functional NK cells which can be used to eradicate this and potentially other hematological malignancies. Our group also showed that CB-NK mediate a specific tumor cell death against MM which is transmissible between cells, and where lipid-protein vesicle trafficking play a relevant role. These findings provide the rationale for the development of CB-NK based therapeutic strategies in the treatment of hematological malignancies.

---

*Corresponding author Beatriz Martín-Antonio: Department of Hematology, Hospital Clinic, IDIBAPS, Barcelona, Spain; Josep Carreras Leukaemia Research Institute, Barcelona, Spain; Tel: +34-932274528; Fax: +34-932275484; E-mail: bmartina@clinic.ub.es

Atta-ur-Rahman (Ed)

In this paper, the strengths and the weaknesses of clinical trials using NK cells are discussed. In addition, some guidelines for the development of future trials are suggested.

Keywords: Stem cell transplantation (SCT), natural killer (NK) cells, cell death, lipids, cell-to-cell communication, autophagy, multiple myeloma (MM).

A BRIEF HISTORY OF HEMOPOIETIC STEM CELL TRANSPLANTATION (SCT)

SCT is an established procedure to treat patients with hematological malignancies and in addition, for patients with metabolic disorders and disorders of the immune system. SCT can be either autologous (auto-SCT) using hemopoietic progenitor cells of the patient, or allogeneic (allo-SCT), where the cell source comes from a donor usually Human Leukocyte Antigen (HLA)-compatible. The main benefit of the allo-SCT is exerted by donor lymphocytes, which mediate a graft *vs.* leukaemia effect that decreases the incidence of the disease recurrence. These lymphocytes will also be responsible for mediating a serious complication termed graft *vs.* host disease (GVHD), which decreases the survival of the allo-SCT [1, 2]. GVHD, which can be either acute or chronic, increases the transplant related mortality (TRM). The main current challenge of the allo-SCT is to increase the graft *vs.* leukemia effect without developing GVHD. Nowadays, statistical data show a continuous improvement in the clinical outcome of the SCT. However, still it is necessary to find the best approach to keep reducing the morbidity and mortality of this procedure and the relapse of the malignant disease.

From the first SCT performed in humans in 1959 till nowadays many changes in the protocols and collaborations between institutions made possible better outcomes decreasing both relapse incidence and TRM. In more detail, the first human bone marrow transplant was performed by E.D. Thomas in 1959 in letally irradiated patients with acute leukaemia providing concept of hematological reconstitution, however these patients died later of relapse [3]. Afterwards, in 1965, G. Mathé described long term engraftment, chimerism, tolerance and anti-leukemic effect in sibling bone marrow transplant [4]. In 1970, results from 203 transplants were published showing disappointing results with only three patients alive, being graft failure, GVHD and relapse the major causes of death. These results determined an extraordinary decrease in the number of SCT performed in the institutions. However, the discovery of the HLA system, which allowed the selection of HLA matched bone marrow donors, new protocols for conditioning regimen, immune suppression and GVHD prevention, better management of early complications and infections, development of new sources of haematopoietic stem cells, and development of bone marrow registries and cord blood banks to look for matched donors allowed the change to better outcomes, and nowadays SCT is a standard procedure to treat patients with hematological malignancies [5]. International collaborations between institutions have been a key factor in the success of this procedure and have led to the development of guidelines available for all the institutions. Most important, analysis of all the data generated in different institutions allowed obtaining important conclusions to modify protocols.

Three different sources of hemopoietic progenitor cells can be used for SCT, either bone marrow (BM) progenitor cells, G-CSF mobilized peripheral blood progenitor cells (PBPC) or cord blood (CB) progenitor cells. Statistical data from 634 different institutions in Europe from 1990-2010 show a different preferential stem cell source used depending on the type of transplant, being peripheral blood (PB) used in 99% of the auto-SCT and in 71% of the allo-SCT procedures. CB, which represents a 6% of the total, was primarily used in allo-SCT. Since 2005 the number of transplants has increased by 37% and by 9% for allo-SCT and auto-SCT, respectively. The most important change with the years is the increase in the number of unrelated donors in comparison to HLA-identical sibling donors, which could be in part due to the creation of donor registries to search for unrelated matched donors and to the impressive number of altruistic donors [6].

CORD BLOOD: A FEASIBLE OPTION TO USE AS CELL SOURCE FOR SCT

Only 30% of patients in which an allo-SCT is indicated will have an HLA-matched donor either from a sibling or from an unrelated origin. For some patients of racial/ethnic minorities the probability is much lower, and in many cases it will not be possible to find a matched donor, and a CB will be the only option. The main practical advantages of using CB are the relative ease of procurement, the absence of risk for mothers and donors, the reduced likelihood of transmitting infections, and the ability to store fully tested and HLA-typed CB in the frozen state, available for immediate use [7].

Thanks to the creation of cord blood banks and to the increase in number of donors registered in Bone Marrow Donor Worldwide (BMDW; www.bmdw.org) the number of matched unrelated donors has increased considerably. Thus, currently the number of allogeneic transplants from unrelated donors is higher than from relatives [6].

Statistical data of large retrospective studies to find out the best source of stem cells have shown that in adults with malignant disease in HLA-identical transplants, PB compared to BM cause higher GVHD incidence and the same overall survival. In the case of CB, the number of cells is critical for the engraftment; unfortunately, CB units have a lower stem cell number than PB and BM. Since the first CB transplant (CBT) performed in France in 1988 in a child with Fanconi Anemia, over 30,000 CBT have been performed and substantial improvements have happened. In children, CBT shows similar survival in comparison to other sources of stem cells. In adults, low cell dose leads to delays in immune reconstitution causing higher incidence of viral infections and increasing the TRM. Changes in protocols have improved this disadvantage and statistical data from 2002-2006 on 1,525 adults with acute leukemia comparing mismatched CB, PB and unrelated BM show same disease free survival. These results support the use of CB for adults with acute leukaemia when there is no HLA-matched unrelated adult donor available, and when a transplant is required urgently [8, 9].

Additional strategies for future improvement of the engraftment of CBT in adults are still needed. Studies infusing two CB units (double CBT) became popular; however, recent results have questioned the benefit of double CBT [10], and other alternatives to increase the expansion or the homing to the bone marrow were explored, such as the use of notch ligand Delta-1 which demonstrated expansion of short-term repopulating cells and an improved time of engraftment [11]. Also the *in vitro* expansion of progenitor cells from one of two CB units with mesenchymal stem cells reported a 30-fold expansion of hemopoietic stem cell count in patients and improved median time of engraftment [12]. Another technique is increasing the homing of progenitor cells into the bone marrow by fucosylation of progenitor cells from CB [13].

NEW CELL THERAPY STRATEGIES TO IMPROVE THE OUTCOME OF ALLO-SCT

Infusion of different types of immune cells expanded *in vitro* is becoming a new tool to improve the outcome of SCT. The aim of this approach is to prevent the development of GVHD after allo-SCT and to increase the graft *vs.* leukemia effect mediated by donor T cells. In this sense, infusion of *in vitro* expanded mesenchymal stem cells into the patient, irrespective of the HLA of the donor [14], and regulatory (reg)-T cells [15, 16], have shown a beneficial effect for patients with steroid-resistant acute GVHD. Mesenchymal stem cells produce more robust results and several bone marrow transplants groups are already infusing them into the patients. For reg-T cells, the beneficial effect has been shown only in murine BM transplantation models and results from clinical trials in patients are not yet available. To increase the graft *vs.* leukemia effect after SCT, natural killer (NK) cells seem to be a promising option. NK cells are cells of the innate immune system with a high cytotoxic activity against both, viral infected and tumor cells.

From now on, we will discuss the main characteristics of NK cells, the discovery of their beneficial properties in the context of SCT and some results of clinical trials performed with NK cells. In addition, based on both the biology of these cells and the biology of tumor cells some suggestions will be made to try to improve the outcome of clinical trials using NK cells. Finally, we will present a new characteristic cytotoxicity of NK cells described by Shah *et al.* [17] and by our group [18], which could open new cell therapy options to treat hematological malignancies.

MORPHOLOGY AND IMMUNOPHENOTYPE OF NK CELLS

NK cells are a subset of cells of the innate immune system. These cells are large granular lymphocytes which constitutively express CD16, CD56, and the T cell associated antigens CD2 and CD7. In PB, we have two different subsets of NK cells: *1)* 90% of NK cells are mature cytotoxic NK cells. They express CD16 bright and CD56 dim; *2)* 10% of the NK cell population is immature, and they are cytokine producers. They are CD16dim, CD56 bright and CD25+. In addition, NK cells express some subunits of the CD3 complex, including the ε and ζ chains. This results in the reactivity of NK cells for CD3 in paraffin-section immunohistochemical analysis because this assay typically uses antibodies reactive with the CD3 ε chain. However, by flow cytometry NK cells are CD3- because the anti-CD3 antibodies used in this assay recognize the fully CD3 assembled complex that NK cells lack. There is some overlapping in the immunophenotype between NK cells and cytotoxic T cells which are usually CD8 bright. On one side, 30% to 50% of NK cells express low levels of CD8; and in the other side, cytotoxic T cells often express CD56 and CD57 [19]. CD57 is a senescence or exhaustion marker which is expressed by two thirds of NK cells in PB. However, in CB, NK cells are immature and they do not express CD57.

NK cells recognize target cells by an array of activating and inhibitory receptors present on their surface. The first known mechanism of NK cell mediated cytotoxicity was the "missing self" recognition, which is mediated by the inhibitory "Killer Immunoglobulin-like Receptors" (KIRs) on NK cells. Inhibitory KIRs interact with the HLA-I (A, C1/C2 and Bw4) molecules present in all nucleated cells. "Missing self" recognition occurs due to the down-regulated or absence of HLA-I expression in virus infected cells or tumor cells. Inhibitory KIRs share a common signaling motif in their cytoplasmic regions, which is an Immunoreceptor Tyrosine-based Inhibitory Motif (ITIM). When ITIM-bearing receptors engage their ligands by phosphorylation of the tyrosine residue, NK cell responses are suppressed and when these receptors are not engaged by their ligands NK cells are activated [20]. The inhibitory KIR-HLA-I interaction inhibits NK cells of attacking the own healthy cells and when this interaction does not happen NK cells become activated (Fig. **1**).

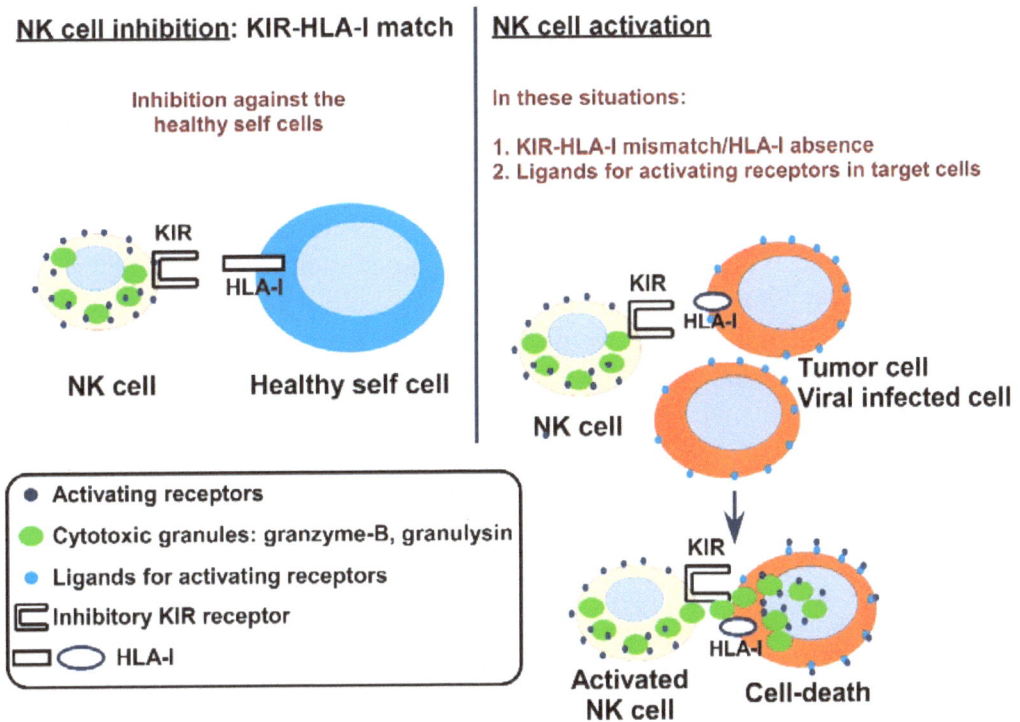

Figure 1: Basic mechanism of natural killer (NK) cells to remove virus infected cells and tumor cells: NK cells have inhibitory KIR receptors which interact with HLA-I in cells of the human body. These interactions inhibit NK cells from attacking the own healthy cells. In some cases, tumor cells or viral infected cells down-regulate the HLA-I expression as a way to evade the immune response mediated by T cells. In this case, inhibitory receptors on NK cells stop being inhibited by the HLA-I and NK cells become activated against such cells.

NK cells also become activated by the interaction of activating receptors on NK cells with ligands induced upon tumor transformation, viral infection and cell stress. There is a vast array of activating receptors to initiate effector functions, and with the exception of CD16 (Immunoglobulin (Ig)-G Fc Receptor III-2), which is the Fc receptor that binds Ig-G mediating antibody dependent cellular cytotoxicity, it is necessary an interplay and cooperation among them. Some of the ligands for these receptors remain unknown. In Table 1 some families of these receptors and their ligands in tumor cells are summarized.

The activating KIRs: This family includes KIR2DS1, KIR2DS2, KIR2DS3, KIR2DS4, KIR2DS5 and KIR3D1. They are highly homologous to their inhibitory KIR counterparts in the extracellular domain, but are characterized by a short cytoplasmic tail which lacks the ITIM. These receptors interact with DAP-12, an adaptor-signaling molecule, carrying an Immunoreceptor Tyrosine-based Activation Motif (ITAM) that can induce NK cell activation.

The C-type lectin-like receptors: include activating (CD94/NKG2C, NKG2D and NKP80) and inhibitory receptors (CD94/NKG2A). CD94/NKG2C, like the activating KIRs, is associated with DAP-12. The inhibitory CD94-NKG2A recognizes HLA-E, which is present in all human cells being responsible for inhibiting the cytotoxicity of NK cells against the own healthy cells in the human body. Of interest, NKP80 by interacting with Activation-Induced C-type Lectin (AICL) exerts an autonomous control of NK cells against excessive inflammatory response. NK cells contain intracellular stores of AICL in the Golgi complex. After inflammatory stimuli, AICL is exposed on the NK cell surface interacting with NKP80 and causing self-NK cells mediated cytolysis. NKG2C expression and lack of NKG2A expression on NK cells have been described to be associated to a memory-like NK cell phenotype [21-23].

The natural cytotoxicity receptors (NCR): include NKP30, NKP44 and NKP46. They are activating receptors that interact with ligands overexpressed on tumor cells. These receptors are type I transmembrane proteins belonging to the Ig superfamily and are composed of one or two extracellular Ig-like domains, which are responsible for ligand binding. Of note, NKP44 is expressed only on activated NK cells. In addition, they contain a transmembrane domain that interacts with signaling adaptor protein containing ITAM. Ligands for these receptors include molecules present in viral infected cells, in tumor cells and also structures like heparin or heparan sulfate present in healthy or tumor cells.

The SLAM (Signaling Lymphocyte Activating Molecule)-related receptors: include SLAMF1 (CD150), 2B4 (CD244), NTD-A (Ly108), CD48, CD84, Ly9 (CD229), and CRACC (CD319). They transmit activating signals to mediate NK cytotoxicity of both, normal hemopoietic cells and lymphoma cells. However, in the absence of SLAM-associated protein family adaptors, they mediate inhibitory signals to suppress the activation mediated by other activating receptors.

Table 1. **Main NK cell receptors, their function, either inhibitory or activating and their ligands on target cells are indicated [24-29].**

Class	Receptor	Ligands in Tumor Cells or Normal Cells
Inhibitory	KIR2DL1	HLA-C2
Inhibitory	KIR2DL2/L3	HLA-C1/C2, some HLA-B
Inhibitory	KIR2DL4	HLA-G
Inhibitory	KIR2DL5	Not described
Inhibitory	KIR3DL1	HLA-Bw4
Inhibitory	KIR3DL2	HLA-A3, A11
Inhibitory	KIR3DL3	Not described
Inhibitory	CD94/NKG2A	HLA-E
Activating	KIR2DS1	HLA-C2
Activating	KIR2DS2	HLA-A11
Activating	KIR2DS3	Not described
Activating	KIR2DS4	HLA-A11, some C1 and C2
Activating	KIR2DS5	Not described
Activating	KIR3DS1	Allotype *014 binds to HLA-Bw4
Activating	CD94/NKG2C	HLA-E
Activating	NKG2D	MICA/B, ULBPs
Activating	NKP30	BAG-6, B7-H6, heparin and heparan sulfates
Activating	NKP44	NKP44L, PCNA, heparin and heparan sulfates
Activating	NKP46	Heparin and heparan sulfates
Activating	NKP80 (KLRF1)	AICL on NK cells (responsible for autonomous control of NK cells during inflammation)
Activating	DNAM1 (CD226)	PVR (CD155), Nectin-2 (CD112)
Activating	2B4 (CD244)	CD48

Upon recognition, NK cells eliminate rapidly (within 30-60 minutes) virus infected cells and tumor cells by two different mechanisms, either granule dependent pathways or death receptor pathways. The granule dependent pathway happens when NK cells after contacting the target cell release granules containing perforin and proteases called granzymes. Perforin (a pore-forming protein) will make transmembrane channels into the target cell to allow the entrance of granzymes into target cells which will lead to cell death. Cytotoxic granules are polarized into the target cell as a result of the formation of an immunological synapse. Granzymes will kill target cells through proteolysis of different proteins in the target cell. There are different types of granzymes (A, B, H, K, M, C, and F) and each one activates different cell death pathways. The mode of action of granzyme B is the most understood. Granzyme B leads to apoptotic cell death through the activation of the caspase cascade either directly, or through release of cytochrome C from the mitochondria that activates the caspase pathway. Granzyme A activates a caspase independent cell death pathway. For the other granzymes there are many questions to be solved in humans as most studies have been performed in mice [30]. Granulysin is another cytotoxic molecule released by NK cell granules which leads to ER stress and cell death [31]. Another type of NK-mediated cell death is the death receptor pathway which is activated by the TNF receptors death family. These receptors include FasL (CD95L) in NK cells and FAS (CD95) in target cells. Interaction of FASL with FAS allows the formation of a death-inducing signaling complex that includes Fas-associated death domain protein (FADD), caspase-8, and caspase-10. Activation of caspase-8 results either in the direct activation of the other effector caspases (-3 and -7) or in the proteolysis of Bid which signals through mitochondria leading to release of cytochrome C and subsequent caspase activation [32].

NK CELLS: INTERPLAY BETWEEN INHIBITORY KIRS AND THE HLA-I. "HLA-I LICENSING" TO INDUCE SELF-TOLERANCE

In the context of SCT, NK cells have an important role, as they are the first immune population that reconstitutes after this procedure, in contrast to B and CD8+ T cells, which do not emerge till a few months later and to CD4+ T cells which can return to normal levels only after years [33]. Thus, during this period of immune recovery after SCT, NK cells are crucial mediating cytotoxicity against both tumor cells and viral infections before T-cell immunity is fully recovered.

In PB there is a high diversity of NK cell clones, some of them might lack inhibitory KIRs for self-HLA-I and potentially they could attack normal self-cells; however during development there is a maturation and self-tolerance process to ensure that this does not occur. This process has been termed HLA-I-dependent "licensing". NK cells are "licensed" by self HLA-I molecules to enable them to monitor other cells for the quality and quantity of their HLA-I expression. This process resembles to the HLA-I-dependent "education" which happens to T cells in the thymus during T-cell development, and requires specific interaction of the inhibitory KIRs with host HLA I molecules. Inhibitory KIRs have a positive role in this process as licensing pairs an inhibitory KIR with its cognate self-HLA-I ligand for functional development. Licensing results in two types of self-tolerant NK cells: 1) licensed NK cells will be the cells that express an inhibitory KIR for self-HLA-I, will show enhanced activity and will be able to kill cells which do not express HLA-I molecules, 2) unlicensed NK cells will be the NK cells, which do not express inhibitory KIRs for self-HLA-I. They do not become licensed to kill, and have no need to be inhibited by HLA-I because they are not functionally competent [34, 35] (Fig. **2A**).

In the allo-SCT context, NK cell HLA-I dependent "licensing" also occurs, and the donor HLA genotype shapes this process. Once full donor chimerism is achieved, the hemopoietic cells are from the donor and express donor HLA-I. Therefore, a stable donor-like NK cell education pattern emerges after KIR-HLA-I ligand mismatched allo-SCT. This donor-like NK cell education pattern is achieved within the first year after allo-SCT and lasts for at least 3 years, enabling NK cells to respond to occasional aberrant leukemic cells that have lost HLA-I and suggesting a sustained graft *vs.* leukemia effect after HLA-I mismatched allo-SCT. For instance, the inhibitory KIR2DL1 recognizes HLA-C2. Therefore, according to the HLA-I licensing, KIR2DL1 mono-KIR cells will become hyporesponsive in a HLA-C1 environment to avoid chronic damage, and responsive in a HLA-C2 environment (Fig. **2B**). This was confirmed after mismatched allo-SCT, showing that KIR2DL1 mono-KIR cells were less responsive in C1/C1 than in C1/C2 or C2/C2 subjects. In addition, the inhibitory CD94-NKG2A receptor played a major role in the NK cell education in an independent and additive fashion with regard to KIRs [36]. For activating KIRs, HLA-licensing has also been described; for example, KIR2DS1 binds to HLA-C2. Therefore, NK cells positive only for KIR2DS1 will be dangerous in a HLA-C2 environment. To avoid this situation, they will become functional only in a HLA-C1 environment (C1/Cx). In contrast, in a HLA-C2/C2 environment they are hyporesponsive to target cells, presumably through a tolerating effect of exposure to C2. This has been confirmed after HLA-mismatched allo-SCT, where KIR2DS1-positive donors conferred higher protection from relapse when donors were homozygous or heterozygous for HLA-C1 (C1/C1 or C1/Cx) than when donors were homozygous for HLA-C2, by killing AML target cells that were HLA-C2 homozygous [37, 38].

The interest of inhibitory KIRs for allo-SCT arose when Ruggeri *et a*l. in 2002 showed that in acute myeloblastic leukemia (AML) the mismatch combination inhibitory KIR (donor)-HLA-I (patient) correlated with good prognosis. In this

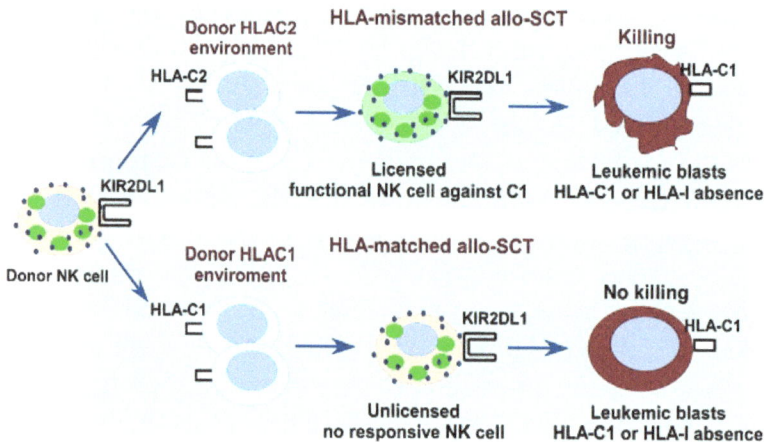

Figure 2. A. HLA-I licensing of NK cells during development: In PB there are different NK cell clones with different combination of inhibitory KIR receptors. NK cell clones with inhibitory KIRs that have a self HLA-I ligand will mature to become functional and HLA-licensed to kill cells which either do not present the same HLA-I or do not express HLA-I. On the contrary, NK cell clones that do not express any inhibitory KIR, potentially they could attack normal self-cells. These cells during development do not get the license to kill cells expressing HLA-I, and remain as hyporesponsive NK cells. **B.** HLA-licensing after allo-SCT: HLA-licensing after allo-SCT is shaped by donor HLA genotype. In a HLA-mismatched allo-SCT, NK cells are educated according to the HLA-I of the donor. For instance, KIR2DL1 mono KIR cells are usually inhibited by HLA-C2; in a HLAC2 environment, they will be licensed to kill HLA-C1 cells, or cells lacking HLA-I which will be the residual leukemic blasts. However, after HLA-matched allo-SCT, the same KIR2DL1 mono KIR cells if they are educated in a HLA-C1 environment these cells will not be HLA-I-licensed against C1 cells, and will not be able to kill leukemic blasts expressing HLA-C1 or blasts lacking HLA-I expression.

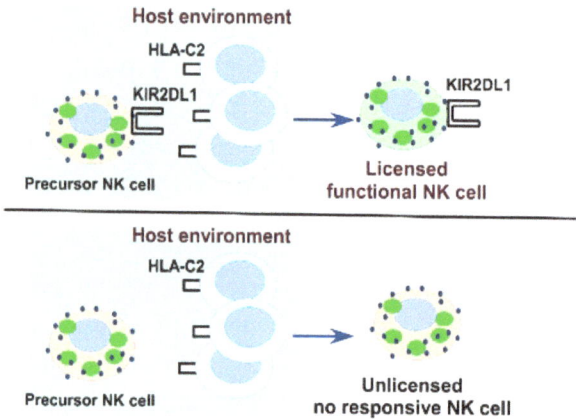

context, NK cells are not inhibited and can exert their cytotoxic function [39]. In haploidentical SCT, *Ruggeri et al.* observed that allo-reactive NK cells in the graft *vs.* host direction can decrease GVHD incidence, and graft rejection in both, AML and ALL patients. Furthermore, they also reported that allo-reactive NK cells in the graft *vs.* host direction increase the graft *vs.* leukemia effect in AML patients, showing a beneficial anti-leukemic effect of NK cells mostly in myeloid malignancies. This allo-reactivity occurs when there is "missing self" recognition, which happens when the donor expresses a specific KIR without the corresponding KIR ligand in the receptor, leading to a KIR (donor)/KIR ligand (recipient) mismatch. When "missing self" recognition occurs, NK cells lyse leukemia blasts, recipient dendritic cells (DCs) and recipient T cells, which translates into a reduction of relapse, prevention of GVHD, and prevention of graft rejection, respectively. The "missing self" recognition led to the proposition of the "missing ligand model" as a powerful algorithm to predict a potent anti-leukemia effect and, as a consequence, a favorable transplant outcome in allo-SCT [39, 40].

However, in ALL, the GVHD reduction and prevention of graft rejection, was not accompanied by an increase of the graft *vs.* leukemia effect observed in AML [41]. Additional studies in ALL have shown this graft *vs.* leukemia effect of allo-reactive NK cells in children [42]. However, in adults there are conflicting results and the impact of the "missing ligand model" is mostly applied only in AML but not in ALL [43]. This effect could be explained in part because unlike ALL, some types of AML cells avoid the T cell immune response, by down-regulating HLA-I and then eliminate the expected graft *vs.* leukemia effect after allo-SCT. However, NK cells become activated due to this HLA-I down-regulation, exerting a graft *vs.* leukemia effect. This is the reason why K562 cell line is the classical tumor cell target used to analyze the functional *in vitro* activity of NK cells. K562 cells were isolated from a patient with chronic myeloid leukemia in terminal blast crises and do not express HLA-I. It could be considered that the "missing ligand model" might be applied just for NK-mediated killing of DCs and T cells of the patient leading to a reduced GVHD and reduced graft rejection, respectively. On the contrary, the graft *vs.* leukemia effect, could be mediated by the KIR-HLA down-regulation, which does not occur in all ALL patients. To support this, clinical studies in ALL confirmed that low HLA-I expression levels in blasts of the patient conferred a beneficial effect mediated by allo-reactive NK cells [44].

In a clinical context, many studies performed in the last years involving NK cells, have not always been successful. Some studies involved NK cell infusion as an immune-therapy option, and some studies performed analysis of the clinical outcomes based on the KIR-HLA in donor and recipient but they did not involve NK cell infusion. Some of these unsuccessful studies might not have considered all the parameters needed for a proper analysis, like for instance, studies which analyzed all patients in the same group instead of grouping by the different hematological malignancies (either myeloid or lymphoid) [45]. To try to give an explanation to the lack of an improvement in clinical outcomes, it is necessary to analyze them deeply and see which parameters should be taken into account. Some questions to consider would be: 1) was the NK cell effect analyzed only in myeloid malignancies? 2) Was the KIR-HLA mismatch considered? 3) And in cases of studies involving NK cell infusion, were NK cells activated previously to the infusion in the patient?

We have reviewed published studies using NK cell infusion as an immunotherapy option. Most of these studies were performed in hematological malignancies, but also there are some studies in non-hematological malignancies such as breast, ovarian and lung cancer. Results are summarized in Table **2**. In more detail, most of these studies used NK cells from haploidentical donors, and considered the KIR-HLA ligand mismatch. NK cells were activated *in vitro* and/or *in vivo* with IL-2, and administered after immunosuppressive treatment based on cyclophosphamide and fludarabine regimens. Patient disease, disease status, number of NK cells infused and number of NK cell infusions differed from one study to another. However, all published studies agree that NK cell infusion is a safe and well-tolerated procedure and not associated with engraftment failure or acute and chronic GVHD. Looking in more detail to these studies, we can observe four general protocols to infuse NK cells:

1) *Protocols infusing NK cells as a consolidation therapy*: three studies have used NK cell infusion as a consolidation therapy for patients with acute leukemia. Rubnitz *et al.* [46] treated 10 children with favorable and intermediate-risk AML in first complete remission; after a median follow up of four years, these patients remain in remission (personal communication). Curti *et al.* [47] showed a clear clinical benefit in elderly patients with high risk AML that at the time of the infusion were in either complete remission (CR) or very early molecular relapse. Both studies [46] and [47], concluded that NK cell therapy could be a promising strategy for consolidation therapy in two different populations who are not candidates for receiving an SCT, children with favorable and intermediate risk AML and elderly patients.

2) *Protocols infusing NK cells in refractory patients*: Bachanova *et al.* [48] compared the outcomes of 3 cohorts of refractory AML patients. All of them received NK cells from haploidentical donors, a preparatory immunosuppressive

Table 2. Immunotherapy based on NK cell infusion for hematological and non-hematological malignancies.

N patients. Disease	NK source	KIR-HLA mismatch	Treatment before NK infusion	NK activation	NK selection	Median number of infused NK (x10^6/Kg)	GVHD/ engraftment failure	Outcome (Reference)
10. AML in CR	Haplo	Yes	Flu/Cy	IL-2 after NK infusion	TCD, CD56+ selection	29	No/No	After 964 days all patients still in remission [46].
13. AML (6 CR, 5 active, disease, 2 MR)	Haplo	Yes	Flu/Cy	IL-2 in culture and after NK infusion	TCD, CD56+ selection	2.74	No/No	1 out of 5 patients with active disease achieved CR for 6 months. The rest had no clinical benefit. 2 patients in MR achieved CR, for 9 and 4 months. 3 out 6 patients in CR were disease free after 34, 32 and 18 months. The other 3 relapsed after NK infusion [47].
57. Refractory AML	Haplo	In 50%	Flu/Cy	IL-2 in culture and after NK infusion	*Cohort 1:* TCD. *Cohort 2:* TCD and CD56+ selection. *Cohort 3:* TCD, BCD, and CD56+ selection	*Cohort 1:* 9.6 *Cohort 2:* 3.4 *Cohort 3:* 26	No/No	Cohort 1 and 2: 9 patients achieved remission (21%) for 2.3 months. DFS at 6 months was 5%. Cohort 3: 15 patients achieved remission (53%), of them 6 patients underwent SCT. The median remission was 11.2 months. DFS at 6 months was 33% [48].
6. Refractory NHL (1 MZL, 1FL, 2, DLBCL, 2 transformed)	Haplo	In 50%	R/Flu/Cy	IL-2 in culture and after NK infusion	TCD	21	No/No	2 patients achieved CR (MZL, FL) and 2 patients PR (1 DLBCL, 1 transformed lymphoma). Three of them underwent SCT, the fourth one maintained a PR for 4 months [49].
10. Relapsed MM	Haplo	Yes	Flu/Mel/Dx	IL-2 in culture and after NK infusion	TCD	1.7	No/No	5 patients (50%) achieved CR or near CR, 2 patients PR, 1 SD and 2 patients PD [50].
13. 6 NHL, 3 MCL, 1 DLBGL, 1 FL, 1 ALCL, 5 MM, 2 HL	Haplo	NA	No treatment	IL-2 in culture	TCD	NA	No/No	7 patients (3 MCL, 2 MM, 1 FL, 1 HL) achieved CR for 23.9 months, 6 patients relapsed [51].
3. Refractory acute leukemia (1 AML, 2 ALL)	Haplo	Yes	TBI/Thio/Flu/OKT3 or OKT3/Mel	IL-2 in culture and after NK infusion	TCD, CD56 enriched	16.2	No/No	All patients achieved complete remission. The AML patient died of early relapse at day +80. ALL patients died of TTP and atypical viral pneumonia on days +45 and +152 [52].
30. 19 myeloid disease, 11 lymphoid disease	14 identical sibling, 16 haplo	NA	Flu/Alemtuzumab	NA	CD56 enriched	9.9	No/2	Patients with myeloid disease had a 50% 1-year survival. Patients with lymphoid disease had a 29% 1-year survival [53].
16. 8 AML, 5 ALL, HL, 1 sarcoma	Haplo	In 11	MA conditioning including ATG or OKT3	None	TCD CD56+ selected	12	3/4	7 patients relapsed (4/8 AML, 1/4 ALL, 1/2 HL, 1/1 sarcoma) of which 6 died. 3 patients died of graft failure, 3 of GVHD and 1 of transplant related neurotoxicity. Survival rates were 44-+ 12% at 1 year and 25 +- 11% at 2 and 5 years [54].

Table 2: contd...

41. 3 CR AML, 29 relapsed AML, 7 relapsed ALL, 1 refractory MDS, 1 refractory DLBCL	Haplo	In 7	Flu/Bu/ATG	IL-15 and IL-21 in culture	TCD	200	1/5	All 3 patients in CR relapsed at 6, 24 and 49 months. One MDS achieved CR. Among 29 patients with refractory AML, 21 achieved a CR (72%), for refractory ALL/lymphoma 4/8 achieved CR (50%). 17 patients showed disease progression; cumulative incidence of disease progression in AML was 38% *vs.* 75% in ALL/lymphoma. Cumulative EFS and OS rates for refractory AML were 31 and 35% and for ALL/lymphoma were 0%. TRM was 27% [55].
15. Advanced NSCLC	Haplo	At least in 50%	None	IL-15 and hydrocortisone in culture	CD56+ selection	4.15	No/No	After 22 months 2 patients were in PR and 6 patients with SD. Median PFS and OS were 5.5 and 1.5 months, respectively [56].
20. 14 ovarian, 6 breast cancer	Haplo	Yes	Flu/Cy Fly/Cy/TBI	IL-2 in culture and after NK infusion	TCD	21.6	No/No	Four patients with PR (all ovarian), 12 with SD (8 ovarian, 4 breast) and 3 with PD (1 ovarian, 2 breast) [57].

Abbreviations. AML: acute myeloid leukemia; ALL: acute lymphoblastic leukemia; NHL: non Hodgkin lymphoma; MZL: marginal zone lymphoma; FL: follicular lymphoma; DLBCL: diffuse large B cell lymphoma; MM: multiple myeloma; ALCL: anaplastic large cell lymphoma; MCL: mantle cell lymphoma; MDS: myelodysplastic syndrome; NSCLC: non-small cell lung cancer; Haplo: haploidentical; NK: natural killer cell; CR: complete remission; PR: partial response; SD: stable disease; PD: progressive disease; MR: molecular relapse; Flu: fludarabine; Cy: cyclophosphamide; TBI: total body irradiation; R: Rituximab; Mel: melphalan; Thio: thiotepa; Dx: dexamethasone; ATG: antithymocyte globulin; Bu: busulfan; MA: myeloablative; TCD: CD3 T-cell depleted; BCD: CD19 B-cell depleted; IL2DT: interleukin-2 diphtheria toxin fusion protein; KIR: killer immunoglobulin-like receptor; GVHD: graft *versus* host disease; TRM: transplant related mortality; EFS: event free survival; OS: overall survival; NA: not available.

regimen, and the same protocols to activate NK cells. Differences between the cohorts were the NK cell selection technique, the median number of infused NK cells and the use of IL-2 diphtheria toxin fusion protein (IL2DT) to deplete host reg-T cells, which was applied in the third cohort of patients. This study showed that the use of IL2DT led to higher NK cell expansion, improved complete remission rate and disease-free survival with no increased toxicity or complications; further, they suggested that the detection of NK cells 7 days after the infusion may serve as a biomarker for clinical response. The same authors [49] reported partial responses in 4 out of 6 refractory lymphoma patients, however, they could not detect NK cells in PB. They assumed that lymphodepletion was incomplete and transient, as they showed persistence of recipient T cells immediately before NK cell infusion.

3) *Protocols infusing NK cells in the auto-SCT context*: two strategies have been reported, one group administered NK cells before the auto-SCT and the other group after the procedure. In the first scenario, Shi *et al.* [50] analysed the impact of NK cell infusion for relapsed MM followed by a delayed rescue with auto-SCT. Some patients had previously received single or tandem auto-SCT. In spite of this, there was a 50% of complete or near complete responses. They failed to demonstrate *in vivo* NK cell expansion, due to persistent recipient T cells probably because of insufficient immunosuppressive treatment. In the second scenario, Klingemann *et al.* [51] administered NK cells 90 days after the auto-SCT without previous immunosuppressive treatment, and as other groups, they could not find the presence of donor NK cells in the recipient PB.

4) *Protocols infusing NK cells in the allo-SCT context*: Koehl *et al.* [52] published one of the first studies with encouraging results. In this report, three children with refractory acute leukemia achieved CR after haploidentical SCT and NK cell infusion, one patient died because of relapse and the other two died in CR. Rizzeri *et al.* [53] suggested that repeated infusions can increase NK cell function and improved duration of responses and overall survival. However, Stern *et al.* [54] did not find relapse rates lower than those of the historical controls treated with haploidentical SCT without NK cell infusion. Choi *et al.* [55] used IL-15 and IL-21 in culture to expand and activate NK cells. They infused escalating doses of NK cells with a median total dose of 2.0×10^8 cells/kg being well tolerated. They analysed the outcome of 41 patients, 38 patients had refractory disease being AML the most frequent disease, when compared to an historical cohort, the study cohort showed less disease progression. However, no differences were observed in the probability of engraftment, grade 2 to 4 acute GVHD, moderate to severe chronic GVHD and TRM.

In the context of non-haematological malignancies, Iliopoulou *et al.* [56] used IL-15 instead of IL-2 to activate and expand NK cells with good results; and then administered NK cells in combination with chemotherapy to 15 patients with advanced non-small cell lung cancer showing potential clinical benefit. Geller *et al.* [57] in patients with recurrent ovarian and breast cancer, added total body irradiation (200 cGy) to fludarabine and cyclophosphamide in order to increase host immune suppression. However, they failed to demonstrate NK cell expansion on day +14, and they could not differentiate the contribution of the NK cells from the chemotherapy regimen in responding patients. It is not clear from these studies whether the lack of NK cell detection in PB is because they migrate to the tumor sites or it is due to a lack of NK engraftment.

Despite all the efforts made in these studies, different aspects have been answered and some remain to be solved in order to use NK cells as an immunotherapy option, such as: 1) should NK cells be infused in patients in remission or with active disease? These studies have shown that NK cells might be a better choice for consolidation therapy. In refractory patients, unsuccessful results might be improved with different strategies like eliminating the immunosuppressive role mediated by reg-T cells. 2) Should NK cells be infused before or after the SCT? In the auto-SCT these studies have shown similar results, with complete response rates of 50% and 53% when NK cells were infused before and after the auto-SCT, respectively. In the allo-SCT, NK cells have been always infused after the procedure. 3) What are the optimal NK cell dose and the number of infusions? These studies have shown that 10^8 cells/kg and repeating serial NK cell infusions are safe and well tolerated 4) How can we improve lymphodepleting treatment in order to reduce the appearance of reg-T cells? Or how can we reduce reg-T cell expression in order to facilitate NK cell activity? These studies have shown that the use of IL2DT is a good choice; however, no other alternatives have been tested. 5) How can we improve NK cell activation and expansion? It seems from these reports that the use of IL15 and IL21 is better than IL2. 6) Is it necessary to consider always KIR-HLA ligand mismatch? Results are not totally clear depending on the malignant disease. However, this effect can be bypassed in some occasions with cytokines like IL2 [34]. Thus, in the case of NK cells previously expanded *in vitro* with cytokines, the KIR-HLA ligand will not be the most important parameter and there will be other parameters more relevant that we will discuss in the following sections. 7) Could other NK cell sources provide with better results? In this sense, CB is a feasible option to obtain NK cells. However, in a CB unit the number of immune cells is low and furthermore they are still naïve and immature. Shah *et al.* [17] and our group [18] have developed a technique to efficiently expand NK cells from CB, which are highly functional. This technique and this type of NK cells will be further discussed in the next sections.

Before moving onto the next section we need to mention pre-clinical studies performed with the NK-92 cell line. This cell line is a type of human NK cells that have been used as a tool to study the properties of NK cells. In the clinical setting, there is only one clinical study that used these cells to treat hematological malignancies (ClinicalTrials.gov Identifier: NCT00990717), but results are not available yet. However, the use of NK-92 cells in pre-clinical studies have provided with some clues to combine NK cells with different drugs in clinical trials. For instance: *1)* Resveratrol, which is an anti-oxidant polyphenol, enhances perforin expression and the cytotoxicity of NK-92 cells against K562, HepG2, and A549 cells [58]. *2)* Radiotherapy by increasing the expression of matrix metalloproteinases reduces the expression of NKG2D ligands on cancer cells, and as a consequence, cancer cells evade the NK cell immune response. Combined treatment of radiation with NK cells and matrix metalloproteinases inhibitors has increased the expression of NKG2D ligands impacting on a higher susceptibility of cancer cells to NK-92 cells [59] *3)* Coloooxib, a nonsteroidal anti-inflammatory drug, by inducing the expression of NKG2D ligands in colon cancer cells, increases tumor cell susceptibility to NK-92 cells [60]. *4)* Calcitriol (1α,25-Dihydroxyvitamin D3) a drug used to increase blood Calcium levels, by down-regulating miR302c and miR-520-c up-regulates NKG2D ligands and enhances the susceptibility of both haematological and solid tumor cell lines to NK-92 cells [61]. *5)* The anti-CD99 antibody by inducing heat shock protein-70 expression on tumor cells enhances NK-92 cytotoxicity against B and T-leukemia cell lines [62]. All these results could be considered when performing clinical trials with NK cells.

ENVIRONMENT AND IMMUNOLOGICAL MEMORY ALSO AFFECT NK CELL PHENOTYPE AND CYTOTOXICITY

In addition to the influence of NK cell receptors in the NK-mediated cytotoxicity, there are supplementary factors impacting in the NK phenotype and cytotoxicity. Thus, mass cytometry analysis for NK cells has shown a very high degree of NK cell diversity, with an estimated 6,000 to 30,000 phenotypic populations within an individual; while genetics largely determined the inhibitory receptor expression, the activating receptors profile was heavily environmentally influenced. Therefore, NK cells may maintain self-tolerance through strictly patterns, while activating and co-stimulatory receptors may widely and flexibly respond to pathogens and tumor cells [63]. Examples of this NK cell education to secure tolerance include the role played by transcription factors in combination with cytokines, such as c-Myc and IL-15, which impact in KIR expression [64]; and the IL2 effect, which causes a less prominent licensing process in NK cells, suggesting that in cases of infection a higher number of NK cells will be more susceptible to become activated [34]. Thus, NK cell education process is more complex than initially thought, implying a connection between the environment surrounding NK cells and signaling from activating and inhibitory receptors.

A particular field related to the activating, inhibitory and co-stimulatory receptors that NK cells have, which has arisen as a field of study in cancer therapy, are the "immune checkpoints". These checkpoints are composed of inhibitory and stimulatory ligand-receptors pairs present in tumor cells, antigen presenting cells and immune cells and they can be critical mediators for tumor cells to evade the immune response. They were first described in T cells, as T cells in order to get a full activation need a costimulatory signal which is mediated by an interplay of these "immune checkpoints". Different immune check points are currently being studied such as lymphocyte activation gene-3 (LAG-3), the cytotoxic T lymphocyte antigen-4 (CTLA-4), programmed death-1 (PD-1), and T cell immunoglobulin and ITIM domain (TIGIT). In particular, PD-1 and CTLA-4 after engaging their ligands in tumor cells act as co-inhibitory receptors on immune cells. Therefore, inhibiting these inhibitory signals will increase the anti-tumor effect mediated by immune cells. The potential beneficial effect of blocking these checkpoints has been shown in clinical studies in melanoma using monoclonal antibodies for CTLA-4 (Ipilimumab) and PD-1 (nivolumab) with successful results [65]. PD-1 engages PD-1 Ligand (PDL1) expressed by many tumors and by antigen-presenting cells [66, 67]. Cytokines like IFN-γ lead to PDL1 up-regulation which in theory could enhance the effect of these monoclonal antibodies. In fact, preclinical studies have shown that IFNγ-inducing cancer vaccines (TEGVAX) up-regulate PDL1 expression in tumor cells and as a consequence increases the response to monoclonal antibodies for PDL1 [68]. In this sense, even NK cells are potent IFN-γ producers it has not been shown that they can increase PDL1 expression on tumor cells. However, it has been described that NK cells from patients with hematological malignancies such as multiple myeloma (MM) express PD-1 whereas normal NK cells do not [69] and that the use of monoclonal antibodies against PDL1 enhances NK cell cytotoxicity in MM patients [70]. Therefore, in some cases, the therapeutic use of NK cells could be affected by overexpression of co-inhibitory immune checkpoints on the surface of tumor cells. In this sense, a therapeutic option would be to combine NK cells with blocking monoclonal antibodies against the co-inhibitory signaling pathway involved.

To add a bit more of complexity to NK cell phenotype and cytotoxicity, it has been described that NK cells may exert a type of memory after viral infections like CMV, inducing a NK cell response similar to the characteristic immunological memory response developed by T cells after infection. NKG2C and NKG2A receptors are involved in this NK cell maturation process. This memory-like NK cell population has a mature phenotype with CD56dim, KIRs, NKG2C and CD57 expression, lack of NKG2A expression, and they are potent IFNγ producers. After allo-SCT, this memory NK cell population for CMV has also been detected, and characterized by the same phenotype and high IFNγ production [21, 22, 71]. Furthermore, when these memory NK cells are transplanted from CMV seropositive donors they expand not only in recipients undergoing CMV reactivation but also in seropositive recipients in absence of detectable viremia [23].

Regarding the environment, NK cells and tumor cells are not alone; other types of immune and non-immune cells with constant cytokine secretion, which affect both NK cytotoxicity and tumor cell survival, surround them. NK cell function can be affected by stimulation with different cytokines or other soluble factors and different isoforms of the receptors [20, 39, 72-76]. A good example of the influence of soluble factors is given by the proinflammatory cytokine macrophage migration inhibitory factor (MIF). MIF is released, after pathogenic or inflammatory stimuli, by pituitary cells, T lymphocytes, macrophages, epithelial and endothelial cells of different tissues. It promotes and amplifies inflammatory reactions [77, 78]. However, over-expression of MIF in cancer cells contributes to their immune escape by down-regulating NKG2D expression which impairs NK cell cytotoxicity toward tumor cells [79]. Another cytokine of interest is the immunosuppressive IL-10. NK cells by producing IL-10 can regulate the adaptive immune response mediated by T-cells, in particular this effect has been mostly studied after viral infection. By doing this, NK cells regulate massive CD8 T cell responses and ameliorate the immunopathology that could be associated to anti-viral CD8 T cell responses [80]. In the context of hematological malignancies and allo-SCT there are few studies available about the impact of IL10 production by NK cells, and most of them are in murine models. For instance, murine donor treatment with α-1-Antitrypsin (a serine protease inhibitor which alters cytokine production) results in enhanced IL10 production with expansion of DCs, reg-T cells, and NK cells, which results in lower GVHD and enhanced graft *vs.* leukemia activity, showing that NK cells might be useful decreasing GVHD without losing the graft *vs.* leukemia activity [81], and indicating that IL10 production by NK cells could not impact negatively after allo-SCT. Moreover, NK cell function is also controlled by interactions with other cells, like DCs, where there is a positive feedback loop among them. Thus, NK cells induce DC maturation, and NK cell-induced DC activation is dependent on TNF-α/IFN-γ secretion and NKp30. On its turn, *in vitro* IL-12, IL-18, IL-15, and IFN-α/β production by DCs enhance NK cell IFN-γ production, proliferation, and cytotoxic potential [82].

CHARACTERISTICS OF TUMOR CELLS AND IMPLICATIONS FOR CHEMO-THERAPY RESISTANCE AND ALLO-SCT

Above the complex network contributing to the NK cell activity has been described, but additionally, there are different types of tumor cells with intrinsic characteristics that make them susceptible to be eliminated through different cell death mechanisms once they become in contact with NK cells.

There are different cell death pathways, like the classic apoptotic cell death, and other forms of cell death like autophagy, necrosis or necroptotic cell death. Different modes of cell death can co-operate depending on the type of cell and the environment [83, 84]. This heterogeneity provides a differential susceptibility to diverse drug treatments impacting on chemo-therapy responses depending on the disease and even on each individual patient. Understanding the factors that contribute to cell death for each type of tumor cell is therefore critical for the development of novel therapies to eliminate cancer cells and to circumvent chemo-resistance. It is necessary to know all these variables to find out whether specific types of hematological malignancies will be susceptible to NK-mediated cytotoxicity.

In particular, necroptosis, which is a type of programmed necrotic cell death, and autophagy, a process of self-cannibalization to generate nutrients and energy, are two types of cell death associated to damaged associated molecular patterns (DAMPs) release. DAMPs are molecules released or exposed by dead, dying, or stressed non-apoptotic cells. These molecules will bind to pathogen recognition receptors, which will activate maturation of DCs, with IL1β production. IL1β will induce T cell proliferation, leading to an anti-tumor response. This type of reaction is termed immunogenic cell death (ICD) (Fig. **3**) [85, 86]. In the allo-SCT context, DAMPs could play a beneficial role increasing the graft *vs.* leukemia effect mediated by T lymphocytes. Furthermore, in the context of auto-SCT, also an increased T cell activity due to DAMPs will remove potential minimal residual disease, improving the clinical outcome of the patient. Some of these DAMPs include calreticulin surface expression, secretion of ATP, and release of high mobility group Box1 protein (HMGB1). On the other hand, the environment, such as the redox status, can change the immunogenic role of DAMPs to a tolerogenic role, contributing to tumor growth, proliferation, and metastasis. For example, HMGB1 will be either immunogenic or tolerogenic, depending on reactive oxygen species (ROS) production from apoptotic cells that will oxidize HMGB1. Oxidized HMGB1 cannot activate DCs, thus being tolerogenic [87-89]. Different types of cell death impact in the amount of DAMPs release.

1) Apoptosis is considered intrinsically tolerogenic because phagocytosis of apoptotic cells by DCs fails to induce maturation and causes tolerance to tumor antigens by generating helpless CD8+ T cells. Apoptotic cells release a low quantity of DAMPS, and with decreased activity.

2) Necroptosis leads to rapid plasma membrane permeabilization and to release of cell contents and exposure of high amount of DAMPs, inducing maturation of DCs and subsequent anti-tumor T cell response.

3) Autophagy is a process of self-cannibalization where lysosomes are the main organelles to carry out this degradation. This self-cannibalization can be either unselective degrading bulk cytoplasm contents to obtain nutrients and energy, or selective by degradation of organelles such as mitochondria (mitophagy), lipids (lipophagy), endoplasmic reticulum (reticulophagy), ribosomes (ribophagy), peroxisomes (pexophagy) [90, 91]. Whereas excessive autophagy leads to tumor cell death, a controlled autophagy can increase tumor cell survival, particularly in some type of tumor cells like MM. The role of DAMPs in autophagy is not clear. For instance, hypericin-photodynamic therapy is an oxidative chemotherapy treatment that triggers autophagic cell death with ROS release and endoplasmic reticulum (ER) stress. This treatment is associated to a release of DAMPs and to an improved ICD. However, on the contrary to expectations, induced autophagy by this oxidative treatment, only regulates calreticulin surface expression, it is not involved in the ATP release, and it is only partially involved in the ICD observed [92].

The ICD effect should be taken into account when infusing NK cells after allo-SCT. Serial additional NK cell infusions after SCT could be done when the T cell immunity starts to recover in the patient. In this context, DAMPs release by dying cells after contacting with NK cells could improve the ICD mediated by T cells.

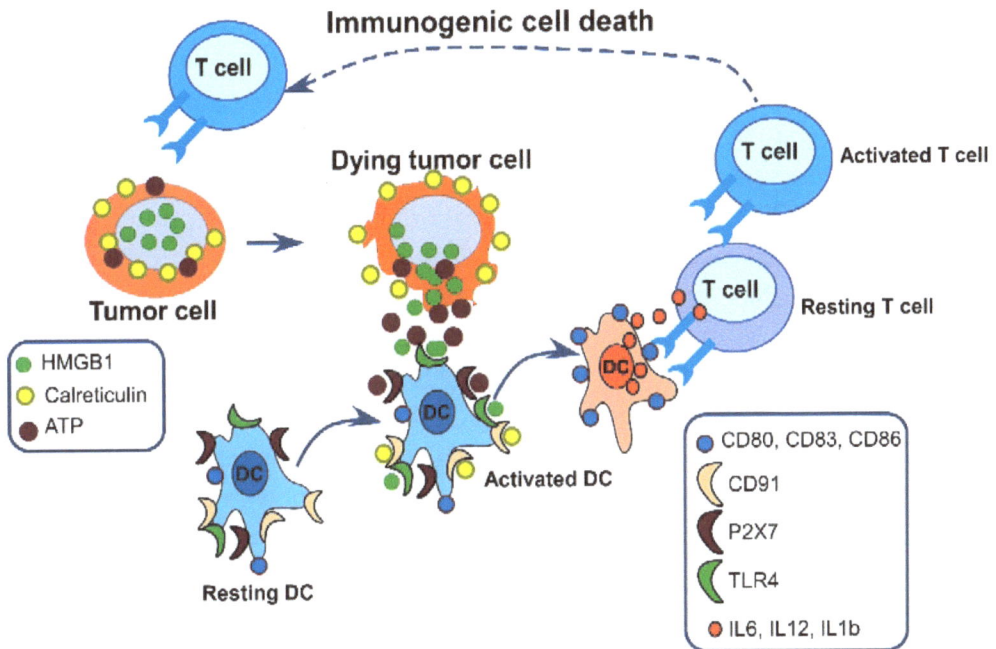

Figure 3: Activation of immunogenic cell death (ICD) by damaged associated molecular patterns (DAMPs): DAMPs are molecules which are released or exposed by dead, dying, or stressed non-apoptotic cells. Some of these DAMPs include the release of the nuclear protein HMGB1, surface exposure of calreticulin, and ATP release. Once released, these DAMPs will bind to pathogen recognition receptors in dendritic cells (DCs) activating their maturation which involves expression of costimulatory molecules (CD80, CD83 and CD86) and IL1β, IL6 and IL12 release that together will induce T cell activation and proliferation leading to an enhanced anti-tumor response.

Multiple Myeloma: A Special Role for Autophagy

MM is a neoplastic plasma cell disorder characterized by clonal proliferation of malignant plasma cells in the bone marrow, monoclonal protein in the blood or urine, and associated organ dysfunction. Plasma cells synthesize large quantities of Ig which are folded in the ER. An excess of Ig synthesis causes a failure in this folding process leading to the release of unfolded or misfolded Ig. Major intracellular protein degradation pathways including the ubiquitin-proteasome system and autophagy normally degrade these Ig. An imbalance between the amount of unfolded or misfolded protein in the ER lumen and the capacity of the ER machinery to refold these proteins results in ER stress and cell death [93, 94]. Proteasome inhibitors are potent anti-MM agents with remarkable clinical efficacy [90] which block the protein degradation process in MM cells increasing ER stress and leading to cell death [95, 96]. The ubiquitin-proteasome system and autophagy regulate each other, where the suppression of one enhances the other as a compensatory mechanism. Thus, an excess of proteasome inhibitor-induced ER stress could also increase autophagy. In some cases this can lead to drug resistance and eventual progression of disease [97-99]. In fact, there is a variable percent of MM patients who do not response to proteasome inhibitors treatment (bortezomib) when used either as a first line treatment (19%) or in relapsed MM patients (50%) [100, 101]. However, the role of autophagy in cancer is not that clear. There is evidence supporting that autophagy can support tumor cells maintenance and that excessive autophagy can lead to cell death. In MM, malignant plasma cells need both protein degradation systems, autophagy and the ubiquitin-proteasome system, to deal with the high Ig production they have. Thus, there is a basal level of autophagy, which confers a state of "autophagic stress", and keeps Ig synthesis and long-lived humoral immunity in MM cells [102]. This autophagy level could be the responsible for MM patients who do not response to proteasome inhibitors treatment. However, this is not clear as bortezomib by causing accumulation of misfolded Ig leads to autophagic cell death. To avoid autophagic cell death, MM cells use caspase-10 to set a threshold for autophagy that, if ruptured leads to excessive autophagy, and cell death [103, 104].

CORD BLOOD DERIVED NK CELLS (CB-NK), A NEW SOURCE OF NK CELLS WITH CHARACTERISTIC CYTOTOXICITY

When NK cells establish the contact with target cells, they kill them by the vesicular delivery of cytolytic molecules such as granzyme-B and granulysin.

This will activate different cell death pathways, which can be caspase-3 dependent, or caspase-3 independent. The caspase-3 independent cell death is *via* ER stress, or lysosomal cell death through release of lysosomal proteases [30, 31, 105-107]. Furthermore, NK cells can release exosomes expressing NK cell activating receptors with cytotoxic activity against tumor cells [108, 109]. We herein present a type of NK cells derived from CB, and termed CB derived NK cells (CB-NK). CB-NK can be expanded *in vitro* to obtain a high number to use in the clinical context, after the expansion they are highly activated and because they are from CB they do not present the high HLA-I restriction that allogeneic donors have. With these conditions we solve some of the problems previously found with NK cells from PB (PB-NK). In addition, CB-NK present specific cytotoxic properties that we will discuss in the following sections.

Generation of Highly Functional NK Cells in Enough Number to be Infused into a Patient

Shah *et al.* [17] have shown a clinically applicable strategy for the generation of highly functional CB-NK which can be used to eradicate MM and potentially other hematological malignancies. These cells can be generated starting either from NK cells isolated from CB, or from CB-mononuclear cells (MNC). Then, they are expanded *in vitro* for 14 days with artificial antigen presenting cells (aAPCs) and adding IL2 exogenously. aAPCs are K562-based aAPCs expressing 41BB ligand, CD64, CD86 and membrane bound IL-21 (Clone9.mbIL21) (Fig. **4**). This expansion technique eliminates the problem of the limited number of NK cells in an unmanipulated CB unit. Thus, starting with 20×10^6 CB-MNCs (approximately 10% of a CB unit), this culture system would allow for the generation of approximately 1.4×10^9 NK cells for infusion, or 1.9×10^7 NK cells/kg for an average 70 kg adult. This number, according to the safety dose study performed by Choi *et al.* [55] would allow for several doses of NK cell therapy to enhance anti-tumor efficacy. Additionally, if the expansion is started from CB-MNC, the final NK product is relatively pure (99%), with only 6×10^4 CD3+cells/kg, thus reducing the potential for GVHD; the amount achieved is over 18 fold higher than the growth seen with CD56+ selected cells expanded with IL-2 alone. This type of NK cells present a characteristic cytotoxicity in comparison to PB-NK, show an activated phenotype with high expression of NCR receptors and are not exhausted as shown by their positive expression for T-bet and Eomesodermin (two transcription factors which are integral to NK cell function). These CB-NK are currently being used in a clinical trial for MM patients (clinicaltrials.gov Identifier: NCT01729091).

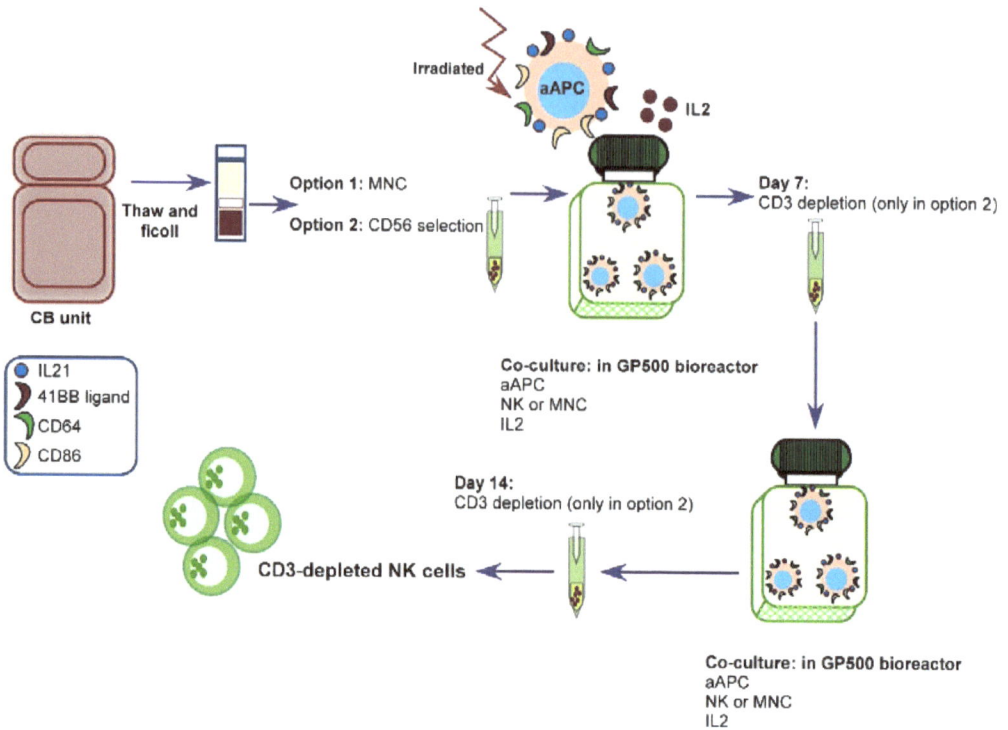

Figure 4: Clinical expansion of NK cells from cord blood (CB). Cord blood-derived NK cells (CB-NK) can be generated starting either from CB-mononuclear cells (MNC) or with NK cells. First, CB cells are ficolled to get the MNC, and then they are either added directly to GP500 bioreactor or subjected to CD56+ selection using magnetic beads. These CD56+ cells will be added to the GP500 bioreactor. Then, they are expanded *in vitro* for 7 days with artificial antigen presenting cells (aAPCs) and adding IL2 exogenously every other day. aAPCs are K562-based aAPCs expressing 41BB ligand, CD64, CD86 and membrane bound IL-21 (Clone9.mbIL21). They are co-cultured in a 2:1 aAPC:MNC or NK cells ratio. On day 7, fresh Clone9 cells are added again in the same ratio and re-cultured in the same conditions for an additional 7 days. On day 7 and day 14, cells are CD3-depleted, only in case the expansion was started with MNC.

Characteristic Cytotoxicity of CB-NK

Shah *et al.* and our group have studied the cytotoxic properties of CB-NK against MM. We have shown a mechanism of transmissible cell death which starts with cytotoxic material transferred from CB-NK to MM cells and secondarily transferred between neighboring MM cells, amplifying the initial CB-NK cytotoxicity achieved. Even these experiments were performed with MM cell lines and not with primary MM cells they give us many clues about the CB-NK cytotoxicity mechanism. This transfer involves both lipid and protein transfer (Figs. **5** and **6**). Such cytotoxicity leads to lysosomal cell death of MM cells which

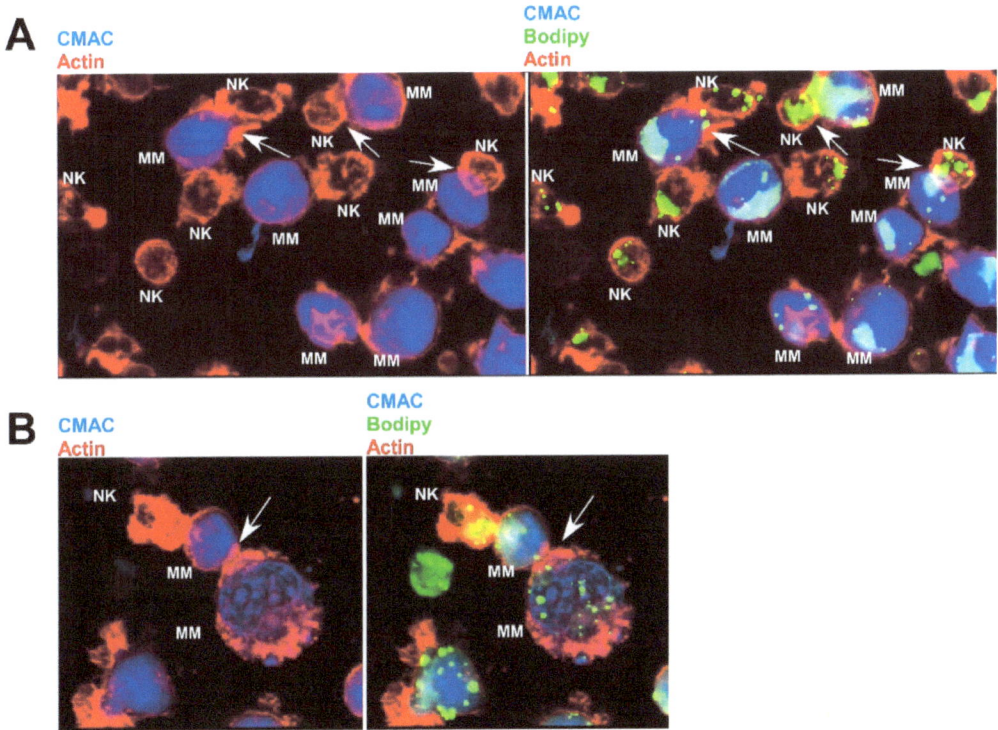

Figure 5: Transmissible cell death by CB-NK. CB-NK when they become in contact with multiple myeloma (MM) cells transmit cytotoxic material in lipid-protein vesicles to target cells through the immunological synapse. MM cells and CB-NK are stained with cell trackers living reagents to follow the transfer of compounds between cells. MM cells are stained in blue (CMAC) and CB-NK cells in green (Bodipy). Bodipy has lipid affinity. F-actin (in red) indicates when cells perform an immunological synapse. On the left, only CMAC and Bodipy are shown and on the right the three colors are shown together. Arrows indicate transfer of CB-NK content (in green) into MM cells (in blue), indicating lipid/protein trafficking. In **A**. CB-NK can be observed transferring contents into MM cells through the immunological synapse and in **b** we can observe one CB-NK transferring content to one MM cell and this MM cell transferring content to a neighboring MM cell through the immunological synapse. Data obtained from Martín-Antonio *et al.* [18].

contacted with CB-NK and to cell death of MM cells even without having been in contact with NK cells. Lysosomal cell death is due to the intracytoplasmatic release of proteolytic enzymes like cathepsin B. Depending on the magnitude of lysosomal permeabilization, lysosomal cell death features can vary activating either an apoptotic like programmed cell death (partial permeabilization) or necrotic cell death (massive permeabilization) [107]. In addition, this indirect and transmissible CB-NK cytotoxicity involves the transfer between cells of the NK cell receptors NKG2D and NKP30; and leads also to a transmission between MM

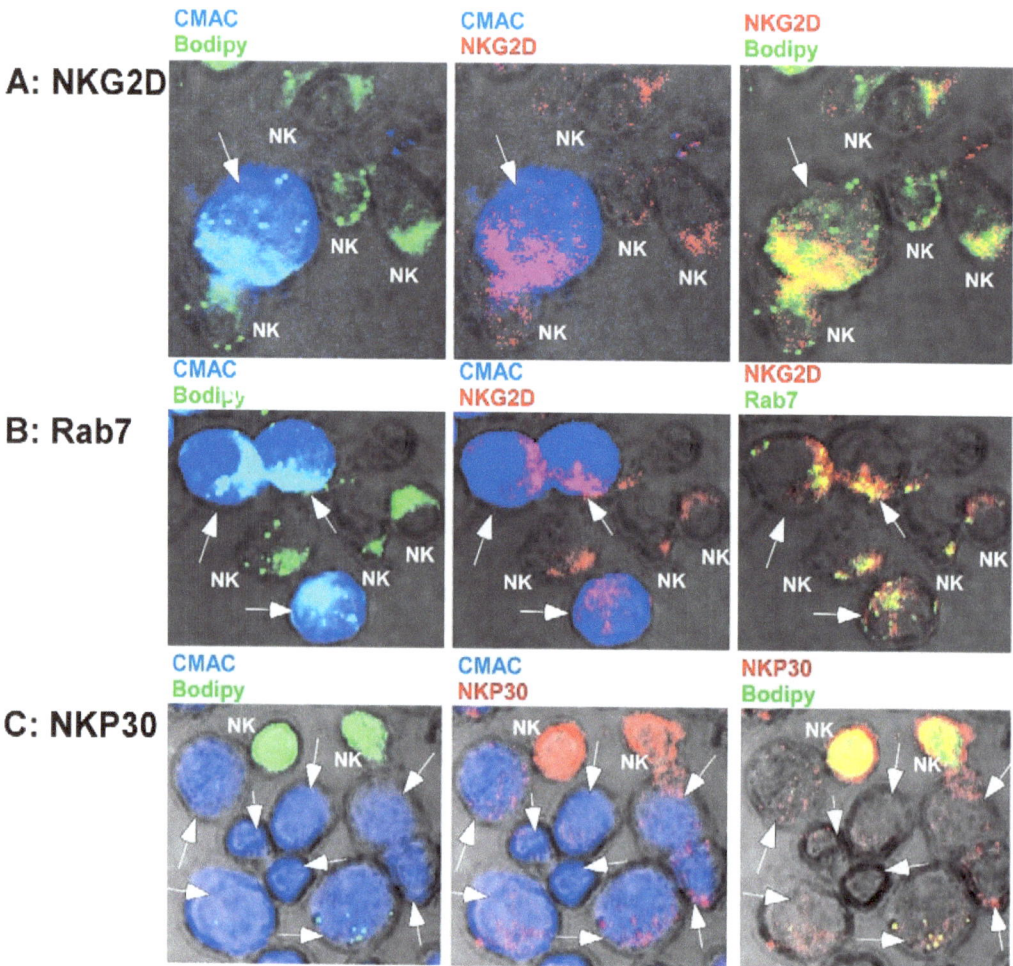

Figure 6: Concomitant lipid-protein trafficking between CB-NK and MM cells. CB-NK transmit cytotoxic material to MM cells containing lipids and proteins. These proteins include at least NK cell receptors NKG2D and NKP30 and this transfer also involves lysosomes, defined by the lysosome marker Rab7. MM cells are stained in blue (CMAC) and CB-NK in green (bodipy). NKG2D, Rab7 and NKP30 are shown in red. **A** shows transfer of NKG2D from one CB-NK to one MM cell. In the third column, co-staining of NKG2D and bodipy (in yellow) indicates concomitant lipid-protein trafficking for NKG2D. **B** shows transfer of NKG2D between two MM cells after it was transferred from CB-NK and co-localization (third column in yellow) with the lysosomal marker Rab7. **C** shows transfer of NKP30 from one CB-NK to one MM cell and then secondarily between MM cells. In the third column, co-staining of NKP30 and bodipy (in yellow) indicates concomitant lipid-protein trafficking for NKP30. Arrows indicate MM cells. Data obtained from Martín-Antonio *et al.* [18].

cells of the initial cytotoxic effects observed (decreased lysosome levels). This cytotoxicity differed between different types of tumor cells in terms of: *1)* granzyme B was necessary for the cytotoxicity *vs.* K562 cells and not *vs.* MM cells, *2)* NKG2D and NKP30 receptors contributed in a higher degree to the cytotoxicity *vs.* MM cells than *vs.* K562 cells; and *3)* lipids, which were transferred between cells played a more relevant role in the cytotoxicity *vs.* MM cells than *vs.* K562 cells [18]. These findings highlight the importance of:

1) There is a differential CB-NK cytotoxicity depending on the tumor cell. Therefore understanding the cell death mechanisms for each malignancy will help to enhance the CB-NK mediated cytotoxicity after allo-SCT; for instance, combining CB-NK with chemotherapy drugs which activate other cell death pathways to achieve a synergistic effect with CB-NK.

2) We observed the importance of lipids, lipid transfer and cell-to-cell communication in CB-NK cytotoxicity. This is a novel finding not previously observed in NK cells. Lipids appeared to play a more relevant cytotoxic role *vs.* MM cells than *vs.* K562 cells, which could be due to the intrinsic characteristics of MM cells. Therefore, lipid transfer from CB-NK to tumor cells appears as a new potential anti-tumor tool to be applied in certain hematological malignancies. The implications of cell-to-cell communication in the transmissible cell death and the novel cytotoxic activity of lipids will be discussed with more detail below.

CELL-TO-CELL COMMUNICATION AND LIPID-BASED THERAPIES AS A NEW CANCER THERAPY OPTION

Cell-to-Cell Communication

To support the CB-NK mediated transmissible cell death that we have observed in our studies, different reports in the last years have shown that cells communicate between them by transferring organelles, including ER/Golgi, endosomes, lysosomes, and mitochondria. Even though this transfer is not a universal phenomenon, it has been described in many different cell types, in the same cell lineage and between different cell lineages. Most reports have shown organelle transfer *in vitro*, but there are also studies showing *in vivo* transfer, for instance, exogenously administered bone marrow-derived mesenchymal stem cells donate mitochondria to injured alveolar epithelial cells of the mouse lung to protect them against acute lung injury. This organelle transfer between cells occurs by means of lipid nanotubes composed of F-actin and resembles to the immunological synapse formation which happens between immune cells and target cells (Fig. **7**) [110].

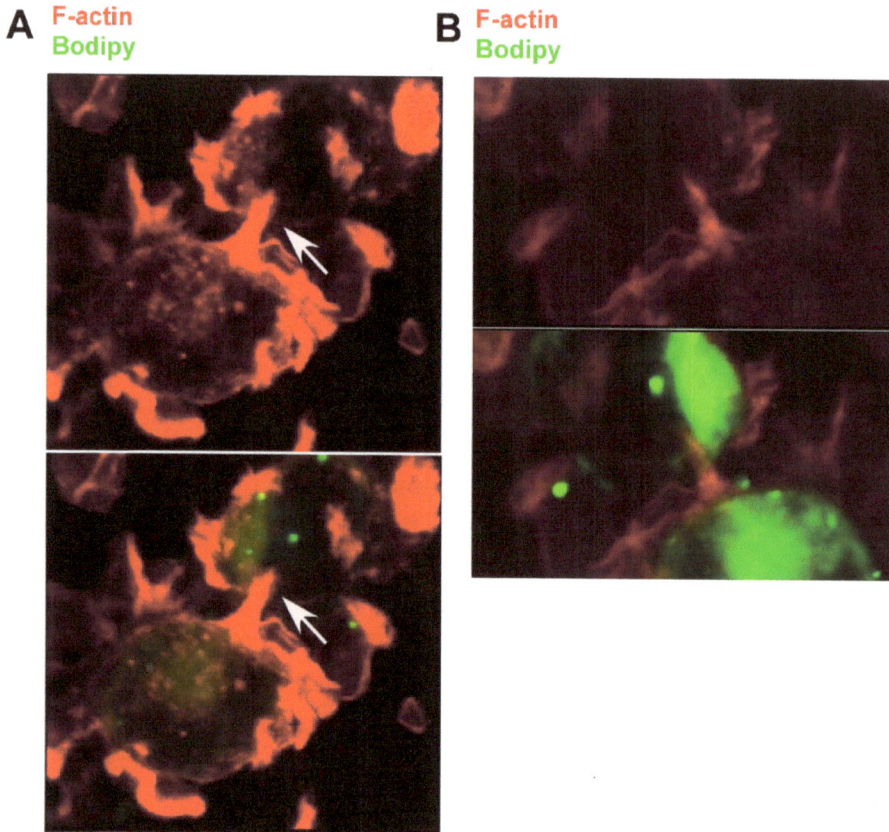

Figure 7: Nanotube formation between cells. MM cell lines were stained in green (bodipy) and left in culture. **A**. Nanotube formation between two cells can be visualized by staining with F-actin (in red). **B** shows a more detailed image of the nanotube formation observed from a different perspective.

The synapse formed between immune cells and target cells is one of the ways that immune cells have to mount an effective immune response. This synapse is a focal point for both exocytosis and endocytosis. Immune signaling triggers massive reorganization of F-actin, microtubule cytoskeletons and secretory lysosomes at the immunological synapse, and both Golgi and recycling endosomes polarize to the synapse moving along microtubules to direct the secretion and endocytosis to the target cell [111].

Usually, organelle transfer happens from a healthy donor cell to an injured receptor cell to enhance cell survival of the injured cell [112, 113]. Cell stress might be a major determinant of this intercellular organelle transfer because the normal transfer rate doubles when cells are stressed, and further, the stressed cells are the ones that sprouts the nanotube to receive the organelles from the donor cell

[110]. If these observations are applied to the transmissible cytotoxicity that CB-NK mediate against MM cells, it could be considered that MM cells become stressed after contacting with CB-NK and therefore, they will increase the synapse formation with neighboring MM cells. However, instead of transferring healing, they will transfer the cytotoxicity transmitted originally from CB-NK. In this transmissible cytotoxicity, lipids are crucial, since in our studies it was observed: *1)* concomitant lipid-protein trafficking between CB-NK and MM cells, and secondarily between MM cells, including lysosomes and NK cell receptors (Fig. **6**); and that *2)* lipids play a relevant role in the CB-NK cytotoxicity against MM cells [18]. It has not been tested whether this transfer would also happen with other sources of NK cells such as PB-NK, but based on the description of this phenomenon for different cell types it should not be discarded the possibility and it should be studied in the future.

Some of the properties that lipids have to activate different types of cell death, and their role as a cancer therapy option will be mentioned. We will specially focus on MM and how lipids, CB-NK and different chemotherapy drugs might be combined to improve the outcome of MM patients.

Anti-Tumor Activities of Lipids and Multiple Myeloma

Lipids are emerging as new cell regulators as cells exchange information *via* release of exosomes or lipid vesicles. Lipids interact extensively with other organelles and are involved in immune responses. Moreover, interactions between lipids with membranes and the endocytic system affect the membrane physical properties impacting in cell-to-cell communication. Lipids are also appearing as new targets for cancer therapies, as they cause apoptosis in cancer cells [114, 115], and other types of cell death like lysosomal cell death and necroptosis by destabilizing lysosomes [116]. Interestingly, they can both inhibit and induce autophagy; and they can cause ER stress similar to what proteasome inhibitors do.

The ER is the main site for lipid membrane synthesis and new protein folding, assembly, and secretion. Disruption of ER homeostasis caused by an excess of lipid accumulation, leads to accumulation of misfolded/unfolded proteins in the ER, which will cause ER stress and eventually cell death. However, the relationship between ER stress and lipid metabolism is bidirectional. While some lipids, like saturated fatty acids, can cause ER stress leading to cell death, also ER stress can lead to altered cholesterol biosynthesis. Furthermore, the levels of lipids can change the function of the ER. Studies have suggested that in liver cells of obese patients, the ER can shift from being the major site of protein synthesis to carrying out lipid synthesis; and resulting eventually in impaired ER calcium retention, protein misfolding and ER stress [117-124].

Regarding autophagy, lipids can either activate or inhibit the autophagy process by controlling different fundamental aspects. First, lipids undergo autophagic degradation in lysosomes (lipophagy) [125]. Also, they regulate signaling cascades converging onto the mammalian TOR (mTOR) pathway, which suppresses autophagy, or lipids like phosphatidylethanolamine by binding to Atg8/LC3 mediate the formation of the phagophores, which are the crucial vesicles to start the autophagy process. Furthermore, lipids can control membrane dynamics by directly affecting the physicochemical properties of lipid bilayers independently of protein effectors. For instance, cholesterol-enriched microdomains play a role in chaperone-mediated autophagy [126]. There are many different types of lipids and each lipid has different cytotoxicity for each type of cancer cell.

In particular, ceramide, which belongs to the sphingolipid family, is involved in many different cell death pathways [127]. Ceramide can trigger mitochondrial-cell death regulated pathways, ER stress, and lysosomal cell death. Ceramide can be metabolized into different pathways impacting into tumor cell survival. Therefore, cancer cells by impairing sphingolipid metabolism, reduce proapoptotic ceramide generation and gain survival advantages against therapy. Proapoptotic ceramide can be degraded into the mitogenic sphingosine-1-phosphate or metabolized into glucosylceramide by the glucosylceramide synthase (GCS) enzyme, impacting on ceramide levels (Fig. **8**). Higher expression of GCS increases chemoresistance (to adriamycin, vinblastine, vincristine, etoposide and doxorubicin) of drug-sensitive cells in breast cancer, colon cancer, leukemia and melanoma cells [114]. In the case of AML, ceramide levels are lower and the activity of GCS is higher in chemo-resistant patients than in chemo-sensitive patients. In chronic myeloid leukemia, the drug resistant K562 cells also express higher level of GCS, and pharmacological inhibition of GCS sensitizes K562 cells to adriamycin. In the case of MM, thalidomide requires ceramide to induce its anti-tumor effects [128]. Furthermore, tamoxifen inhibits ceramide degradation into glucosylceramide in cancer cells and leads to lysosomal cell death. Ceramide can also mediate autophagic cell death by targeting mitochondria to autophagolysosomes, leading to inhibition of mitochondrial function [129]; and necroptotic cell death in lung cancer cells [130]. Therefore, combining therapies with tamoxifen and ceramide looks an option, and in fact, the combination of both has shown higher anti-tumor efficacy [131, 132].

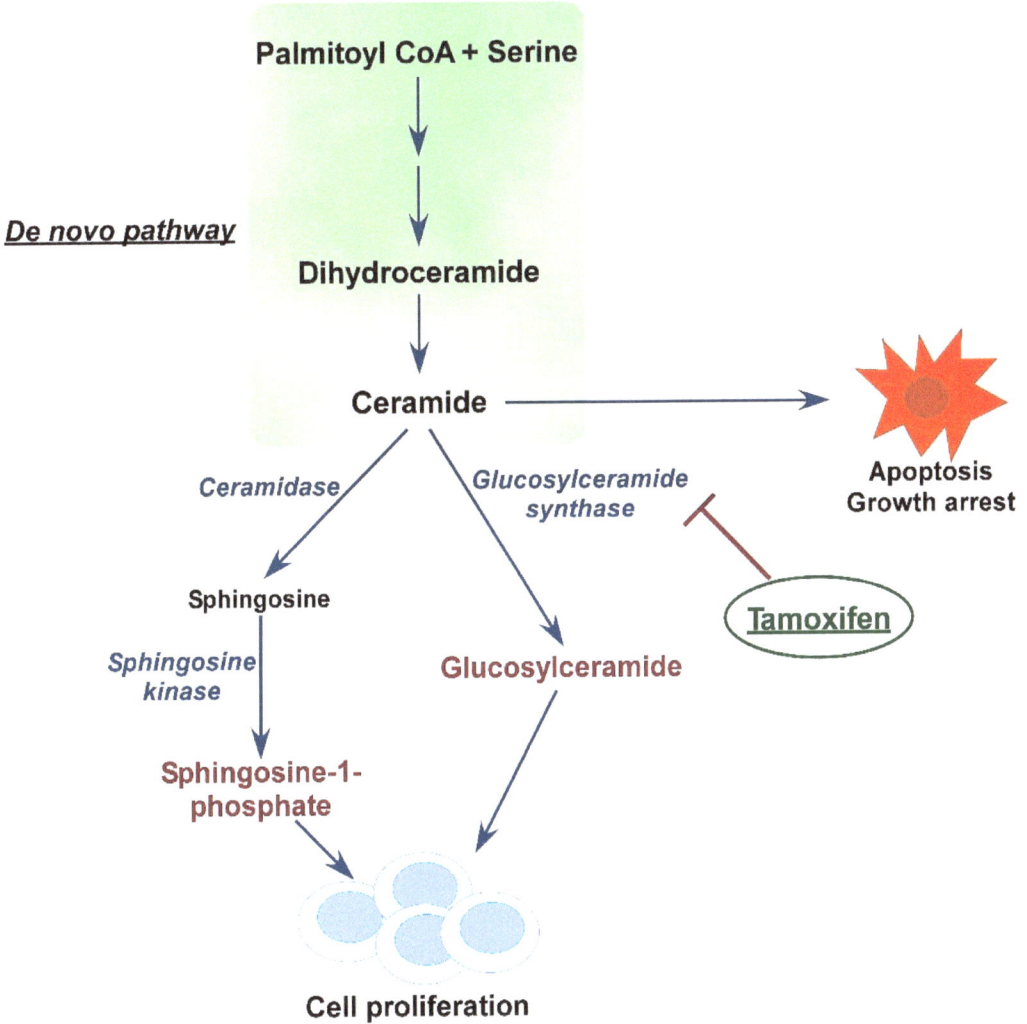

Figure 8: Metabolic pathways which affect ceramide levels. Ceramide after being synthesized has anti-tumor activities causing apoptosis in tumor cells. Ceramide can be either degraded to sphingosine-1-phosphate by the enzyme sphingosine kinase or metabolized into glucosylceramide by the enzyme glucosylceramide synthase. Both compounds will induce tumor cell proliferation. Tamoxifen exerts anti-tumor activities by inhibiting glucosylceramide synthase and decreasing glucosylceramide levels.

In our studies, some similarities between the CB-NK mediated cytotoxicity *vs.* MM cells and the ceramide and tamoxifen mediated cytotoxicity was shown. It was observed that *1)* CB-NK have a higher amount of lipids in comparison to non-expanded PB-NK. *2)* CB-NK mediate a lysosomal cell death *vs.* MM cells not observed for PB-NK. *3)* CB-NK lead to an increased expression of ER stress related markers (DNA-damage-inducible transcript 3: DDIT3) in MM cells. *4)*

CB-NK decrease glucosylceramide levels in MM cells. All these features resemble to the cell death mechanisms that tamoxifen and ceramide mediate. Additionally, we have confirmed the synergistic effect of tamoxifen and ceramide against MM cell lines (unpublished data) confirming the relevance of lipids as therapeutic targets in MM. These features, which were not observed in K562 cells, could be explained by the intrinsic characteristics of MM cells.

As previously mentioned, MM is characterized by the presence of malignant plasma cells that present a basal level of autophagy to survive to deal with the high amount of proteins that they produce, but in the other side, an excessive autophagy can lead to cell death. Proteasome inhibitors, like bortezomib, by blocking the ubiquitin proteasome system cause excessive autophagy and cell death, but sometimes this can lead to refractoriness. To take advantage of the specific characteristics of MM cells, two different scenarios can be considered. In one scenario, there would be refractory patients who do not response to bortezomib treatment. In this case, a tempting option would be to treat MM patients with autophagy inhibitors to try to lower autophagy levels and then try again bortezomib and watch whether patients show any response. In fact, in a very recent study it has been shown that autophagy inhibition can revert bortezomib resistance [133]. There are already two clinical trials on-going (clinicaltrials.gov Identifiers: NCT00568880 and NCT01438177) using both drugs together and even a proposal to correlate autophagy levels with clinical outcome (NCT01594242). In a second scenario, for patients who receive first line treatment it could be tried to combine drugs to reach excessive autophagy levels leading to cell death. In this case, it could be considered combining ceramide, which causes autophagic cell death, with proteasome inhibitors, or with tamoxifen to achieve an additive effect. Furthermore, it could be considered treating MM patients with CB-NK, which seem to induce a similar type of cell death, in combination with either proteasome inhibitors or tamoxifen to enhance their cytotoxicity. However these hypotheses still need to be confirmed *in vivo* in mice studies.

CONCLUSION

Results of SCT have improved during last years. However, clearly better clinical outcomes are still needed. It has been suggested that NK cell therapy could contribute to further improvement of SCT results obtained in hematological malignancies. However, this has not been confirmed by several clinical trials, indicating the need to perform changes and to consider new options. NK cell *in vitro* expansion with cytokines can bypass the KIR-HLA ligand mismatch solving one of the limitations when looking for a donor. In addition, the response to such therapy may be different depending on the hematological malignancy treated that could be due to the intrinsic characteristics of the tumor cells. It is necessary to understand which cell death pathways are activated in tumor cells, in order to combine drugs activating the same cell death pathway and lead to a synergistic effect. We propose the approach of using CB-NK, which can be easily expanded from a CB unit and obtain enough number to infuse into a patient in serial infusions. CB-NK mediate a characteristic cytotoxicity depending on each target cell. They can transmit cell death by transferring lipids and other cytotoxic molecules. Deciphering what are the main lipids involved and to search drugs that induce the same cytotoxic effect for each hematological malignancy may lead to a stronger CB-NK mediated cytotoxicity. And hopefully, the use of these powerful cells may contribute to further improvement of the SCT results in hematological malignancies.

ACKNOWLEDGEMENTS

We acknowledge Institute of Health Carlos III (project: PI14/00798) and Josep Carreras Leukaemia Research Institute for providing funding for NK studies. BMA was the recipient of a "Sara Borrell" fellowship from Institute of Health Carlos III.

CONFLICT OF INTEREST

The author(s) confirm that this chapter contents have no conflict of interest.

ABBREVIATIONS

aAPCs	=	Artificial antigen presenting cells
AICL	=	Activation-Induced C-type Lectin
ALCL	=	Anaplastic large cell lymphoma
AML	=	Acute myeloblastic leukemia
ALL	=	Acute lymphoblastic leukemia
Allo-SCT	=	Allogeneic stem cell transplantation
ATG	=	Antithymocyte globulin
Auto-SCT	=	Autologous stem cell transplantation

BCD CD19	=	B-cell depleted
BM	=	Bone marrow
Bu	=	Busulfan
CAR	=	Chimeric antigen receptor
CB	=	Cord blood
CBT	=	Cord blood transplant
CB-NK	=	Cord blood derived natural killer cells
CMV	=	Cytomegalovirus
CR	=	Complete remission
CTLA-4	=	Cytotoxic T lymphocyte antigen-4
Cy	=	Cyclophosphamide
DAMPs	=	Damaged associated molecular patterns
DCs	=	Dendritic cells
DLBCL	=	Diffuse large B cell lymphoma
Dx	=	Dexamethasone
EFS	=	Event free survival
ER	=	Endoplasmic reticulum
FADD	=	Fas-associated death domain protein
FL	=	Follicular lymphoma
Flu	=	Fludarabine
GCS	=	Glucosylceramide synthase

GVHD	=	Graft *vs.* host disease
Haplo	=	Haploidentical
HLA	=	Human leucocyte antigen
HMGB1high	=	Mobility group Box1 protein
ICD	=	Immunogenic cell death
Ig	=	Immunoglobulin
IL2DT	=	IL-2 diphtheria toxin fusion protein
ITAM	=	Immunoreceptor tyrosine-based activation motif
ITIM	=	Immunoreceptor tyrosine-based inhibitory motif
KIRs	=	Killer cell immunoglobulin-like receptors
LAG-3	=	Lymphocyte activation gene-3
MA	=	Myeloablative
MCL	=	Mantle cell lymphoma
MDS	=	Myelodysplastic syndrome
Mel	=	Melphalan
MIF	=	Macrophage migration inhibitory factor
MM	=	Multiple myeloma
MNC	=	Mononuclear cells
MZL	=	Marginal zone lymphoma
MR	=	Molecular relapse
NA	=	Not available

NCR   = Natural cytotoxicity receptors

NHL   = Non Hodgkin lymphoma

NK   = Natural killer

NSCLC   = Non-small cell lung cancer

OS   = Overall survival

PB   = Peripheral blood

PB-NK   = NK cells from PB

PBPC   = Peripheral blood progenitor cells

PD   = Progressive disease

PD-1   = Programmed death-1

PDL1   = PD-1 Ligand

PR   = Partial response

Reg   = Regulatory

R   = Rituximab

ROS   = Reactive oxygen species

SLAM   = Signaling lymphocyte activating molecule

SCT   = Stem cell transplantation

TRM   = Transplant related mortality

SD   = Stable disease

TBI   = Total body irradiation

TCD CD3   = T-cell depleted

TIGIT   = T cell immunoglobulin and ITIM domain

Thio   = Thiotepa

REFERENCES

[1]   Copelan EA. Hematopoietic stem-cell transplantation. N Engl J Med. 2006; 354(17): 1813-26.
[2]   Martín-Antonio B, Granell M, Urbano-Ispizua A. Genomic polymorphisms of the innate immune system and allogeneic stem cell transplantation. Expert Rev Hematol. 2010; 3(4): 411-27.
[3]   Thomas ED, Lochte HLJ, Lu WC, Ferrebee JW. Intravenous infusion of bone marrow in patients receiving radiation and chemotherapy. N Engl J Med. 1957 257(11): 491-6.
[4]   Mathé G, Amiel JL, Schwarzenberg L, Cattan A, Schneider M. Adoptive immunotherapy of acute leukemia: experimental and clinical results. Cancer Res. 1965; 25(9): 1525-31.
[5]   Gluckman E. A brief history of HSCT. In: Apperley J, Carreras E, Gluckman E, Masszi T, editors. The EBMT Handbook Hematopoietic Stem Cell Transplantation: European School of Haematology; 2012.
[6]   Passweg JR, Baldomero H, Gratwohl A, Bregni M, Cesaro S, Dreger P, *et al*. The EBMT activity survey: 1990-2010. Bone marrow transplantation. 2013; 48(7): 1161-2267.
[7]   Ballen KK, Gluckman E, Broxmeyer HE. Umbilical cord blood transplantation: the first 25 years and beyond. Blood. 2013; 122(4): 491-8.
[8]   Eapen M, Rocha V, Sanz G, Scaradavou A, Zhang MJ, Arcese W, *et al*. Effect of graft source on unrelated donor haemopoietic stem-cell transplantation in adults with acute leukaemia: a retrospective analysis. Lancet Oncol. 2010; 11(7): 653-60.
[9]   Shaw BE, Madrigal A. Immunogenetics of allogeneic HSCT. In: Apperley J, Carreras E, Gluckman E, Masszi T, editors. The EBMT handbook Hematopoietic Stem Cell Transplantation. 1. 6 ed: European School of Haematology; 2012.
[10] Wagner JE, Jr., Eapen M, Carter S, Wang Y, Schultz KR, Wall DA, *et al*. One-unit *versus* two-unit cord-blood transplantation for hematologic cancers. N Engl J Med. 2014; 371(18): 1685-94.
[11] Delaney C, Heimfeld S, Brashem-Stein C, Voorhies H, Manger RL, Bernstein ID. Notch-mediated expansion of human cord blood progenitor cells capable of rapid myeloid reconstitution. Nat Med. 2010; 16(2): 232-6.
[12] de Lima M, McNiece I, Robinson SN, Munsell M, Eapen M, Horowitz M, *et al*. Cord-blood engraftment with *ex vivo* mesenchymal-cell coculture. N Engl J Med. 2012; 367(24): 2305-15.
[13] Robinson SN, Thomas MW, Simmons PJ, Lu J, Yang H, Parmar S, *et al*. Fucosylation with fucosyltransferase VI or fucosyltransferase VII improves cord blood engraftment. Cytotherapy. 2014; 16(1): 84-9.
[14] Le Blanc K, Frassoni F, Ball L, Locatelli F, Roelofs H, Lewis I, *et al*. Mesenchymal stem cells for treatment of steroid-resistant, severe, acute graft-*versus*-host disease: a phase II study. Lancet. 2008; 371(9624): 1579-86.
[15] Edinger M, Hoffmann P. Regulatory T cells in stem cell transplantation: strategies and first clinical experiences. Curr Opin Immunol. 2011; 23(5): 679-84.
[16] Di Ianni M, Falzetti F, Carotti A, Terenzi A, Castellino F, Bonifacio E, *et al*. Tregs prevent GVHD and promote immune reconstitution in HLA-haploidentical transplantation. Blood. 2011; 117(14): 3921-8.
[17] Shah N, Martin-Antonio B, Yang H, Ku S, Lee DA, Cooper LJN, *et al*. Antigen presenting cell-mediated expansion of human umbilical cord blood yields log-scale expansion of natural killer cells with anti-myeloma activity. Plos One. 2013; 8(10): e76781.
[18] Martin-Antonio B, Najjar A, Robinson SN, Chew C, Li S, Yvon E, *et al*. Transmissible cytotoxicity of Multiple Myeloma cells by NK cells mediated by vesicle trafficking. Cell Death Differ. 22. 2015; 1: 96-107.
[19] Morice WG. The immunophenotypic attributes of NK cells and NK-cell lineage lymphoproliferative disorders. Am J Clin Pathol. 2007; 127(6): 881-6.

[20]   Lanier LL. Up on the tightrope: natural killer cell activation and inhibition. Nat Immunol. 2008; 9(5): 495-502.

[21]   Foley B, Cooley S, Verneris MR, Pitt M, Curtsinger J, Luo X, *et al*. Cytomegalovirus reactivation after allogeneic transplantation promotes a lasting increase in educated NKG2C+ natural killer cells with potent function. Blood. 2012; 119(11): 2665-74.

[22]   Lopez-Vergès S, Milush JM, Schwartz BS, Pando MJ, Jarjoura J, York VA, *et al*. Expansion of a unique CD57+NKG2Chi natural killer cell subset during acute human cytomegalovirus infection. Proc Natl Acad Sci U S A. 2011; 108(36): 14725-32.

[23]   Foley B, Cooley S, Verneris MR, Curtsinger J, Luo X, Waller EK, *et al*. Human cytomegalovirus (CMV)-induced memory-like NKG2C(+) NK cells are transplantable and expand *in vivo* in response to recipient CMV antigen. J Immunol. 2012; 189(10): 5082-8.

[24]   Moretta L, Pietra G, Montaldo E, Vacca P, Pende D, Falco M, *et al*. Human NK cells: from surface receptors to the therapy of leukemias and solid tumors. Front Immunol. 2014; 5(87).

[25]   Sivori S, Carlomagno S, Pesce S, Moretta A, Vitale M, Marcenaro E. TLR/NCR/KIR: Which One to Use and When? Front Immunol. 2014; 5(105).

[26]   Marcus A, Gowen BG, Thompson TW, Iannello A, Ardolino M, Deng W, *et al*. Recognition of tumors by the innate immune system and natural killer cells. Adv Immunol. 2014; 122: 91-128.

[27]   Kruse PH, Matta J, Ugolin S, Vivier E. Natural cytotoxicity receptors and their ligands. Immunol Cell Biol. 2014; 92(3): 221-9.

[28]   Parham P, Moffett A. Variable NK cell receptors and their MHC class I ligands in immunity, reproduction and human evolution. Nat Rev Immunol. 2013; 13(2): 133-44.

[29]   Klimosch SN, Bartel Y, Wiemann S, Steinle A. Genetically coupled receptor-ligand pair NKp80-AICL enables autonomous control of human NK cell responses. Blood. 2013; 122(14): 2380-9.

[30]   Ewen CL, Kane KP, Bleackley RC. A quarter century of granzymes. Cell Death Differ. 2012; 19(1): 28-35.

[31]   Saini RV, Wilson C, Finn MW, Wang T, Krensky AM, Clayberger C. Granulysin delivered by cytotoxic cells damages endoplasmic reticulum and activates caspase-7 in target cells. J Immunol. 2011; 186(6): 3497-504.

[32]   Chavez-Galan L, Arenas-Del Angel MC, Zenteno E, Chavez R, Lascurain R. Cell death mechanisms induced by cytotoxic lymphocytes. Cell Mol Immunol. 2009; 6(1): 15-25.

[33]   Tomblyn M, Chiller T, Einsele H, Gress R, Sepkowitz K, Storek J, *et al*. Guidelines for preventing infectious complications among hematopoietic cell transplantation recipients: a global perspective. Biol Blood Marrow Transplant. 2009; 15(10): 1143-238.

[34]   Kim S, Poursine-Laurent J, Truscott SM, Lybarger L, Song YJ, Yang L, *et al*. Licensing of natural killer cells by host major histocompatibility complex class I molecules. Nature. 2005; 436(7051): 709-13.

[35]   Anfossi N, André P, Guia S, Falk CS, Roetynck S, Stewart CA, *et al*. Human NK cell education by inhibitory receptors for MHC class I. Immunity. 2006; 25(2): 331-42.

[36]   Haas P, Loiseau P, Tamouza R, Cayuela JM, Moins-Teisserenc H, Busson M, *et al*. NK-cell education is shaped by donor HLA genotype after unrelated allogeneic hematopoietic stem cell transplantation. Blood. 2011; 17 (3): 1021-9.

[37]   Miller JS, Blazar BR. Control of acute myeloid leukemia relapse--dance between KIRs and HLA. N Engl J Med. 2012; 367(9): 866-8.

[38]   Venstrom JM, G. P, T.A. G, Chewning JH, Spellman S, Haagenson M, *et al*. HLA-C-dependent prevention of leukemia relapse by donor activating KIR2DS1. N Engl J Med. 2012; 367(9): 805-16.

[39]   Bertaina A, Locatelli F, Moretta L. Transplantation and innate immunity: the lesson of natural killer cells. Ital J Pediatr. 2009; 35: 1-5.

[40]   Moretta L, Locatelli F, Pende D, Marcenaro E, Mingari MC, Moretta A. Killer Ig-like receptor-mediated control of natural killer cell alloreactivity in haploidentical hematopoietic stem cell transplantation. Blood. 2011; 17(3): 764-71.

[41]   Ruggeri L, Capanni M, Urbani E, Perruccio K, Shlomchik WD, Tosti A, *et al*. Effectiveness of donor natural killer cell alloreactivity in mismatched hematopoietic transplants. Science. 2002; 295(5562): 2097-100.

[42]   Pende D, Marcenaro S, Falco M, Martini S, Dondero ME, Montagna D, *et al*. Anti-leukemia activity of alloreactive NK cells in KIR ligand-mismatched haploidentical HSCT for pediatric patients:

evaluation of the functional role of activating KIR and redefinition of inhibitory KIR specificity. Blood. 2009; 113(13): 3119-29.

[43]   Cooley S, Weisdorf DJ, Guethlein LA, Klein JP, Wang T, Le CT, *et al*. Donor selection for natural killer cell receptor genes leads to superior survival after unrelated transplantation for acute myelogenous leukemia. Blood. 2010; 116(14): 2411-9.

[44]   Feuchtinger T, Pfeiffer M, Pfaffle A, Teltschik HM, Wernet D, Schumm M, *et al*. Cytolytic activity of NK cell clones against acute childhood precursor-B-cell leukaemia is influenced by HLA class I expression on blasts and the differential KIR phenotype of NK clones. Bone marrow transplantation. 2009; 43(11): 875-81.

[45]   Brunstein CG, Wagner JE, Weisdorf DJ, S. C, Noreen H, Barker JN, *et al*. Negative effect of KIR alloreactivity in recipients of umbilical cord blood transplant depends on transplantation conditioning intensity. Blood. 2009; 113(22): 5628-34.

[46]   Rubnitz JE, Inaba H, Ribeiro RC, Pounds S, Rooney B, Bell T, *et al*. NKAML: a pilot study to determine the safety and feasibility of haploidentical natural killer cell transplantation in childhood acute myeloid leukemia. J Clin Oncol. 2010; 28(6): 955-9.

[47]   Curti A, Ruggeri L, D'Addio A, Bontadini A, Dan E, Motta MR, *et al*. Successful transfer of alloreactive haploidentical KIR ligand-mismatched natural killer cells after infusion in elderly high risk acute myeloid leukemia patients. Blood. 2011; 118(12): 3273-9.

[48]   Bachanova V, Cooley S, Defor TE, Verneris MR, Zhang B, McKenna DH, *et al*. Clearance of acute myeloid leukemia by haploidentical natural killer cells is improved using IL-2 diphtheria toxin fusion protein. Blood. 2014; 123(25): 3855-63.

[49]   Bachanova V, Burns LJ, McKenna DH, Curtsinger J, Panoskaltsis-Mortari A, Lindgren BR, *et al*. Allogeneic natural killer cells for refractory lymphoma. Cancer Immunol Immunother. 2010; 59(11): 1739-44.

[50]   Shi J, Tricot G, Szmania S, Rosen N, Garg TK, Malaviarachchi PA, *et al*. Infusion of haplo-identical killer immunoglobulin-like receptor ligand mismatched NK cells for relapsed myeloma in the setting of autologous stem cell transplantation. Brit J Haematol. 2008; 143(5): 641-53.

[51]   Klingemann H, Grodman C, Cutler E, Duque M, Kadidlo D, Klein AK, *et al*. Autologous stem cell transplant recipients tolerate haploidentical related-donor natural killer cell-enriched infusions. Transfusion. 2013; 53(2): 412-8; quiz 1.

[52]   Koehl U, Sorensen J, Esser R, Zimmermann S, Gruttner HP, Tonn T, *et al*. IL-2 activated NK cell immunotherapy of three children after haploidentical stem cell transplantation. Blood Cell Mol Dis. 2004; 33(3): 261-6.

[53]   Rizzieri DA, Storms R, Chen DF, Long G, Yang Y, Nikcevich DA, *et al*. Natural killer cell-enriched donor lymphocyte infusions from A 3-6/6 HLA matched family member following nonmyeloablative allogeneic stem cell transplantation. Biol Blood Marrow Transplant. 2010; 16(8): 1107-14.

[54]   Stern M, Passweg JR, Meyer-Monard S, Esser R, Tonn T, Soerensen J, *et al*. Pre-emptive immunotherapy with purified natural killer cells after haploidentical SCT: a prospective phase II study in two centers. Bone marrow transplantation. 2013; 48(3): 433-8.

[55]   Choi I, Yoon SR, Park SY, Kim H, Jung SJ, Jang YJ, *et al*. Donor-derived natural killer cells infused after human leukocyte antigen-haploidentical hematopoietic cell transplantation: a dose-escalation study. Biol Blood Marrow Transplant. 2014; 20(5): 696-704.

[56]   Iliopoulou EG, Kountourakis P, Karamouzis MV, Doufexis D, Ardavanis A, Baxevanis CN, *et al*. A phase I trial of adoptive transfer of allogeneic natural killer cells in patients with advanced non-small cell lung cancer. Cancer Immunol Immunother. 2010; 59(12): 1781-9.

[57]   Geller MA, Cooley S, Judson PL, Ghebre R, Carson LF, Argenta PA, *et al*. A phase II study of allogeneic natural killer cell therapy to treat patients with recurrent ovarian and breast cancer. Cytotherapy. 2011; 13(1): 98-107.

[58]   Lu CC, Chen JK. Resveratrol enhances perforin expression and NK cell cytotoxicity through NKG2D-dependent pathways. J Cellular Physiol. 2010; 223(2): 343-51.

[59]   Heo W, Lee YS, Son CH, Yang K, Park YS, Bae J. Radiation-induced matrix metalloproteinases limit natural killer cell-mediated anticancer immunity in NCI-H23 lung cancer cells. Mol Med Rep. 2015; 11(3): 1800-6.

[60]   Kim SJ, Ha GH, Bae JH, Kim GR, Son CH, Park YS, *et al.* COX-2- and endoplasmic reticulum stress-independent induction of ULBP-1 and enhancement of sensitivity to NK cell-mediated cytotoxicity by celecoxib in colon cancer cells. Exp. Cell Res. 2015; 330(2): 451-9.

[61]   Min D, Lv XB, Wang X, Zhang B, Meng W, Yu F, *et al.* Downregulation of miR-302c and miR-520c by 1,25(OH)2D3 treatment enhances the susceptibility of tumour cells to natural killer cell-mediated cytotoxicity. Br J Cancer. 2013; 109(3): 723-30.

[62]   Husak Z, Dworzak MN. CD99 ligation upregulates HSP70 on acute lymphoblastic leukemia cells and concomitantly increases NK cytotoxicity. Cell Death Dis. 2012; 3: e425.

[63]   Horowitz A, Strauss-Albee DM, Leipold M, Kubo J, Nemat-Gorgani N, Dogan OC, *et al.* Genetic and environmental determinants of human NK cell diversity revealed by mass cytometry. Sci Transl Med. 2013; 5(208): 208ra145.

[64]   Cichocki F, Hanson RJ, Lenvik T, Pitt M, McCullar V, Li H, *et al.* The transcription factor c-Myc enhances KIR gene transcription through direct binding to an upstream distal promoter element. Blood. 2009; 113(14): 3245-53.

[65]   Ott PA, Hodi FS, Robert C. CTLA-4 and PD-1/PD-L1 blockade: new immunotherapeutic modalities with durable clinical benefit in melanoma patients. Clin Cancer Res. 2013; 19(19): 5300-9.

[66]   Keir ME, Butte MJ, Freeman GJ, Sharpe AH. PD-1 and its ligands in tolerance and immunity. Annu Rev Immunol. 2008; 26: 677-704.

[67]   Peggs KS, Quezada SA, Allison JP. Cancer immunotherapy: co-stimulatory agonists and co-inhibitory antagonists. Clin Exp Immunol. 2009; 157(1): 9-19.

[68]   Fu J, Malm IJ, Kadayakkara DK, Levitsky H, Pardoll D, Kim YJ. Preclinical evidence that PD1 blockade cooperates with cancer vaccine TEGVAX to elicit regression of established tumors. Cancer Res. 2014; 74(15): 4042-52.

[69]   Benson DM, Jr., Bakan CE, Mishra A, Hofmeister CC, Efebera Y, Becknell B, *et al.* The PD-1/PD-L1 axis modulates the natural killer cell *versus* multiple myeloma effect: a therapeutic target for CT-011, a novel monoclonal anti-PD-1 antibody. Blood. 2010; 116(13): 2286-94.

[70]   Ray A, Das DS, Song Y, Richardson P, Munshi NC, Chauhan D, *et al.* Targeting PD1-PDL1 immune checkpoint in plasmacytoid dendritic cell interactions with T cells, natural killer cells and multiple myeloma cells. Leukemia. 2015.

[71]   Della Chiesa M, Falco M, Muccio L, Bertaina A, Locatelli F, Moretta A. Impact of HCMV Infection on NK Cell Development and Function after HSCT. Front Immunol. 2013; 4(458).

[72]   Bottino C, Castriconi R, Moretta L, Moretta A. Cellular ligands for activating NK receptors. Trends Immunol. 2005; 26: 221-6.

[73]   Moretta L, Moretta A. Unravelling natural killer cell function: triggering and inhibitory human NK receptors. EMBO J. 2004 23: 255-9.

[74]   Bryceson YT, March ME, Ljunggren HG, Long EO. Synergy among receptors on resting NK cells for the activation of natural cytotoxicity and cytokine secretion. Blood. 2006; 107(1): 159-66.

[75]   Veillette A, Dong Z, Latour S. Consequence of the SLAM-SAP signaling pathway in innate-like and conventional lymphocytes. Immunity. 2007; 27(5): 698-710.

[76]   Watzl C, Long EO. Natural killer cell inhibitory receptors block actin cytoskeleton-dependent recruitment of 2B4 (CD244) to lipid rafts. Exp Med. 2003; 197(1): 77-85.

[77]   Lue H, Dewor M, Leng L, Bucala R, Bernhagen J. Activation of the JNK signalling pathway by macrophage migration inhibitory factor (MIF) and dependence on CXCR4 and CD74. Cell Signal. 2011; 23(1): 135-44.

[78]   Shi X, Leng L, Wang T, Wang W, Du X, Li J, *et al.* CD44 is the signaling component of the macrophage migration inhibitory factor-CD74 receptor complex. Immunity. 2006; 25(4): 595-606.

[79]   Krockenberger M, Dombrowski Y, Weidler C, Ossadnik M, Hönig A, Häusler S, *et al.* Macrophage migration inhibitory factor contributes to the immune escape of ovarian cancer by down-regulating NKG2D. J Immunol. 2008; 180(11): 7338-48.

[80]   Lee SH, Kim KS, Fodil-Cornu N, Vidal SM, Biron CA. Activating receptors promote NK cell expansion for maintenance, IL-10 production, and CD8 T cell regulation during viral infection. J Exp Med. 2009; 206(10): 2235-51.

[81]   Marcondes AM, Karoopongse E, Lesnikova M, Margineantu D, Welte T, Dinarello CA, *et al.* alpha 1-Antitrypsin (AAT) modified donor cells suppress GVHD but enhance the GVL effect: a role for mitochondrial bioenergetics. Blood. 2014; 124(18): 2881-91.

[82]   Walzer T, Dalod M, Robbins SH, Zitvogel L, Vivier E. Natural-killer cells and dendritic cells: "l'union fait la force". Blood. 2005; 106(7): 2252-8.

[83]   Inoue H, Tani K. Multimodal immunogenic cancer cell death as a consequence of anticancer cytotoxic treatments. Cell Death Differ. 2014; 21(1): 39-49.

[84]   Nikoletopoulou V, Markaki M, Palikaras K, Tavernarakis N. Crosstalk between apoptosis, necrosis and autophagy. Biochim Biophys Acta. 2013; 1833(12): 3448-59.

[85]   Ghiringhelli F, Apetoh L, Tesniere A, Aymeric L, Ma Y, Ortiz C, *et al.* Activation of the NLRP3 inflammasome in dendritic cells induces IL-1beta-dependent adaptive immunity against tumors. Nat Med. 2009; 15(10): 1170-8.

[86]   Dudek AM, Garg AD, Krysko DV, De Ruysscher D, Agostinis P. Inducers of immunogenic cancer cell death. Cytokine Growth Factor Rev. 2013; 24(4): 319-33.

[87]   Hou W, Zhang Q, Z. Y, Chen R, Zeh Iii HJ, Kang R, *et al.* Strange attractors: DAMPs and autophagy link tumor cell death and immunity. Cell Death Dis. 2013; 4(e966).

[88]   Krysko O, Løve Aaes T, Bachert C, Vandenabeele P, Krysko DV. Many faces of DAMPs in cancer therapy. Cell Death Dis. 2013; 4(e631).

[89]   Kaczmarek A, Vandenabeele P, Krysko DV. Necroptosis: the release of damage-associated molecular patterns and its physiological relevance. Immunity. 2013 38(2): 209-23.

[90]   Sonneveld P, Goldschmidt H, Rosiñol L, Bladé J, Lahuerta JJ, Cavo M, *et al.* Bortezomib-based *versus* nonbortezomib-based induction treatment before autologous stem-cell transplantation in patients with previously untreated multiple myeloma: a meta-analysis of phase III randomized, controlled trials. J Clin Oncology. 2013; 31(26): 3279-87.

[91]   Rabinowitz JD, White E. Autophagy and metabolism. Science. 2010; 330(6009): 1344-8.

[92]   Garg AD, Dudek AM, Ferreira GB, Verfaillie T, Vandenabeele P, Krysko DV, *et al.* ROS-induced autophagy in cancer cells assists in evasion from determinants of immunogenic cell death. Autophagy. 2013; 9(9).

[93]   Aronson LI, Davies FE. DangER: protein ovERload. Targeting protein degradation to treat myeloma. Haematologica. 2012; 97(8): 1119-30.

[94]   Walter P, Ron D. The unfolded protein response: from stress pathway to homeostatic regulation. Science. 2011; 334(6059): 1081-6.

[95]   Mimura N, Fulciniti M, Gorgun G, Tai YT, Cirstea D, Santo L, *et al.* Blockade of XBP1 splicing by inhibition of IRE1alpha is a promising therapeutic option in multiple myeloma. Blood. 2012; 119(24): 5772-81.

[96]   Tagoug I, Jordheim LP, Herveau S, Matera EL, Huber AL, Chettab K, *et al.* Therapeutic Enhancement of ER Stress by Insulin-Like Growth Factor I Sensitizes Myeloma Cells to Proteasomal Inhibitors. Clin Cancer Res. 2013; 19(13): 3556-66.

[97]   Milani M, Rzymski T, Mellor HR, Pike L, Bottini A, Generali D, *et al.* The role of ATF4 stabilization and autophagy in resistance of breast cancer cells treated with Bortezomib. Cancer Res. 2009; 69(10): 4415-23.

[98]   Wang XJ, Yu J, Wong SH, Cheng AS, Chan FK, Ng SS, *et al.* A novel crosstalk between two major protein degradation systems: Regulation of proteasomal activity by autophagy. Autophagy. 2013; 9(10): 1500-8.

[99]   Choi AM, Ryter SW, Levine B. Autophagy in human health and disease. N Engl J Med. 2013; 368(19): 1845-6.

[100]   San Miguel JF, Schlag R, Khuageva NK, Dimopoulos MA, Shpilberg O, Kropff M, *et al.* Bortezomib plus melphalan and prednisone for initial treatment of multiple myeloma. N Engl J Med. 2008; 359(9): 906-17.

[101]   Richardson PG, Sonneveld P, Schuster MW, Irwin D, Stadtmauer EA, Facon T, *et al.* Bortezomib or high-dose dexamethasone for relapsed multiple myeloma. N Engl J Med. 2005; 352(24): 2487-98.

[102]   Pengo N, Scolari M, Oliva L, Milan E, Mainoldi F, Raimondi A, *et al.* Plasma cells require autophagy for sustainable immunoglobulin production. Nat Immunol. 2013; 14(3): 298-305.

[103]   Lamy L, Ngo VN, Emre NC, Shaffer AL, 3rd, Yang Y, Tian E, *et al.* Control of autophagic cell death by caspase-10 in multiple myeloma. Cancer cell. 2013; 23(4): 435-49.

[104]   Carroll RG, Martin SJ. Autophagy in multiple myeloma: what makes you stronger can also kill you. Cancer cell. 2013; 23(4): 425-6.

[105]   Okada S, Li Q, Whitin JC, Clayberger C, Krensky AM. Intracellular mediators of granulysin-induced cell death. J Immunol. 2003; 171(5): 2556-62.

[106]   Aporta A, Catalán E, Galán-Malo P, Ramírez-Labrada A, Pérez M, Azaceta G, *et al*. Granulysin induces apoptotic cell death and cleavage of the autophagy regulator Atg5 in human hematological tumors. Biochem Pharmacol. 2014 87(3): 410-23.

[107]   Zhang H, Zhong C, Shi L, Guo Y, Fan Z. Granulysin induces cathepsin B release from lysosomes of target tumor cells to attack mitochondria through processing of bid leading to Necroptosis. J Immunol. 2009; 182(11): 6993-7000.

[108]   Lugini L, Cecchetti S, Huber V, Luciani F, Macchia G, Spadaro F, *et al*. Immune surveillance properties of human NK cell-derived exosomes. J Immunol. 2012; 189(6): 2833-42.

[109]   Corrado C, Raimondo S, Chiesi A, Ciccia F, De Leo G, Alessandro R. Exosomes as intercellular signaling organelles involved in health and disease: basic science and clinical applications. Int J Mol Sci. 2013; 14(3): 5338-66.

[110]   Rogers RS, Bhattacharya J. When cells become organelle donors. Physiology (Bethesda). 2013; 28(6): 414-22.

[111]   Griffiths GM, Tsun A, Stinchcombe JC. The immunological synapse: a focal point for endocytosis and exocytosis. J Cell Biol. 2010; 189(3): 399-406.

[112]   Islam MN, Das SR, Emin MT, Wei M, Sun L, Westphalen K, *et al*. Mitochondrial transfer from bone-marrow-derived stromal cells to pulmonary alveoli protects against acute lung injury. Nat Med. 2012; 18(5): 759-65.

[113]   Yasuda K, Khandare A, Burianovskyy L, Maruyama S, Zhang F, Nasjletti A, *et al*. Tunneling nanotubes mediate rescue of prematurely senescent endothelial cells by endothelial progenitors: exchange of lysosomal pool. Aging (Albany NY). 2011; 3(6): 597-608.

[114]   Huang WC, Chen CL, Lin YS, Lin CF. Apoptotic sphingolipid ceramide in cancer therapy. J Lipids. 2011; 2011(565316).

[115]   Ikonen E. Cellular cholesterol trafficking and compartmentalization. Nat Rev Mol Cell Biol. 2008; 9(2): 125-38.

[116]   Bröker LE, Kruyt FA, Giaccone G. Cell death independent of caspases: a review. Clin Cancer Res. 2005; 11(9): 3155-62.

[117]   Maxfield FR, Tabas I. Role of cholesterol and lipid organization in disease. Nature. 2005; 438(7068): 612-21.

[118]   Basseri S, Austin RC. Endoplasmic reticulum stress and lipid metabolism: mechanisms and therapeutic potential. Biochem Res Int. 2012; 2012: 841362.

[119]   Singh R, Kaushik S, Wang Y, Xiang Y, Novak I, Komatsu M, *et al*. Autophagy regulates lipid metabolism. Nature. 2009; 458(7242): 1131-5.

[120]   Liu K, Czaja MJ. Regulation of lipid stores and metabolism by lipophagy. Cell Death Differ. 2013; 20(1): 3-11.

[121]   Fu S, Yang L, Li P, Hofmann O, Dicker L, Hide W, *et al*. Aberrant lipid metabolism disrupts calcium homeostasis causing liver endoplasmic reticulum stress in obesity. Nature. 2011; 473(7348): 528-31.

[122]   Zhang X, Zhang K. Endoplasmic reticulum stress-associated lipid droplet formation and type II diabetes. Biochem Res Int. 2012; 2012: 247275.

[123]   Bozza PT, Magalhães KG, Weller PF. Leukocyte lipid bodies - Biogenesis and functions in inflammation. Biochim Biophys Acta. 2009; 1791(6): 540-51.

[124]   Saka HA, Valdivia R. Emerging roles for lipid droplets in immunity and host-pathogen interactions. Annu Rev Cell Dev Biol. 2012; 28: 411-37.

[125]   Meng Q, Cai D. Defective hypothalamic autophagy directs the central pathogenesis of obesity *via* the IkappaB kinase beta (IKKbeta)/NF-kappaB pathway. J Biol Chem. 2011; 286(37): 32324-32.

[126]   Dall'Armi C, Devereaux KA, Di Paolo G. The role of lipids in the control of autophagy. Curr Biol. 2013; 23(1): R33-45.

[127]   Li Y, Li S, Qin X, Hou W, Dong H, Yao L, *et al*. The pleiotropic roles of sphingolipid signaling in autophagy. Cell Death Dis. 2014; 5(e1245).

[128]   Yabu T, Tomimoto H, Taguchi Y, Yamaoka S, Igarashi Y, Okazaki T. Thalidomide-induced antiangiogenic action is mediated by ceramide through depletion of VEGF receptors, and is antagonized by sphingosine 1 phosphate. Blood reviews. 2005; 106(1): 125-34.

[129]   Sentelle RD, Senkal CE, Jiang W, Ponnusamy S, Gencer S, Selvam SP, *et al.* Ceramide targets autophagosomes to mitochondria and induces lethal mitophagy. Nat Chem Biol. 2012; 8(10): 831-8.

[130]   Saddoughi SA, Gencer S, Peterson YK, Ward KE, Mukhopadhyay A, Oaks J, *et al.* Sphingosine analogue drug FTY720 targets I2PP2A/SET and mediates lung tumour suppression *via* activation of PP2A-RIPK1-dependent necroptosis. EMBO Mol Med. 2013; 5(1): 105-21.

[131]   Morad SA, Levin JC, Tan SF, Fox TE, Feith DJ, Cabot MC. Novel off-target effect of tamoxifen--inhibition of acid ceramidase activity in cancer cells. Biochim Biophys Acta. 2013; 1831(12): 1657-64.

[132]   Morad SA, Madigan JP, Levin JC, Abdelmageed N, Karimi R, Rosenberg DW, *et al.* Tamoxifen magnifies therapeutic impact of ceramide in human colorectal cancer cells independent of p53. Biochem Pharmacol. 2013; 85(8): 1057-65.

[133]   Chen S, Zhang Y, Zhou L, Leng Y, Lin H, Kmieciak M, *et al.* A Bim-targeting strategy overcomes adaptive bortezomib resistance in myeloma through a novel link between autophagy and apoptosis. Blood. 2014; 124(17): 2687-97.

CHAPTER 4

Nanoparticles in Health and Disease: An Overview of Nanomaterial Hazard, Benefit and Impact on Public Health Policy - Current State and Outlook

Stanislav Janousek*, Dagmar Jirova, Kristina Kejlova and Marketa Dvorakova

Center of Toxicology and Health Safety, National Institute of Public Health, Srobarova 48, 100 42, Prague, Czech Republic

Abstract: Nanotechnology is a new interdisciplinary platform for medical research offering a novel experience in the disease treatment at the nanoscale level where most of the biological molecules functionate. In this chapter we address the application of nanomaterials in human life and medical practice which may anticipate a great impact on public and individual health. Nanoparticles and nanotherapeutics have recently been regulated by a conventional regulatory framework. The European Commission has worked out an Action Plan for Europe "Nanoscience and Nanotechnologies" and has called on the Member States for its implementation. Namely, the regulatory agency for responsibility, supervision, protection and promotion of human (animal) health in the European Union (EMA) provides regulatory guidance and authorization for the safety of nanomedicines. Interdisciplinary approaches are recommended for the application of scientific results in practice with simultaneous strict adherence to the Community legislative requirements on nanoproduct safety. The Report recommends additional specialized expertise, together with adaptation of existing methodologies and development of new methods for the evaluation of nanoproduct quality, safety, efficacy and risk management. In this chapter we will focus on liposome-, nanocrystal, virosome-, polymer therapeutic-, nanoemulsion-, and nanoparticle-based approaches to nanotherapeutics, which represent the most successful and commercialized categories within the field of nanomedicine. In addition, we will inform about generic nanotherapeutics and pitfalls of similar colloidal-based nanoformulations. We will pay attention to topics such as nanoparticles and nanomaterials in hematological and malignant disorders. Finally, we will discuss consumer nanoproduct safety (or risks) as well as future directions in nanomaterial commercialization, *i.e.* what are the forthcoming human (animal) health safety concerns and how relevant is the potential negative impact on the environment, life cycle and living systems when nanoproducts (and their use) are expected to be extended.

Keywords: Nanomaterials, nanotherapeutics, nanoparticle safeties and toxicities, public health policy, consumer products, ecotoxicology.

*Corresponding author Stanislav Janousek: Center of Toxicology and Health Safety, National Institute of Public Health, Srobarova 48, 100 42, Prague, Czech Republic; Tel: +420267082563; Fax: +420267311188; E-mail: stanislav.janousek@szu.cz

Atta-ur-Rahman (Ed)

INTRODUCTION

Nanoparticles and nanotherapeutics which are molecular assemblies of functional chemistries are able to overcome biological barriers, accumulate preferentially in target location and specifically recognize and destroy pathophysiologically destitute and proliferative cells. Moreover, nanoparticles display unique physicochemical properties such as size, shape and surface area. The ability of nanotherapeutics to enhance modus of action while reducing side effects, or periods of the treatment, has resulted in different biopharmaceutical profiles of nanotherapeutics and their medicinal applications. In addition, the use of nanoparticles in consumer products has the potential to enhance quality of life but also raises concerns regarding human health safety. Nevertheless, extended evaluations are needed to assure the safety and efficacy of both types of nanoproducts.

The area of nanoparticle-based medicine has recently received particular attention as it holds the promise to revolutionize medical treatment with more potent, less toxic and more sophisticated therapeutics. From the second half of the twentieth century, a great deal of research has been devoted to the development of nanomaterials as well as to nanoparticle delivery systems. Since most of the biological processes occur at the nanoscale, one can expect that nanoparticulate technology has a promising future in this respect, *i.e.* in developing novel preventive, diagnostic and nanotherapeutic agents [1]. On the other hand, there have also been some criticisms about the discrepancy between early promises or expectations of nanotherapeutics and their practical use in the disease treatment [2]. The field of nanomedicine is defined by the Medical Standing Committee of the European Science Foundation as "the comprehensive monitoring, control, construction, repair, defense and improvement of all human biological systems, working from the molecular level using engineered devices and nanostructures, ultimately to achieve medical benefit" [3]. In the past, drug loaded nanoparticles were commonly used in the treatment of a number of diseases such as metabolic disorders, autoimmune diseases, inflammatory disorders, neurodegenerative diseases or malignancies. For instance, since the 1930s for treatment of iron-deficiency anaemia in patients with chronic kidney disease, when oral iron supplements cannot be used or fail to provide therapeutic effect, parenteral iron formulations in the forms of nano-colloidal systems have been utilized [4, 5]. Another example is liposome encapsulated formulations that were initially used to study biological membranes in the mid-1960s and were being administered as early as the 1990s [6, 7]. Despite these and other drug histories, nanotherapeutics are viewed as a novel and emerging discipline, probably due to a great progress that has currently been made in this field [8]. In addition to the beneficial outcomes, engineered nanoparticles have recently created a serious worldwide concern due to their potential toxicities, both for humans or animals [3]. Over a thousand consumer products are currently listed as containing NPs, including sunscreens, paints, semiconductors and cosmetics, with expected production volume on global markets exceeding $1 billion by 2015 [9]. In general, pharmaceuticals and personal care compounds such as chemotherapy drugs, antihistamines, stimulants, antimicrobials, analgesics, antibiotics, fragrances and various cosmetic additives, have commonly been found in the environment, especially from wastewater discharges [10-13]. Moreover, organisms in the environment may be exposed to a number of pollutants from different compound groups concurrently. Even though the environmental concentrations of individual pollutants (including natural pollutants) might be too low to exert an effect on their own, the presence of several similarly acting compounds is expected to induce effects through combined toxicity at concentrations below their individual No Observed Effect Concentrations [14, 15]. Besides the classic pollutants like the polychlorinated biphenyls, polyaromatic hydrocarbons, alkylphenols, biocides, *etc.*, other pollutants of concern including nanoconstructs are constantly emerging, and have been found to co-occur [16]. In addition, trophic transfer relates to the accumulation of hazardous chemicals by an organism following consumption of another organism previously exposed to them [17]. To date, cytotoxic and genotoxic data for most manufactured nanomaterials (NM) have not been established at a rate corresponding to their use and the mechanisms underlying nanomaterial toxicity are not fully understood [18]. Moreover, it was demonstrated that the physico-chemical properties of NPs modulate their dynamic interaction with biomolecules and cellular organelles. The formation of a biomolecular corona covering the surface of NPs determines their functionalities in biological fluids. The protein corona may reduce cell uptake, limit the penetration of NPs into the cell [19, 20], disrupt the folding process of proteins and initiate an inflammatory response [21] or may moderate cellular damage [22-24]. However, although much progress has been made in explaining toxicity of NPs, a satisfactory integration of various experimental observations is extremely challenging.

The properties exhibited by nanoscale materials are considered to be different to those exhibited by their larger counterparts. In this respect, NPs potentially cross more easily portals of entry into the body such as the gastrointestinal tract, lung epithelium, skin, gills and internal barriers, such as the blood– brain barrier, placenta or blood–testis barrier. NM with different characteristics (size, coating, shape) may also differ considerably in the extent of their transportation across these barriers. The types of particles that can be transferred across barriers have not been systematically investigated, and to date very little is known about the parameters that influence differences in the extent of transportation [25, 26]. Such information is crucial to determine if NM have reached the systemic circulation, and if so, to what extent their bioavailability is increased, and whether, for instance, systemic effects, including cardiovascular or immunological toxicity, should be considered. In contrast to many soluble chemicals, NM generally tend to disappear rapidly from the blood by being taken up into tissues. The apparent very short blood-plasma half-life is usually in sharp contrast with the apparent long whole body or tissue half-life [27, 28].

Nanomaterial, Nanoparticle, Nanomedicine and Nanotechnology Definition(s)

Usually, nanomaterials are defined as chemical molecules that have surfaces with at least one dimension smaller than 100 nm; moreover, engineered nanomaterials are supposed to have the same specific physicochemical characteristics and are manufactured intentionally [28]. Definitions of nanotechnology, nanofabrication, nanomaterials, nanomedicine and even nanoparticle may vary widely in scientific literature. Basically, nanotechnology covers nanomaterials, nanomedicine, nanofabrication and others. However, the term "nanomaterials" is usually considered as nanoparticles, nanosheets, nanotubes and other forms of nanoscale structures. Nanotechnology is a vast topic which includes diverse fields such as medicine, biotechnology, energy storage, organic chemistry, semiconductors and others. Fig. **1** presents the nanotechnology in this respect. The term "nanotechnology" was specified by professor Norio Taniguchi in 1974 [29]. At nanoscale the material

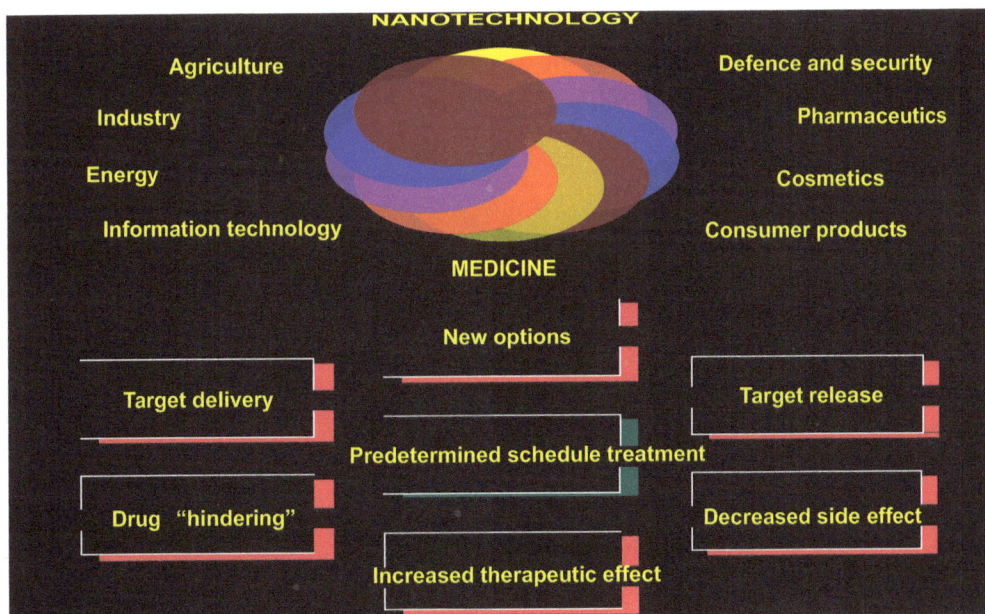

Figure 1: Nanotechnology - the interdisciplinary platform.

properties change dramatically. Importantly, these properties are not exhibited by micro or macro scale [30]. Although the concept and terminology of nanomaterials was introduced in the late eighties of the last century, nanomaterials have been used for a long time. Generally, any material can be classified as nanomaterial, if it has following key properties [31]: Firstly, dimensions (at least one) should be in the range of 1-100 nm; secondly, processes involved in design must display fundamental control on attributes of molecular-scale structures (both physical and chemical) and thirdly, a larger structure can be prepared from them. It is necessary to note that some scientists restrict the definition of NM on working with molecules and devices between 1 and 100 nm. However, others widen these parameters to 1- 1,000 nm as we will discuss later [32].

Nanomedicine refers to any application of nanomaterials for medical purposes, ranging from diagnostic to therapeutic applications such as the use of nanoscale materials for diagnosis, monitoring, control, prevention, and treatment of diseases. Nanomedicine holds great promise to revolutionize medicine across disciplines and specialties. Beyond the typical complications associated with drug development, the fundamentally different and novel physical and chemical properties of some nanomaterials compared to materials on a larger scale (*i.e.*, their bulk counterparts) can create a unique set of opportunities as well as safety concerns. In scientific publications, the terms of nanodrugs, nanotherapeutics, nanomedicines, and nanopharmaceuticals are sometimes used interchangeably. Although a consensus definition has not been reached by the various scientific and international regulatory communities, a construct is loosely classified as a nanomedicine if it has at least one dimension in the nanoscale range up to 1,000 nm and exhibits properties dependent upon those dimensions. The regulatory communities are challenged by the confusion of definitions for nanomaterials and nanotechnology-enabled products as well as by limited standard nomenclature and reference materials. Nanoproducts are developed but should not be released to market until safety testing is completed; however, appropriate regulations cannot be developed until a weight of evidence and a convergence of data support conclusions on nanoproduct safety are preceded. Nanopharmaceutical regulators must thus often apply existing regulations to nanoscale materials and devices with complex and frequently novel properties. Nanotechnology integrates knowledge from physics, mathematics, materials and pharmaceutical science, chemistry, biology and engineering. Medical research encompasses primarily human biology and bioengineering, research tools and methods, therapeutics and diagnostics. Research challenges include the measurement of physical, chemical, and functional properties; synthesis, reproducibility, and scale-up; and *in vivo* assessment, tracking, and imaging of materials with nanoscale size or features.

Nanomaterials used in medicine are becoming increasingly more and more sophisticated. Researchers have developed techniques to create complex multifunctional conjugates with special attach coatings, targeting molecules, prodrugs, tracking moieties, imaging agents and other functionalities. As this next generation of sophisticated nanomedicines are developed, novel methods for characterization of the nanomedicines' physical, chemical and biological properties (including size, morphology, charge, purity, *etc.*) and techniques for characterization of performance (including protein binding, cellular uptake, drug release, and metabolism) have to be developed. As sophisticated nanomedicines enter clinical trials, there is a need for physicochemical characterization data obtained through several methods and for seeking a detailed understanding of how the unique properties of nanomedicines influence their biological performance. This type of thorough understanding can be achieved only through an integrated approach where physicochemical characterization informs biological performance testing and *vice versa*. Kostoff has claimed that nanotechnology is best defined by the capacity to artificially construct and manipulate structures at nanoscale and nanoscale's novel properties [33]. For comparison, an H atom has the size of 0.1 nm in diameter, a lysosome is between 200 and 500 nm, an *Escherichia coli* bacterium is about 2 µm in length, and most of eukaryotic cells have a size between 8 and 30 µm in diameter or more [34]. The size of proteins is in a range between 3 and 90 nm; therefore, many enzymes, signaling molecules and receptors are functioning in the nanoscale range [34]. As mentioned, nanotherapeutics used for medical applications may operate up to 1,000 nm.

One reason for the extended definition up to 1,000 nm is that during the life-cycle of a nanomaterial when the particles are released from the agglomerates (or aggregates), their agglomerates or aggregates may exhibit the same properties as the unbound particles [35]. The term of particle aggregation is used for the tight fusing and binding the particles together. The agglomeration of nanoparticles is used for more loosely bound particles. However, in the field of medicine nanoparticle properties such as bonding or sticking the particles together, or their substantial combination exhibiting a coincident localization, should be considered more widely. The properties exhibited by materials at a "nano" scale are known to be immensely different to those exhibited by their larger counterparts. Depending on how the nanoparticles aggregate or agglomerate, they form larger masses. Concerns arise when aggregates or agglomerated groupings of nanoparticles begin to be biologically active (*e.g.* in the sense of their fate in organism). The agglomeration (or the aggregation) of NM is driven by molecular forces and the interaction of NPs with proteins or body fluid may result in significant changes in their grouping. When an amount of nanoparticles enters the body or the cell and they clump together, they may form a quantitatively novel mass even larger than the previous agglomerates (or aggregates). Moreover, the number (particle) size distribution should cover the fact that nanomaterials consist most typically of many particles present in various sizes in a dynamic particular distribution. Without specifying the number (particle) size distribution, it would be difficult to determine if a specific material complies with the definition in case that some particles are below 100 nm while others are not. Such approach is in line with the opinion that the particle distribution should be presented as the distribution based on the number concentration (*i.e.* the particle number). Therefore, the definition of what is a nanomaterial should include particles in agglomerates (or aggregates), whenever the constituent particles are in the size range 1 -100 nm (or up to 1,000 nm), regardless of the fact how tightly or loosely the particles are bound. Finally, agglomerated or aggregated nanoparticles may even exhibit the same properties as the unbound particles in spite of their ontogeny.

Moreover, there can be a great discrepancy between the measurement of the specific surface area and the number size distribution from one nanomaterial to another. It has been claimed that it should be specified which results for number size distribution prevail actually. It would not be relevant to demonstrate that a material is not a nanomaterial when the specific surface area of nanoparticles is out of the definition range. In this case, the definition including other nanomaterial descriptors should be subject to a further description to ensure that it corresponds to our practical needs, *i.e.* whether the number size distribution threshold should be increased or decreased, or whether included materials with internal structure or surface structure are in the nanoscale of different nano components, *i.e.* as is the case of nano-porous or nano-composite materials. This may even result in important consequences whether to exclude such materials from the scope of application of specific legislation and legislative provisions, or not. It may likewise be argued whether to include corroborative materials and its threshold of the number size distribution as nanoproduct characteristics with a size smaller than 1 nm or greater than 100 nm [35]. In the EU, the respective Commission Recommendation defines a "nanomaterial" to be a "natural, incidental or manufactured material containing particles, in an unbound state or as an aggregate or as an agglomerate and where, for 50% or more of the particles in the number size distribution, one or more external dimensions is in the size range 1–100 nm". The Commission Recommendation provides for flexibility in the number size distribution threshold of 50% and lays down that "fullerenes, graphene flakes, and single wall carbon nanotubes with one or more external dimensions below 1 nm should [also] be considered as nanomaterials" [36, 37].

The International Organization for Standardization defines the term "nanomaterial" as "material with any external dimensions in the nanoscale or having internal structure or surface structure in the nanoscale". Similarly, the term "nanoscale" is defined as size range from approximately 1 nm to 100 nm, however, the number size distribution should cover for the fact that nanomaterials most typically consist of many particles present in different sizes in a particular distribution. It means that without specifying the number size distribution, it would be difficult to determine if a specific material complies with the definition where some particles are below 100 nm while others are not. There is no unequivocal scientific basis to suggest a specific value for the size distribution below which materials containing particles in the size range 1 nm - 100 nm are not expected to exhibit properties specific to nanomaterials. The scientific advice was to use a statistical approach based on standard deviation with a threshold value of 0.15 %. Given the widespread occurrence of materials that would be covered by such a threshold and the need to tailor the scope of the definition for use in a regulatory context, the threshold should be higher. Even a small number of particles in the range between 1 nm - 100 nm may in certain cases justify a targeted assessment. Nevertheless, there may be specific legislative cases where concerns for the environment, health, safety or competitiveness warrant the application of a threshold below 50 % [35]. The current adopted Commission Recommendation in the EU is given as follows:

1. Member States, the Union agencies and economic operators are invited to use the following definition of the term "nanomaterial" in the adoption and implementation of legislation and policy and research programs concerning products of nanotechnologies.

2. "Nanomaterial" means a natural, incidental or manufactured material containing particles, in an unbound state or as an aggregate or as an agglomerate and where, for 50 % or more of the particles in the number size distribution, one or more external dimensions is in the size range 1 nm-100 nm. In specific cases and where warranted by concerns for the environment, health, safety or competitiveness the number size distribution threshold of 50 % may be replaced by a threshold between 1 and 50 %.

3. By derogation from point 2, fullerenes, graphene flakes and single wall carbon nanotubes with one or more external dimensions below 1 nm should be considered as nanomaterials.

4. For the purposes of point 2, "particle", "agglomerate" and "aggregate" are defined as follows: (a) "particle" means a minute piece of matter with defined physical boundaries; (b) "agglomerate" means a collection of weakly bound particles or aggregates where the resulting external surface area is similar to the sum of the surface areas of the individual components; (c) "aggregate" means a particle comprising of strongly bound or fused particles.

5. Where technically feasible and requested in specific legislation, compliance with the definition in point 2 may be determined on the basis of the specific surface area by volume. A material should be considered as falling under the definition in point 2 where the specific surface area by volume of the material is greater than 60 m^2/cm^3. However, a material which, based on its number size distribution, is a nanomaterial should be considered as complying with the definition in point 2 even if the material has a specific surface area lower than 60 m^2/cm^3.

6. The definition set out in points 1 to 5 is supposed to be reviewed in the light of experience and of scientific and technological developments. The review should particularly focus on whether the number size distribution threshold of 50% should be increased or decreased.

7. This Recommendation is addressed to the Member States, Union agencies and economic operators [35].

The major characteristics of nanoparticles include:

(a) nanoparticle size, which can be measured using a dynamic light scattering method, or a transmission electron microscope or a scanning electron microscope;

(b) zeta potential, which indicates the surface charge of nanoparticles and can be measured using a zeta potential analyzer;

(c) polydispersity index, which indicates nanoparticle size distribution and can be measured by a dynamic light scattering method;

(d) physical and chemical stability, which indicates the stability of nanoparticles and loaded compounds, respectively;

(e) as to nanopharmaceuticals, encapsulation efficiency is considered as the amount of drug present in the nanoparticles with respect to the total amount of the drug;

(f) loading capacity is related to the total amount of drug, the amount of unbound drug and the weight of nanoparticles.

While in most cases of nanopharmaceuticals, smaller size can be preferable for enhancing absorption of encapsulated compounds into target tissues, it may not always be true in the respect to their efficiency because such nanoparticles also easily may move in and out of the target tissues [38]. Additionally, smaller size requires a larger amount of stabilizers, which may introduce additional properties of drug nanoconstructs [3]. As the size range of the nanoparticles used in medicine is considered to be in the range of 1-1,000 nm, a wide choice for selection of route of administration is thus possible.

Nanomaterial Applications in Medicine

Nanomaterial application in medicine is envisioned to have a great impact on patient treatment. It uses nanosized tools for the diagnosis, prevention, and treatment of disease and encompasses several distinct application areas: drug delivery, drugs and therapies, *in vivo* imaging, *in vitro* diagnostics, biomaterials, and active implants. Over the last two decades, significant progress has been made in this field, resulting in a number of products, including therapeutics and imaging agents, enabling more effective and less toxic therapeutic and diagnostic interventions [39]. An interdisciplinary approach resulted in the development of innovative nanomedicine preparations, which have both diagnostic and therapeutic potential and are believed to have significant potential in enabling personalized nanomedicinal treatments [40, 41]. The major applications of nanotechnology in the biomaterial field are in hard tissue implants, bone substitute materials, dental restoratives, soft tissue implants, and antibiotic materials [42-44]. The use of nanomaterials in medicine requires a molecular level understanding of how nanoparticles (NPs) interact with cells in a physiological and pathophysiological environment. For instance, it has been emphasized that extracellular serum proteins present in blood tend to adsorb onto the surface of NPs, forming a natural "protein corona". Therefore, a critical difference between well-controlled *in vitro* experiments and *in vivo* applications is the presence of a complex mixture of extracellular proteins. Serum proteins adsorb onto the surface of both cationic and anionic NPs, forming a net anionic protein–NP complex. Although these protein-NP complexes have commonly similar diameters and effective surface charges, they show the exact opposite behavior in terms of cellular binding [45]. Biologics such as proteins, peptides, antibody fragments, and nucleic acids bounded on therapeutic nanoparticles may also serve as antigens and the immune response to nanotherapeutics can be elicited by different sources. Nanoparticles can be antigenic themselves, with the immunogenicity of nanoparticles being affected by their size, surface characteristics, charge, hydrophobicity or solubility. Depending on these properties, some nanoparticles can be opsonized by plasma proteins and recognized as foreign bodies, resulting in the activation of complement pathway. Based on [46]: In the classical pathway, the recognizing protein C1q binds to targets (activators) and thereby activates two proteases, C1r and C1s. When C1s is activated, it cleaves and activates the next two proteins of the system (C2 and C4). The domains of these proteins then form a complex C4b2a, which is itself a protease that cleaves and activates the most abundant complement protein C3. C3 is activated to form C3b, which binds back onto the surface of the target. The target becomes coated with clusters of hundreds of C3b molecules, which are gradually cleaved by other proteases (iC3b or C3d/C3dg). C3b is recognized by the complement receptor 1 (CR1) resulting in the target bond to red blood cells. A C3b-coated target, bound to red blood cells circulates in the blood and C3b is being gradually converted to iC3b. As the red blood cells with the bound target pass through the liver or spleen, they come into contact with macrophages, which express the surface receptors for iC3b (CR3 and CR4). The target which is attached now with iC3b, is stripped off from the red blood cell and transferred to the macrophage, which ingests the target and destroys it intracellularly. When iC3b is further broken down to C3d/C3dg, it interacts with the receptor CR2 on B lymphocytes. This interaction can stimulate the synthesis of antibodies against the target. Dendritic cells, which capture foreign materials and present them to the adaptive immune system as antigens also have complement receptors, thus coating of the target with C3 fragments also helps to develop the adaptive immune response (antibodies and cytotoxic T cells) against the target. Once C3 has been activated, the protease C4b2a can then activate the next complement protein C5, forming the fragments C5a and C5b. C5b binds to C6, C7, C8, and C9, and this large protein complex MAC (membrane attack complex) can insert itself into lipid bilayers (in cell membranes). If the target has a cell membrane, the MAC will effectively puncture the membrane, destroying the target. During the activation of C4, C3, and C5, small peptides C4a, C3a, and C5a are released, and these potentially have inflammatory effects. They affect the smooth muscle of blood vessels and cause fluid leakage from the blood into the tissue spaces. They have effects on cytokine and chemokine release, and C5a is also a chemotactic factor, *i.e.*, it attracts cells (*i.e.*, granulocytes). At the site of a wound, these activities would cause the leakage of fluid into the wound, releasing more complement proteins from the blood into the site, to opsonize infectious microorganisms and also cause granulocytes to migrate to the site, where they ingest and kill bacteria. In the lectin pathway, the binding of mannan-binding lectin or ficolins to carbohydrate structures present on a wide range of microorganisms is initiated. Mannan-binding lectin has also been reported to bind to IgA and to galactosyl IgG (IgG-Go). Mannan-binding lectin and ficolins circulate in serum, complexed with serine protease proenzymes which are structurally similar to C1r and C1s with identical domain composition. The mechanisms for MASP activation have yet to be fully determined; but upon MBL of ficolin binding to targets, MASP-2 is auto-activated and cleaves C4 and C2 to form the C3 convertase C4b2a, similar to that in the classical pathway.

The alternative complement pathway is initiated differently from the classical and lectin pathways. To trigger this pathway, C3b has to be deposited on the surface of a target. C3b may be derived from the activation of the other pathways, arising from a non-enzymatic slow turnover of C3, or it may arise because other non-complement proteases can activate C3 to a minor extent. Once one molecule of C3b is bound to the target surface, factor B can bind to it, and is then cleaved by factor D to form C3bBb, which is a protease complex homologous to C4b2a in the other pathways. Once this C3-cleaving enzyme is formed, it leads to the same events which occur in the classical and lectin pathways. Since C3bBb can form more molecules of C3b, the alternative pathway acts as an amplification loop, causing more C3b to be produced and deposited on the target surface. The different pathways of complement respond to different targets. C1q in the classical pathway binds to a very wide range of targets. These include antibody–antigen complexes formed with IgG or IgM, Gram-negative bacteria, some viruses, polyanionic molecules like DNA/RNA, or anionic phospholipid micelles, altered host proteins such as clots or amyloids, and importantly also synthetic or nanomaterials (such as for instance Perspex - carbon nanotubes). It is well known that C1q recognizes mainly charged clusters, but probably also hydrophobic patches when occurred on surfaces.

Moreover, the interaction of drug and carrier may result in significant conformation changes that may increase immunogenicity of used drugs. The complement activation can lead to rapid phagocytosis and clearance by macrophages. The complement activation may provoke life-threatening allergic, anaphylactic and hypersensitivity reactions or an activation of specific humoral and cellular immune responses [47-49]. Nanoparticles are able to increase antigenicity of weak antigens and may serve as adjuvants. Furthermore, they are disposed for lytic functioning. The red blood cell hemolysis is one of such examples. The hemolysis can be immune-mediated through nanoparticle-specific antibody or non-immunogenic through nanoparticle–erythrocyte interaction [49]. The recurring hemolysis induced by nanoparticles may lead to anemia. The nanoparticles with a combination of hydrophobic and hydrophilic areas on the surfaces act as potent disruptors of cell membranes [50] and those with positive surface charge from cationic surface groups of unprotected primary amines increase the erythrocyte damage [51, 52]. The released hemoglobin and cell debris can in turn attach to the nanoparticles, causing rapid clearance and potential immune response [53, 54]. The thrombogenicity is another example of an undesirable outcome of interactions between nanoparticles and blood components, resulting in blood clotting and partial (or complete) vessel occlusions. This may especially happen when the nanoparticles are engineered to have longer circulation time. In the most severe form, the fatal disseminated intravascular coagulation may occur. On the contrary, appropriately modified nanoparticles may lose their previous undesirable characteristics. Greish *et al.* reported that amine-terminated dendrimers triggered disseminated intravascular coagulation, whereas carboxyl- and hydroxyl- terminated dendrimers of similar sizes were well tolerated [55].

Thus to maximize nanoparticle uptake at the target site, the nanoparticles must be able to evade detection by the immune system. Macrophages recognize and clear bacteria by recognizing specific protein patterns on bacterial surface coatings. Therefore nanoparticle interfaces should be designed so as not to bind at the surface of known opsonins such as fibrinogen, IgG and IgA, or components of the complement system (especially C3b, C4b, and iC3b) [56, 57]. For instance, low affinity polymers such as polyethylene glycol are known to reduce non-specific protein binding due to a combination of steric hindrance, polymer flexibility and hydrophilicity [58]. Thus polymer layers are added on NPs to reduce serum adsorption on NPs, namely in particular IgG and complement adsorption as well as to protect NP-induced serum dependent complement activation *via* both the classical and alternative pathways or to become long-circulating through recognition evasion [59-61]. This may also be reached by the use of dysopsonins such as human serum albumin or apolipoproteins which promote prolonged circulation of NPs in the bloodstream [62]. Low affinity polymers and appropriate coating greatly lowers the amount of non-specific adsorption on NPs and improves NP circulation times. Notwithstanding, it is more recently emerging that considerable protein adsorption events remain sometimes too long to allow NPs to meet a target. This may lead to the failure of drug functioning [63, 64]. A specific targeting is one of the most important issue that is intended to increase the specificity of delivery of therapeutics, to maximally reach an intended location and to reduce side effects related to unspecific accumulation in other organs or cellular compartments. A highly developed targeting strategy would exploit and utilize all of the nano-scale drug advantages [56]. Target ligands such as antibodies, small peptides, receptor binding compounds *etc.* are incorporated on the surface of nanoparticles [65] to increase uptake and bioavailability of loaded compounds [59, 60] to maintain their integrity and stability [66-68] or to protect them from degradation by enzymes as well as to prolong their circulation by stabilizing them against opsonization [1, 69].

Nanodrug delivery can either use passive targeting mechanisms, such as the enhanced permeability and retention effect, or active targeting mechanisms, using ligands directed against differentially overexpressed cell surface markers [70]. The passive targeting usually utilizes generic and specific tissue characteristics such as leaky vasculature, higher rates of metabolism or different levels of oxygenation of the tissue, whereas the active targeting is based on specific ligands to actively deliver the drug through biorecognition. A series of clinical studies have substantiated the potential of nanoparticle-based therapeutics with less adverse effects. Nanoparticle (NP) formulations, through passive or active targeting, can release therapeutic payloads [71]. Although most commercialized nanomedicines have been based on a passive-targeting strategy that exploits the enhanced permeability and retention effect (EPR), there is an increasing need for the development of active-targeting technologies. The leading strategy has been the conjugation of ligand molecules that specifically bind to overexpressed receptors [72]. To common types of nanoparticles that have yet been approved (or are in late stages of clinical trials) belongs polymer-drug conjugates, micelles, protein-based carriers, liposomes, polymeric nanoparticles, and inorganic nanoparticles. Most of them are organic based due to a relatively low toxicity and high biocompatibility.

Basically, targeting strategies can be divided into two categories: physical and chemical ones. Particle size has proven to be highly important. Larger particles take advantage of the EPR effect but they are unable to penetrate into the cells in the way smaller particles can. Therefore, structures with smart and dynamic size properties should be applied to overcome this situation. Similarly, as for immunological applications, smaller particles are able to target dendritic cells in lymph nodes better, whereas larger particles better target the dendritic cells in the periphery. The stiffness of a particle was also shown to be useful in targeting and uptake by certain cell types. Some cells are better able to engulf soft particles whereas others (such as macrophages) the harder ones. In chemical targeting, the most well studied strategies involve molecular recognition. However, these are generally limited by the tendency of targeting moieties to enhance delivery to undesirable organs (even more than to the intended tissues). An alternative approach to the specific targeting involves responsive systems recognizing the biochemical environment around the target. In terms of *in vivo* chemical-targeting efforts, antibody conjugation is a good strategy when significant amounts of active binding sites are present after conjugation. Encapsulating pharmaceuticals in nano- or microparticles offers another solution to multiple problems. Nanoparticles can be made from various materials including lipids [73]; inorganic materials [74]; proteins [75]; or polymeric systems [1]. Lipids are widely used and well characterized in carrier systems because their dynamic nature allows clustering of peptides or other ligands, enhancing the affinity of the interaction with target cells [76]. However, this same dynamic nature also makes them less stable than other carriers. Inorganic materials provide the advantage of a relative stability; but this may be associated with their retention in the body thus limiting clinical applications. As regards both the chemical and physical targeting strategies, there are other important issues to be considered when using size nanoconstructs in well controlled delivery systems. For instance, nanotechnology became an attractive choice of molecular delivery systems for prevention of immediate clearance from the kidneys. Studies using different sizes of quantum dots indicate that only the particles with a diameter of 5.5 nm resulted in rapid and efficient urinary excretion [77]. However, there is not one particular size for all applications and several studies have demonstrated opposing results using nanoparticles or microparticles. In addition, methods used for determination of the size of particles differ. Furthermore, reported particle sizes may be misleading due to a high polydispersity of size caused by imperfect methods of preparation. The polydispersity can affect the apparent number and volume diameter and, more importantly, can result in microparticle formulation procedures that produce microparticles in combination with some nanoparticles and *vice versa*. A light-activated release of pharmacological agents is an example of the way how to control systems for specific delivery targeting. Light-activated particles are created using optically active substances that are capable of degrading or releasing their load under UV, visible light, and near infrared (NIR) light. As such example, particle encapsulation and UV-triggered release of drug compounds can be developed using amphiphilic block copolymers composed of poly(ethylene oxide) and poly (2-nitrobenzyl methacrylate) [78]. UV light has the highest power and can break polymeric bonds more easily but can also damage the surrounding tissues. Visible light-responsive systems can incorporate a chromophore that can absorb visible light, which is dissipated locally as heat, but the visible light cannot penetrate well into tissues. NIR light between 750 and 1,000 nm has been shown to penetrate more deeply with minimal risk to surrounding tissues. This optical systems present an opportunity to develop novel stimuli-responsive systems to control delivery of pharmaceuticals in a harmless but still effective manner. Polymers with a high sensitivity have been developed using self-immolative monomers that can potentially sense a single triggering event to degrade the entire particle [79]. *In vitro* studies show that these polymeric nanoparticles can be used to encapsulate small molecules and can be released rapidly, followed by degradation. Although the field of NIR light responsive particles is relatively new, it seems to be one of the most promising systems to control drug delivery. pH responsive systems are another attractive way among the improved drug delivery systems. These systems can be applied for external and internal cellular release [80]. For example, human tumors have been shown to exhibit acidic pH states range from 5.5 to 6.8 To the acidity of tumor microenvironment contribute hypoxia and poor lymphatic drainage as well as elevated rate of glucose uptake, reduced rate of oxidative phosphorylation and lactic acid accumulation in tumor cells. The persistence of high lactate production provides tumor growth advantage in the presence of oxygen (Warburg's effect). NPs have been formulated for pH-dependent drug release by using carriers (polymers) that change their physical and chemical properties, such as by swelling and solubility, based on local pH levels. NPs take these actions by responding to the acidic pH of tumor microenvironments, as opposed to those in oral drug delivery where the elevated pH is often used as trigger. The acidic tumor pH may thus be exploited to achieve high local drug concentrations and to minimize overall systemic exposure. Several studies have successfully developed and demonstrated the potential of using this particular pathological stimuli to deliver encapsulated drugs [81].

Moreover, nanoparticles can also be triggered to release drugs once they are taken up by cells *via* endocytosis [82]. As nanoparticles are taken up by cells, they enter the endocytic pathway. The early endosome begins to acidify within minutes and progressively becomes more acidic as it moves toward the lysosomes where the pH can be as low as 4. In addition, changes in ionic strength causes a proton sponge effect in certain cationic buffering polymers that can allow the endosome to release encapsulated material. Delivery of drugs using pH responsive nanoparticles *via* the endosome and into the cytoplasm is very effective and in some cases seems also to improve drug resistance [83]. A combination of pH responsive nanoparticles with other targeting strategies can help to further target and enhance the therapeutic outcome. Combining cell-penetrating peptides with pH responsive nanoparticles has been successfully used to efficiently enter and then release drug content into nonphagocytic cells [84].

To summarize, both targeted and non-targeted nanoparticles arrive to the cell vicinity *via* the passive targeting or enhanced permeation and retention effect [85], after which the mechanism of the cell internalization could be enhanced by the presence of surface ligands for active targeting [86]. However, it should be kept in mind that the utilization of a targeting ligand for specific cells does not guarantee higher nanoparticle accumulation [87, 88]. Higher targeting efficiency could be obtained by combining strategies, or developing such nanoparticles that are functionalized and based on specific and separate mechanisms of their actions [56, 88]. By programming the degradation of NPs, a prolonged drug release can be achieved, eliminating the need for repetitive dosages and enabling more sustained and consistent drug concentrations [89]. Nanodrug delivery systems can carry one or a combination of therapeutics, including cytotoxic agents, chemo sensitizers, antiangiogenic agents and others [90]. Drug encapsulation within nanoparticles can enhance the bioavailability of drugs administered *via* routes other than intravenous. Both insoluble and soluble drugs can be incorporated within nanoparticulate sols, extending their stability as they travel through the blood, which in turn improves their overall pharmacokinetic half-life [89]. One prominent trait of NM is the fact that, during the lifetime of a given NM, organisms can be exposed to different forms of the nanomaterial due to an interaction with surrounding substances, dissolution or aggregation effects. In general, the effects of NM may be related to: 1) the particles themselves and their coatings (particle effects); 2) ions or molecules released from the particles (chemical effects); 3) molecules formed by the catalytic surface of the particle (nanorelated effects) [36]. As an example, *in vitro* dissolution tests in physiologically relevant media (*e.g.*, lysosomal fluid, gastrointestinal fluid, lung lining fluid) may give indications on the time frame in which mainly particles, both particles and ions, and mainly ions are present [26, 90-94]. If particles are unlikely to dissolve, efforts should mainly focus on long-term particle-related effects, since the particles are likely to accumulate over time [92]. If particles dissolve, both particle and ion-related toxicity should be considered. In that case, accumulation of particles is to be expected to a smaller extent, due to elimination by dissolution of particles to ions. The ion-related effects may be different from the effects of the dispersed particles. If NM distribution and cellular uptake occurs in the particulate form, the ions may be released at different sites than when the NM are distributed as dispersed particles. Hence, the resulting toxicity may be different. NM uptake by cells without its significant clearance from those cells may pose a potential prospective risk that has not yet been met but it should be taken into account in the context of its further risk assessment [36].

The first approved "nanodrug" that had successfully been introduced in the market (in 1995 as a treatment for AIDS-related Kaposi's sarcoma) was Doxil [95]. Doxil liposomes consist of a single lipid bilayer membrane composed of hydrogenated soy phosphatidylcholine and cholesterol with doxorubicin encapsulated in the internal compartment [96]. The mean size of the vesicles is in the range of 80–90 nm [97] and each vesicle can hold a payload of up to 15,000 molecules of doxorubicin [96]. The PEG functionalization makes the particles nearly invisible to the reticuloendothelial system (RES), earning them the distinction of stealth liposomes [96, 97]. While not specific to the Doxil formulation, the primary mechanism for accumulation and distribution throughout target sites is believed to be due to the combination of long circulation time, the microvascularity and the enhanced permeability and retention (EPR) effect [98, 99]. The volume of the distribution of Doxil is only slightly larger than the plasma volume itself, indicating that there is a very little uptake of the liposomes by healthy tissue. Consequently, many studies have indicated that Doxil efficacy is substantially higher than that of free doxorubicin on a mg-to-mg scale [97]. Moreover, despite its significantly longer circulation time than doxorubicin itself, Doxil has dramatically different and less severe side effects than the free drug [98]. Doxil shows a drastic decrease in the cardiotoxicity over doxorubicin, for which cardiotoxicity is the dose-limiting side effect. The two most severe side effects of Doxil are mucositis and palmar plantar erythrodysesthesia which is a toxic effect unique to Doxil (not observed with free doxorubicin) and is attributed largely to the long circulation time of the vesicles and a tendency of stealth liposomes to accumulate at the skin [100]. Myocet is another liposomal doxorubicin nanomedicine whose main difference from Doxil is that it lacks the PEG functionalization on the particle surface. Advantages of the formulation are lower toxicity of this drug and without PEGylation, the circulation time is significantly shorter than observed for Doxil and the liposomes are not "invisible" to the RES. Therefore, Myocet is not associated with palmar plantar erythrodysesthesia, the dose-limiting toxicity of Doxil as mentioned, and it also shows reduced incidence of mucositis when compared [101]. The increased release rate of Myocet is attributed primarily to the lack of PEG coating, which stabilizes the membrane of the liposome and prevents the leakage of the payload [99]. Myocet is marketed in Canada and Europe in combination with cyclophosphamide as a first-line treatment for metastatic breast cancer but has not yet been approved for use in the USA (Phase III clinical trials as first-line treatment for HER2 positive metastatic breast cancer is underway). DaunoXome is a nanotherapeutic which is approved in the USA as a first-line treatment for patients with advanced HIV-associated Kaposi's sarcoma. DaunoXome is a liposomal daunorubicin that relies on a passive-targeting mechanism. Due to the small vesicle size, net neutral charge, and incorporation of cholesterol and a lipid molecule with a high phase-transition temperature, DaunoXome is also able to avoid the RES. Celsion's ThermoDox is very similar to Doxil and Myocet that is composed of doxorubicin encapsulated within the aqueous inner core of bilayer liposomes. The property that makes ThermoDox unique is that upon heating the liposomes to temperatures $\geq 39.5°C$, they release their payloads within seconds [102]. Unlike Doxil, Myocet, DaunoXome and Celsion's ThermoDox, there are nanomedicines such as Abraxane, Rexin_G or Oncaspar that utilize other forms of carriers. Abraxane (marketed by Celgene) utilizes albumin and the active agent is paclitaxel. Because of paclitaxel's hydrophobic nature, it needs a nonpolar carrier to make it clinically effective. Albumin is an ideal carrier as it is the natural carrier of hydrophobic molecules [1, 103]. Table **1** illustrates some nanoparticle platforms associated with nanotherapeutic products. Figs. (**2**) and (**3**) outline polymeric micelles, nanotherapeutic Doxil and mode of their action.

Table 1. Examples of marketed nanotherapeutic products.

Nanoplatform.	Formulation, Payload, Component.	Brand Name, Label.	Indication –Application.
Nanocrystals	Papeliperidon Olazanpin Aprepitant Fenofilbrate Sirolimus	Xeplion Zypandera Emend Tricor Rapamune	Schizophrenia Nausea Hypercholesterolemia Graft rejection
Liposome	Doxorubicine Doxorubicine Doxorubicine Doxorubicine Daunorubicin Cytarabine Vincristine Amphotericin B Mifamurtide Irinotecan/Cisplastin Irinotecan/Floxuridine Cytarabin/Daunorubicin Amikacin Fluorouracil Lurtotecan Morphine Cytarabine Cisplastine Annamycin 9-Nitrocamptothecin Vertepofin Oxiliplatin Morphine	Doxil/Caelyx Myocet Sarcodoxome MCC465 DaunoXome CPX-351 OncoTCS AmBisome Mepact CPX-571 CPX-1 CPX-351 MiKasome 5-FU OSI/NX-211 DepoDur DepoCyt SPI-077 L-Annamycin l-9NC Visudyne MBP-426 DepoFoam	Kaposi's sarcoma Prostate cancer Ovarian cancer Fungal infection Breast neoplasms Colorectal cancer Multiple myeloma Lymphoblastic Myeloblastic leukemia Neoplastic menigitis Various cancer
Nanoemulsions	Cyclosporine Ritonavir	Neoral Norvir	Post-transplant profylaxis HIV/AIDS infection

Table 1: contd…

Virosomes	Inactivated hepatitis A virus Influenza viruses	Epaxal Inflexal	Hepatitis A vaccine Antiviral medicine
Nanoparticles	Paclitaxel Y-ibritumomab tiuxetan Colloidal gold/TNF Silica NPs/Au coated Gold NPs	Abraxane Zevalin Aurimune AuroLase Verigene	Breast neoplasma Lymphoma Solid tumor Neck cancer Diagnostics
Polymer-protein conjugates	Peginterferon α-2a -2b Pegaspargase Certolizumab pegol Pegfilgrastin (PEG-rhGCFS) Methoxy polyethylene glycol epoetin β	Pegasys Pegintron Oncaspar Cimzia Neulasta Mircera	Hepatitis-B -C Acute leukemia Rheumatoid arthritis Chemotherapy induced neutropenia Kidney failure anemia
Iron nanocomplex	Iron(III)-hydroxid Sucrose complex Sodium ferric gluconate Iron(III) isomaltoside Ferumoxytol Iron oxide NPs coated - dextran - carboxydextran	Venofer Ferrlecit Monofer Rienso Combidex Feridex Resovist	Iron deficiency Sideropenia Tumor imaging
Polymeric – micelles	Paclitaxel Paclitaxel 1,3-bis(2-chloroethyl)-1-nitrosourea Doetaxel Estradiol	Genexol PM, NK 105 BiCNU (Carmustine) Taxotere Estasorb	Cancer Glioblastoma Menopause
Polymeric - nanoparticles	-Paclitaxel -Rapamycin - Docetaxel - Tanespimycin/Albumin, Cremophor EL	Abraxane ABI-009 ABI-008 ABI-0010 Taxol	Cancers Breast cancer

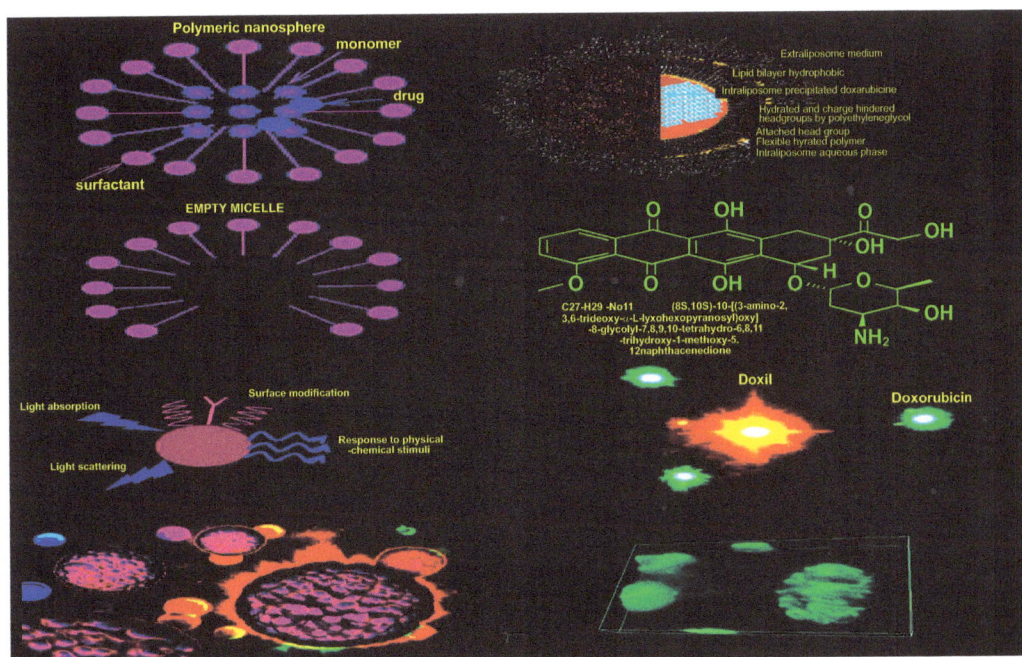

Figure 2: Nanodrug constructs and structures.

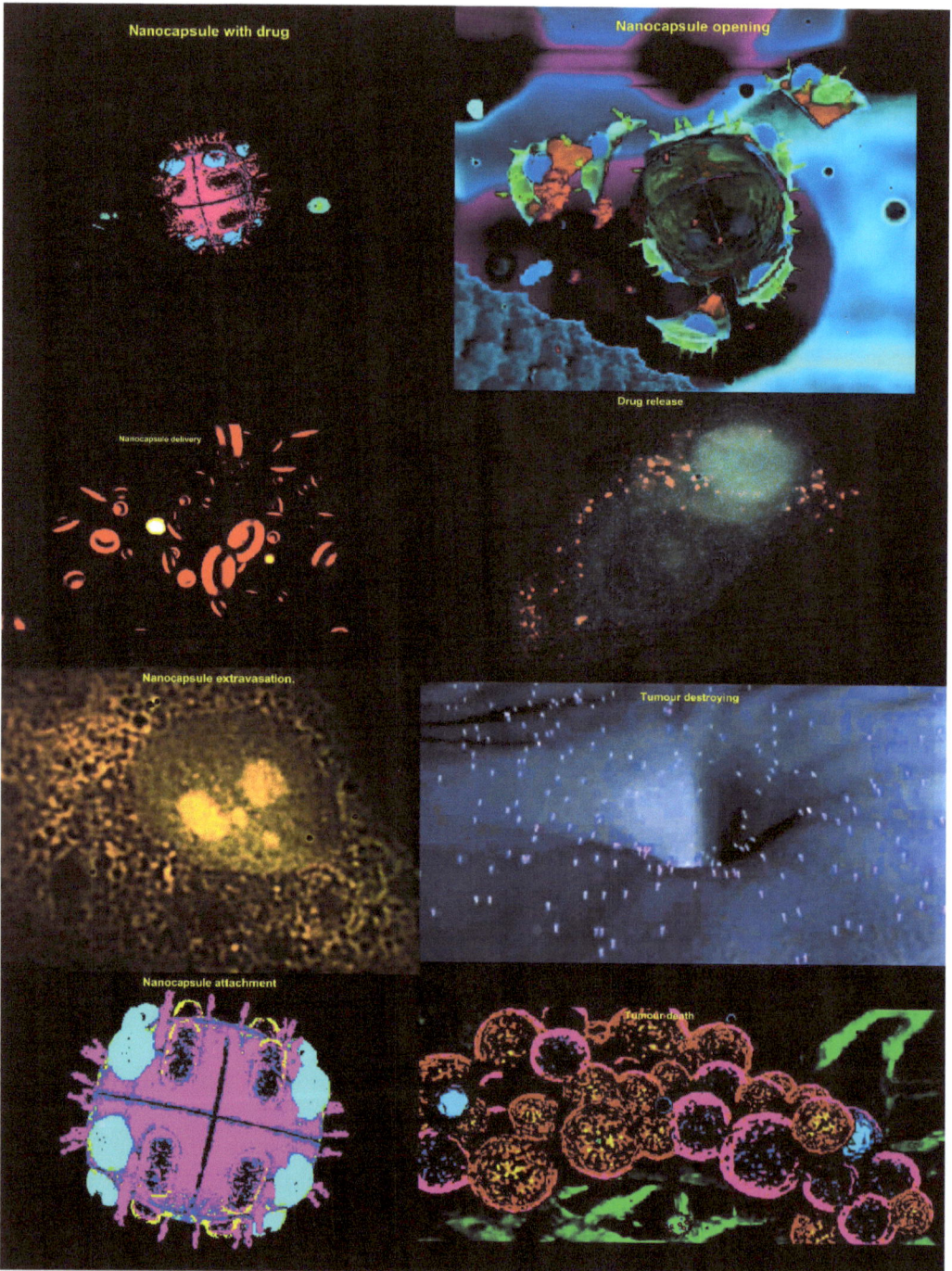

Figure 3: The mode of nanodrug action against tumor cells.

Nanoparticle Bioimaging

Nanoparticles are utilized not only in disease therapy but also as probes for various types of bioimaging, including fluorescence, magnetic resonance imaging, computed tomography, and positron emission tomography. Moreover, "theranostics" is defined as a methodology that combines both therapeutic and diagnostic approaches [104]. By combining therapeutic and diagnostic capability into one single agent, a new protocol is anticipated to shape up a treatment according to the test results. Such combination agents are capable of detecting and treating diseases in a single procedure. Emerging nanotechnology is offering great opportunities to design and generate agents, wherein the detection modality is extensively associated with the treatment procedure. One of the most promising aspect is the possibility to localize NPs in specific sites of destitute tissues and mitigate undesired side effects. It is well known that tumor blood vessels tend to be irregularly dilated and leaky, and due to this feature the nanoparticles can easily extravasate from the blood pool into tumor tissues and are retained due to the poor lymphatic drainage. Namely an advantageous property of NPs involves the high surface area-to-volume ratio, thereby yielding a high loading capacity of therapeutic drug or imaging agent. The greater surface area can be functionalized with stabilizing agents and ligands aimed at cloaking and targeting purposes. Since multiple types of molecules can be incorporated onto a single nanoparticle surface concurrently, the same nanoscale complex can achieve multiple functions. In fact, many nanoparticles already serve as imaging agents, thereby may readily be upgraded to theranostic agents by incorporating the therapeutic utility within. A qualified targeting approach facilitates accumulation of theranostic agents at the targeted region of the interest, thereby raising the local concentration of therapeutic agent, and additionally increasing the target-to-background contrast in the imaging [105]. When modified with the macromolecule to improve biocompatibility, the nanoparticles are very stable and remain in circulation for days, far longer than the few hours typical of many molecular imaging agents [106]. As example, DNA nanostructures can readily be incorporated with components possessing multiple functionalities. Recent advances in predicting the secondary structures of a DNA fragment or interactions between multiple DNA strands, as well as technologies to automatically synthesize predesigned DNA sequences, have a great potential of advanced applications of DNA structures in nanotheranostic and bioimaging. It has been demonstrated that the nanoparticles conjugated with selected aptamers are capable of differentiating among different subtypes of leukemia and in similar, among different cancers of the lung, liver, ovaries, colon, brain, breast and pancreas [107]. An aptamer moiety allows specific target cell recognition and a double-stranded DNA section forms the drug loading.

Moreover, a conventional assembly of DNA nanostructures may exploit the hybridization of a DNA strand to part of its complementary one or the DNA nanostructures may be self-assembled through liquid crystallization of DNA, which occurs at high concentrations of the nucleic acid [108]. Such advances in targeting are now making it possible to deliver some combinations of drugs. A liposome-based combination chemotherapy delivery system of two synergistic chemotherapeutics has been reported [109]. One of them promotes the dynamic rewiring of apoptotic pathway which sensitizes malignant cells to subsequent exposure to the DNA-damaging second one. By incorporating a hydrophobic molecule into the lipid bilayer shell while packaging the hydrophilic inside of the liposomes, control of the drug release in time dependent manners may also be achieved [110]. In addition, advances in the field of the contrasting agents promise a novel next generation of these agents [111]. Nanoparticle probes, nanocantilever, nanowire and nanotube arrays are the subject of intensive research and are expected to solve an accurate localization of tumors and their metastases, *via* nanoparticles loaded with a diagnostic aid, facilitating other therapies, such as radiotherapy, photodynamic therapy and surgery and their limitation [112, 113]. For instance, Perche and Torchillin have suggested that a possible direction for research may be the coupling of ligands of different natures (antibodies, proteins, peptides and chemokines, hormone analogs) to target at least two tumor cell populations, providing more sensitive malignant lesion detection and reducing relapses [114, 115]. Shapira has looked forward to the development of "theragnostic" nanovehicles that carry four major components: a selective targeting moiety, a diagnostic imaging aid for localization of the malignant tumor and its metastases, a cytotoxic small molecule drug(s) or innovative therapeutic biological matter, and a chemosensitizing agent to neutralize drug resistance – the advent of "quadrugnostic" nanomedicine [113]. Heller's group has described the goal of research as "the development of a cancer therapy monitoring/diagnostic platform device". This would provide real-time monitoring of patient blood for cancer cells, cell derived nanoparticulates (such as high molecular weight DNA fragments), and carry out cancer-related genotyping, gene expression and immunochemical analysis [116].

Nanotherapeutics in the EU

Current progress in the development of nanotherapeutics and nanotechnology bioimaging tools allow to improve drug solubility (*e.g.* micelles [117] and nanocrystals [118]), to guide drugs to the desired location of action with increased precision (*i.e.* drug targeting [119, 120], to control the drug's release (*e.g.* nanoparticles [121] and liposomes [122], and/or to enhance the transport across biological barriers (*e.g.* micelles) [123] and nanoparticles [124]. The main goal is to improve drug bioavailability, pharmacokinetics, efficacy, and safety to promote the treatment of diseases which currently cannot be achieved with conventional dosage forms [125]. Naturally, there are the justified expectations of the power of nanotherapeutics, their commercialization and application in practical treatment of many diseases [126]. In the EU, marketed nanoproducts include nanocrystals, liposomes, virosomes, nanoemulsions, polymer-protein conjugates, polymeric drugs, nanocomplexes or nanoparticles. The majority of nanotherapeutics on the market are intended for parenteral administration but nanotherapeutics designed for oral administration are also constructed. It is expected that a large number of on-going preclinical and clinical investigations will shortly result in the approval of nanotherapeutics intended for other than non-parenteral or oral routes such as ocular, pulmonary, nasal, dermal and vaginal ones [127]. The choice of the delivery route, and consequently the barriers to be crossed, are of particular importance for any drug-delivery system [128]. A versatile library of nanomedicines such as carbon nanotubes, organic nanostructures (*e.g.* liposomes, dendrimers, and polymer-based nanocarriers), and inorganic nanoparticles (*e.g.* metal, silica, and semiconductor quantum dots, *etc.*) has already been constructed. These extensive libraries are composed of an assortment of different sizes, shapes, and materials with various chemical and surface properties.

Nanocrystals

Drug nanocrystals are nanoscopic crystals of the parent compound (nanoscopic crystals of a hydrophobic parent drug of size: 50–1,000 nm). Approximately 40% of drug candidates are poorly soluble in water, which hinders their development and clinical applications [129]. One of the strategies employed to overcome this limitation is the reduction of drug crystal size, *i.e.* drug nanonization [129]. This strategy involves the production of drugs by either chemical precipitation or disintegration [130-134]. This is one of the basic approaches of enhancing the oral bioavailability of hydrophobic drugs. Most common preparation methods include media milling, high-pressure homogenization, and nanoprecipitation [129]. Media milling is the oldest method employed in the production of drug nanocrystals. The disintegration of drug powders into nanoparticles is a result of strong shear forces generated by high-speed rotation of a milling chamber charged with milling pearls, dispersion media (*e.g.* water), drug powders, and stabilizers.

Polymer Nanoparticles

Polymeric nanoparticles are nanosized solid particles that consist of natural or synthetic polymers. Two types of nanoparticles can be distinguished:

1) nanospheres, which are matrix systems in which the drug is uniformly dispersed; and 2) nanocapsules, which are reservoir systems in which the drug is located in the core surrounded by a polymer membrane. Drugs are physically entrapped within the nanoparticles, chemically conjugated or adsorbed to the constitutive polymers of the nanoparticle. Polymeric nanoparticles are able to control drug release either by diffusion through polymer matrix or by its degradation. They have been investigated as drug-delivery systems for the site-specific targeting of tumors and for the transport of drugs across biological barriers, particularly the blood–brain barrier [126].

Polymer Nanotherapeutics

Polymeric nanotherapeutics include the family of compounds and the drug delivery technologies that use water-soluble polymers as a common core component. The term specifically refers to polymeric drugs, polymer–drug conjugates, polymer–protein conjugates, polymeric nonviral vectors, and dendrimers. Polymer therapeutics are nanosized (typically of 2–25 nm), but they are quite distinct from polymeric nanoparticles and the polymer molecular weight, polydispersity, architecture, and conjugation chemistry have a significant impact on their characteristic, safety and efficacy [126].

Polymeric Micelles

Polymeric micelles are formed upon the self-assembly of amphiphilic polymers, with the hydrophobic portion of the polymer orientated toward the core and the hydrophilic segments on the outer shell [123]. The micelles are in the size range of 20–80 nm, suitable for the entrapment of hydrophobic drugs [126]. They consist of a hydrophobic core and a hydrophilic shell and are useful drug carriers, due to their tunable size and surface functions, high monodispersity and excellent stability. Polymeric micelles have the ability to form hydrogels and are used for drug encapsulation or drug conjugation [135]. Under the right conditions thermosensitive polymers form a hydrogel at body temperature but are water soluble at lower temperature. This allows them to be injected as a liquid but they form a hydrogel *in situ*, resulting in prolonged drug release of the encapsulated drug [136]. Polymeric micelles/conjugates have been emerging as a highly integrated theranostic nanoplatform for cancer diagnostics and therapy. Of numerous materials, the most successful strategy has been to modify nanocarriers with polymers, which leads to decreases in immunogenicity and antigenicity as well as increases in body-residence time and stability. Furthermore, polymers are capable of shielding the core of nanocarriers from degradation by steric hindrance, reducing kidney clearance by virtue of an increased hydrodynamic size of polymer carrier conjugate, and increasing the solubility of nanocarriers as a result of its hydrophilicity [107]. There has been significant efforts in developing novel block copolymer micelles as nanomedicines to achieve improved delivery of poorly soluble, highly toxic and/or unstable drugs, to increase tissue targeting and/or to improve the efficiency of cytosolic delivery of macromolecular drugs [137, 138]. Typically, such self-assembled micelles are prepared from amphiphilic block copolymers (although other chemistries are being proposed) that spontaneously assemble into polymeric micelles in aqueous media; hydrophobic interactions typically drive this self-association. However, other driving forces, such as electrostatic interactions, may be used to promote micelle formation and enhance micelle stability. The active substance component can be incorporated into the inner core of the block copolymer micelle product by chemical conjugation or physical entrapment. Functional features may also be added to the system by targeting molecule conjugation to the block copolymer, or by the addition of another homopolymer to stabilize the micelle or active substance, modify its release rate and/or increase its loading. Innovative block copolymer micelle products have a carefully designed structure in which the inner core typically serves as a container for active substance component and this is surrounded by an outer shell of hydrophilic polymers. The chemistry of such block copolymer micelles may be designed to ensure high stability after dilution and administration due to a low critical association concentration (the concentration above which monomers spontaneously form micelles), to optimize the pharmacokinetics (targeting), and to better control the drug release kinetics, both in terms of timelines and site. Thus, the dissociation of such block copolymer micelles may be kinetically slow. These properties are different from traditional surfactant micelles used to entrap/solubilize/aid the transport of drugs. Moreover, a block copolymer micelle product can contain multiple components within the core, including the active substance, which in certain cases may be covalently bound. Nonclinical studies involving block copolymer micelles have demonstrated their potential to preferentially accumulate in solid tumors due to microvascular hyperpermeability and impaired lymphatic drainage [137]. The specific physicochemical properties of block copolymer micelles, such as size, surface-charge, composition and stability are important determinants of safety and efficacy in all proposed applications. Several block copolymer micelle products containing anti-tumor agents as the active substance are currently in clinical development and block copolymer micelle products containing proteins are in preclinical development [138, 139].

Dendrimers are globular macromolecular compounds consisting of an inner core, which can be manipulated to alter its shape and size, surrounded by a series of branches with surface functional groups. They can carry a multiple payload of active targeting molecule, diagnostic agent and therapeutic drug, and those with a hydrophobic core and hydrophilic surface groups can form micelles, which can then be designed for site-specific release of their payload, *via* pH and enzyme dependent mechanisms [140, 141]. Polymeric micelle nanotherapeutics have not yet been approved in the EU. Recently, the Committee for Medicinal Products for Human Use of the European Medicines Agency has drafted a reflection paper on the pharmaceutical development and preclinical and early clinical studies of polymeric micelle medicinal products designed to affect the pharmacokinetics, stability, and distribution of incorporated or conjugated drug *in vivo* [142]. However, because of the complexity of polymeric micelle products and limited research, their general advice is to seek product-specific scientific consults when developing such nanoproducts [126].

Liposomes and Nanoliposomes

Liposomes have been classically described as artificially prepared lipid-based vesicles composed of one or more lipid bilayers enclosing one or more aqueous compartments. They include mono- and multi-lamellar liposomes, and some formulations benefit from the addition of sterols, size reduction and surface modification with covalently linked polymers. Typically, active substance(s) are entrapped in the aqueous phase of a liposome, or by incorporation or binding to the lipid components. The early goals for design of parenteral liposomal products was the generation of drug delivery systems to improve solubility of entrapped components, for disease-specific targeting, to control drug release rates and/or to produce a pharmaceutical formulation suitable for clinical use *via* reducing unacceptable toxicity [5, 143, 144]. Liposomes have been in use for the past several decades and are established as drug and imaging agent carriers with proven clinical efficacy [5]. They are artificial phospholipid vesicles 50 nm to 1 μm in size, either unilamellar or multilamellar, with one or more aqueous compartments [5, 136]. The pharmaceutical active compound can be held in the aqueous compartment(s) or lipid layer [115]. During the development of first parenteral liposomal products, a number of physicochemical properties were identified as critical determinants of *in vivo* behavior. Experience with liposomal formulations developed with reference to an innovator is still very limited. A regulatory approach similar to the one established for analogous biological medicinal products that requires the stepwise comparability approach for quality, safety and efficacy between the reference biological medicinal product and the biosimilar has been suggested [145, 146]. Due to its biphasic character, liposomes can serve as carriers for both hydrophilic (in the central aqueous compartment) and hydrophobic (in lipid bilayers) compounds [147] However, the term nanoliposome has been introduced recently to exclusively refer to nanometric size of liposomes [148]. Although, in a broad sense, liposomes and nanoliposomes have the same chemical, structural and thermodynamic properties, the smaller size of nanoliposomes could produce larger interfacial area of encapsulated compounds with biological tissues and thus provide higher potential to increase the bioavailability of encapsulated compounds [3]. Especially for solid tumor treatment, nanoliposomes can accumulate more in tumors because of the enhanced permeation and retention effect [147, 149]. Higher energy input is required to produce nanoliposomes in the aqueous solution [150]. The commonly used methods for nanoliposome synthesis include sonication, extrusion, freeze-thawing, ether injection and microfluidization.

Emulsions and Nanoemulsion

An emulsion is a mixture composed of two immiscible liquids. When oil is dispersed, it will form into droplets through the aqueous phase; this is referred to as oil-in-water emulsion. On the contrary, an aqueous solution dispersed in oil phase is referred to as water-in-oil emulsion [151]. In order to disperse two immiscible liquids and to stabilize the emulsion structure, a surfactant or emulsifier is required, which has an amphiphilic structure with one fragment being hydrophilic and the other one being hydrophobic [152, 153]. Novel self-emulsifying drug delivery systems formulations are isotropic mixtures of oils, surfactants, nutrients (or drugs), usually with one or more cosurfactants or coemulsifiers [154-156]. When oral administration of novel self-emulsifying drug delivery systems is in solid, liquid or semiliquid form, the mixture is dispersed into gastrointestinal fluids, yielding fine oil-in-water emulsions containing hydrophobic compounds upon gentle agitation in the gastrointestinal tract [154]. In nanoemulsions oil nanodroplets are dispersed within an aqueous continuous phase and stabilized with surfactant molecules [129]. Nanoemulsions can be prepared by high-pressure homogenization, microfluidization, ultrasonication, and spontaneous emulsification. Advantages of nanoemulsions include increased drug loading and enhanced bioavailability. Therapeutic nanoemulsion-based products in the form of self-emulsifying drug-delivery systems formulated in gelatine capsules are particularly interesting because nanoemulsions form when the formulation reaches the gastrointestinal tract. Only few formulations have been commercialized due to limitations related to usage of surfactants and cosolvents and a possibility of drug precipitation upon aqueous dilution *in vivo* resulting in unpredictable oral bioavailability [157].

Solid (lipid) Nanoparticles

Solid lipid nanoparticles have a similar structure of the oil-in-water nanoemulsion including a hydrophilic shell and a hydrophobic lipid core, which is solid at room temperature [158]. Solid lipid nanoparticles were developed in the early 1990s as an alternative novel carrier system to traditional nanocarriers such as polymeric nanoparticles, nanoemulsions and nanoliposomes [159]. Solid lipid nanoparticles are usually composed of solid lipids, surfactants and water, with or without cosurfactants. Although solid lipid nanoparticles are one of the most useful lipid based nanocarriers in nutraceutical and pharmaceutical research, some limitations exist, such as low compound loading capacity and leakage during storage. These limitations are overcome by the recently developed nanostructured lipid carriers, an improved, new generation of lipid nanocarriers [160]. The common methods for making solid lipid nanoparticles and nanostructured lipid carriers include high-pressure homogenization, cold homogenization, hot homogenization/ultrasonication, phase inversion and solvent evaporation/emulsification [161].

Virosomes

Virosomes are a drug or vaccine delivery systems composed of viral membranes that are reconstituted with viral lipids and proteins, allowing the fusion with target cells [162]. Virosome production generally includes detergent solubilization of the influenza virus and subsequent reconstitution with two influenza envelope glycoproteins, hemagglutinin and neuraminidase. These glycoproteins have a significant role in structural stability, homogeneity, targeting, receptor-mediated endocytosis, and endosomal escape after endocytosis.

Metal- and Other Inorganic Based Nanomaterials

Basically, three topics related to the biomedical application of metal NPs can be considered: cellular membrane-permeable nanoparticles, self-assembled nanoparticles and nanoparticle-based vaccines. The water dispersibility of nanoparticles can be provided by the attachment of hydrophilic functional groups to the ligands such as PEG, carboxylic acids, sulfonic acids, ammonium salts or zwitterions. Charged nanoparticles show higher water dispersibility than do non-charged nanoparticles. Positively charged nanoparticles can be taken up into cells owing to their high affinity for the cell membrane, whereas negatively charged nanoparticles show longer circulation. These two factors are important to the therapeutic use of nanoparticles and should be carefully balanced. As the net charge of nanoparticles is adjustable by the co-display of anionic and cationic ligands, this approach is also expected to be applicable to pH-responsive cellular uptake. It has also been demonstrated a fine tuning of nanoparticle charge by a co-display of cationic and anionic ligands and these NPs were stable under both high and low pH conditions. The exposure of nanoparticles to media containing electrolytes like sodium chloride cause aggregation as the salt neutralizes the electronic repulsion among nanoparticles. There is a need for the balance between hydrophilicity and hydrophobicity to maintain dispersibility of nanoparticles and allow their cellular uptake. When hydrophobic nanoparticles within the cell membrane penetrate into the cytosol, the surface of the nanoparticle which is hydrophilic allows easier permeation. In addition, the hydrophobicity/hydrophilicity of the nanoparticles may be changed by stimuli and the cell permeation may also be accelerated. Typically several stimuli such as light [163-167] or temperature [168, 169] have been employed. The stimuli used to change the particle properties should be both biocompatible and applied internally rather than externally to allow *in vivo* use. Encapsulation strategies associated with the controlled delivery of active pharmaceutical ingredients by self-assembly strategies, has been successfully applied [170-173]. Due to their superior biocompatibility and well-established strategies for surface modification (*i.e.* metal-thiol bonding), metal-based nanomaterials have been investigated as multifunctional probes. The unique optical and photothermal characteristics of metal nanomaterials enable them not only to be applied as sensing utilities but also to induce photothermal effects for therapeutic purposes because of the localized surface plasmon resonance (LSPR) of metal - nanomaterials can be adjusted by tuning particle morphology. Fig. **4** illustrates networks of gold nanoparticles as biological cell sensors or cell-targeting agents. Metal-based nanorods, nanoshells or nanocages exhibit distinctive optical and thermal properties, which can readily upgrade these nanomaterials to be prospective theranostic agents. Metal nanomaterial provides a versatile platform for simultaneously carrying therapeutics and can be intracellularly released from the nanocarriers by exchange reactions [174]. One of the most important functions of nanoparticles is catalysis, especially with noble metal nanoparticles, which have high catalytic activity for many chemical reactions. Namely Au NPs, have emerged as a promising scaffold for drug and gene delivery systems. Importantly, when considering a nanomaterial to be used in biomedical engineering the biodegradability and biocompatibility of the material should be carefully evaluated. Ceramic NPs are particles fabricated from inorganic compounds with porous

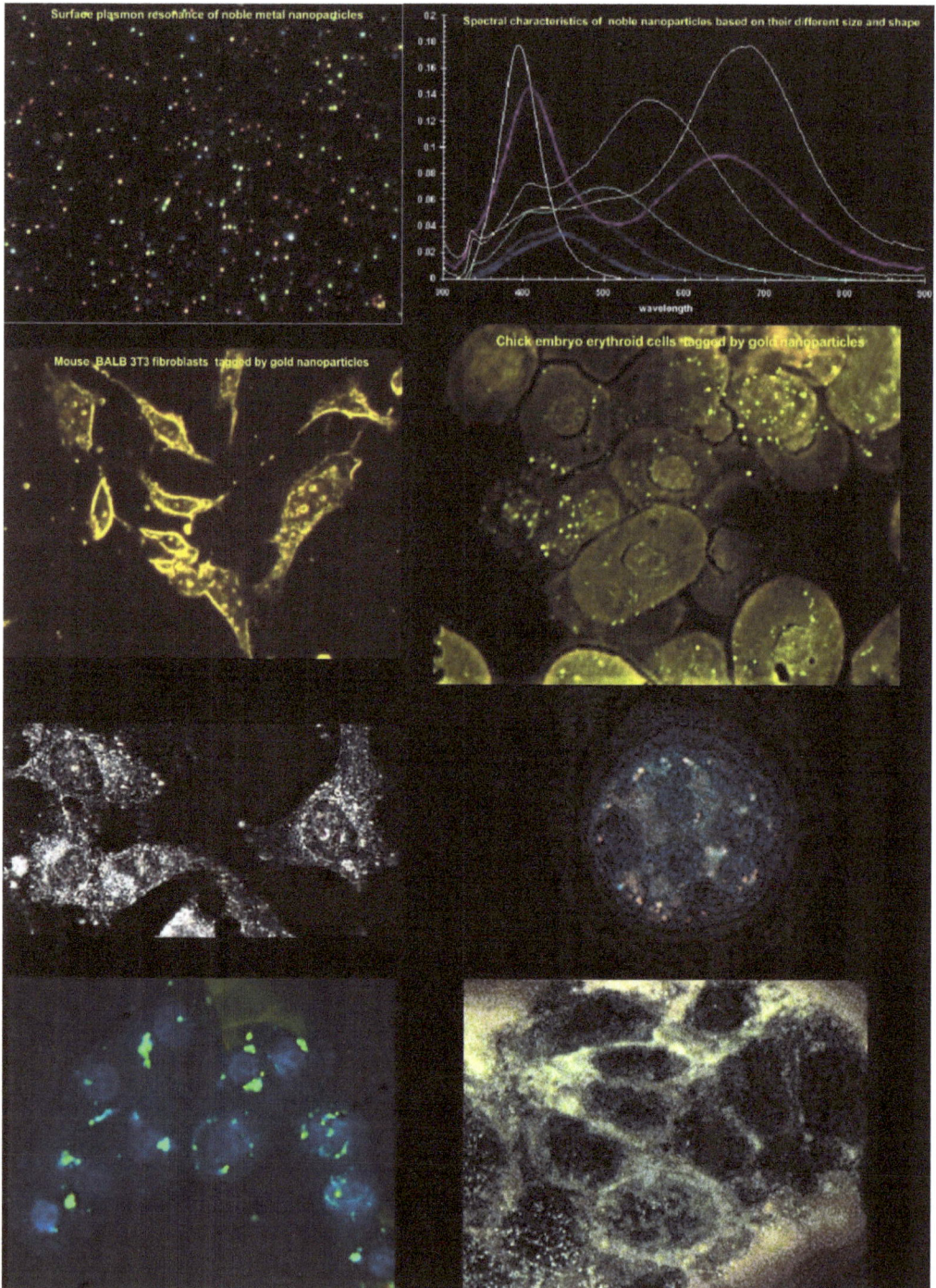

Figure 4: Spectral characteristics and the visualization of noble metal nanoparticles by dark field and fluorescence microscopy.

characteristics, such as silica, alumina and titania [175-177]. Among these, silica NPs have attracted much research attention as a result of their biocompatibility and ease of synthesis, as well as surface modification [178, 179]. Silica-based nanomaterials, unlike many other nanomaterials whose size-dependent properties are commonly observed as their size approaches the nanoscale and as the percentage of atoms at the surface of a material becomes significant, have constant physical properties similar to those of bulk material, *i.e.* except the total surface area which increases as the size decreases. Well-defined tunable nanostructures and well established siloxane chemistry allow to fabricate effectively the desired functionalized surface of these nanoparticles for diagnostic and therapeutic applications [107].

Namely, the remarkable properties of the mesoporous silicon (PSi) and silica (PSiO$_2$) materials as nanodelivery systems are their high surface-to-volume ratio, large surface area (up to 700-1,000 m^2/g) and large pore volume (> 0.9 cm^3/g) [180], possessing a stable and rigid framework with excellent chemical, thermal and mechanical stability. In this respect, the mesoporous materials may act as reservoirs for storing the therapeutic molecules and can thus be tailored *via* different pore size and surface chemistries [181]. In practice, mesoporous silica and silicon nanoparticles differ in their properties and fabrication techniques: Mesoporous silica nanomaterials are synthesized through the "bottom-up" approach, whereas silicon NP materials are produced by the "top-down" one [182-184]. The most common surface treatments of silicon NPs are thermal oxidation and stabilization by thermal carbonization or hydrocarbonization for hydrophilic or hydrophobic surface properties [185-192]. Moreover, mesoporous silica synthesis processes utilize different template systems to direct the silica molecules into a mesoscopically ordered yet amorphous structure containing very unidirectional and uniform pore channel structures. The surface chemistries of these mesoporous silica materials consist of siloxane groups (–Si–O–Si–), with the oxygen on the surface, and of three forms of silanol groups (–Si–OH) [193, 194]. The surface treatment may also affect the loading of the molecules into the pores *via* hydrophobic-hydrophilic interactions and the pore diameters of silicon can vary from few nanometers to micrometers, however in drug delivery applications the mesopores (2–50 nm) are the most used. The mesoporous silica exhibits materials of highly ordered two-dimensional tube-like pore structures with pore diameters typically between 1.5–30 nm. Silicon NPs are readily biodegraded into silicic acid, which is a natural compound of the human body and that can be cleared from the blood through the urine. Silicon NP degradation rate is directly correlated with the particle's size diameter and pore [195] However, in the case of the particles with diameters around 100 nm presenting pores sizes between 5 and 20 nm, the stability of the particles does not strongly depend on the pore size [196]. Moreover, the silicon NPs ranging from 80 to 120 nm are large enough to avoid renal clearance. The biodegradation properties of the silicon NPs provide a safe clearance from the body, and their biodegradability rate, which is often too fast, limits their half-life, thus reducing their *in vivo* delivery efficiency. It is noteworthy that for the control and precise tuning of the drug release profiles it is also possible to use a so-called "gate-keeping" approach, which consists of incorporated responsive polymers or other pH-sensitive compounds attached to the surface of the silicon (or mesoporous silica) structures [197-201].

Mesoporous silica NPs contain hundreds of empty channels (mesopores) arranged in a 2D network of a honeycomb-like porous structure. In contrast to the low biocompatibility of other amorphous silica materials, recent studies have shown that mesoporous silica NPs exhibit superior biocompatibility at concentrations adequate for pharmacological applications [202, 203]. The ability to selectively functionalize the external particle and/or the interior nanochannel surface is advantageous in achieving this goal [204, 205]. The surface of mesoporous silica NPs can be engineered with cell-specific moieties, such as organic molecules, peptides, aptamers and antibodies, to achieve cell type or tissue specificity. Moreover, optical and magnetic contrast agents can be introduced to develop multipurpose drug delivery systems. These strategies demonstrated that the application of target-specific mesoporous silica NPs vehicles *in vitro* s promising; however, mesoporous silica NPs are not biodegradable; consequently, there is a concern that they may accumulate in the human body and cause harmful effects. For further *in vivo* applications, the biocompatibility, biodistribution, retention, degradation and clearance of MSNs must be systematically investigated. Moreover, in the case of silica-based ordered mesoporous materials, the silanol groups that cover the silica walls can interact with physiological fluids leading to the formation of an apatite-like layer similar to that of the natural bone. These bioceramics may regenerate osseous tissues and they are able to act as controlled release systems. Thus it is necessary to consider non biodegradability of nanomaterials on one side (*i.e.* hydrolytic, enzymatic *etc.* stabilities of such nanomaterials) but also the modification of silica walls favoring the adsorption of certain biomolecules.

Carbon Nanomaterials

Based on their bonding structures (sp2 or sp3), carbon nanomaterials can be classified into zero-dimensional fullerene, carbon dot, nanodiamond, one-dimensional carbon nanotube, and two-dimensional graphene. Each of them possesses exceptional physical and unique chemical properties. Graphene and carbon nanotube, on the other hand, exhibit mechanical, electrical, thermal, and biological properties which are suitable for a variety of applications [107]. As example, by virtue of their optoelectronic properties, carbon dot and nanodiamond are able to display natural fluorescence emission.

Single wall nanocarbon tubes (NTs) consist of a single cylindrical carbon layer with a diameter in the range of 0.4–2 nm [206], depending on the temperature at which they have been synthesized. In contrast, multiwall NTs are usually made from several cylindrical carbon layers with diameters in the range of 1–3 nm for the inner tubes and 2–100 nm for the outer tubes [207]. Depending on the type of synthesis, different types of carbon NTs with different properties can be synthesized [208]. The appropriate fabrication technique can be utilized according to the intended application of the carbon nanotubes. In drug delivery, single wall NTs are known to be more efficient than multiwall NTs. This is due to the one-dimensional structure of the single wall NTs and efficient drug-loading capacity because of its ultrahigh surface area [209]. Single wall NTs can also be used for imaging such as single-molecule fluorescence spectroscopy and Raman spectroscopy techniques. Multiwall NTs are known to be more useful than single wall NTs for thermal treatment [210]. This is due to the fact that the multiwall NTs release substantial vibrational energy after exposure to near infrared light. However, the primary drawback of carbon-based nanomaterials appears to be their toxicity. Experiments have shown that carbon NTs can lead to cell proliferation inhibition and apoptosis. Although they are less toxic than carbon fibers, the toxicity of carbon NTs increases significantly when carbonyl, carboxyl and/or hydroxyl functional groups are present on their surface. In order to promote the application of carbon NTs for drug delivery, researchers have functionalized their surface, rendering them benign [211]. Unfortunately, concerns that functionalized carbon NTs may revert back to a toxic state if the functional group is detached has limited the pursuit of using these modified carbon NTs for biomedical applications. One study of human lung tumor cells showed that carbon NPs are even more toxic than multiwall NTs or carbon nanofibers [212].

Nanoirons, Iron Based Magnetic and Superparamagnetic NPs (SPIONs)

Ferromagnetic materials exhibit a long- range ordering phenomenon at the atomic level which causes the unpaired electron spins to line up parallel with each other in a so called domain. Within the domain, the magnetic field is intense, but in a bulk sample the material will usually be unmagnetized because the many domains will be randomly oriented with respect to one another. Magnetic materials are classified by their magnetic behaviors demonstrated in the absence or presence of an externally applied magnetic field. They can be broadly categorized as diamagnetic, paramagnetic, ferromagnetic or ferrimagnetic. The magnetic properties of the materials are governed in part by the existence of multiple magnetic domains, or areas in which the magnetic moments are parallel and the magnetization is regionally uniform. However, as the size of a material is reduced below 100 nm, the same materials can exhibit significantly different properties. The particles become magnetized up to their saturation magnetization, and on removal of the magnetic field, they no longer exhibit any residual magnetic interaction. This property is size-dependent and generally arises when the size of nanoparticles is as low as 10–20 nm. At such a small size, these nanoparticles do not exhibit multiple domains as found in large magnets. They exhibit "a single magnetic domain" and act as a "single super spin" with a high magnetic susceptibility. On application of a magnetic field, these nanoparticles provide a stronger and more rapid magnetic response compared with bulk magnets with negligible remanence (residual magnetization) and coercivity (the field required to bring the magnetism to zero) [213, 214]. Such superparamagnetic behavior is very important for nanoparticle use as drug delivery vehicles. Generally, the transport of drug molecules to their target site can be realized under the influence of an applied magnetic field, but once the applied magnetic field is removed, the particles may easily be dispersed. No residual magnetism is retained by them and hence they are much less tending to agglomerate. Superparamagnetic iron oxide nanoparticles (SPIONs) are thus the most commonly known and used [107]. SPIONs are small synthetic-Fe_2O_3 (maghemite), Fe_3O_4 (magnetite) or a-Fe_2O_3 (hematite) particles with a core ranging from 1 nm to 20 nm in diameter. Moreover, mixed oxides of iron with transition metal ions such as copper, cobalt, nickel, and manganese, also exhibit superparamagnetic properties and may fall into the category of SPIONs because of a long- range ordering phenomenon at the atomic level which causes the unpaired electron spins to line up parallel with each other in a so called domain. Within this domain, the magnetic field is intense, but in a bulk sample the material will usually be unmagnetized because the many domains will be randomly oriented with respect to one another.

For instance, magnetic cobalt ferrite ($CoFe_2O_4$) nanoparticles can be prepared by hydrolysis of non-aqueous cobalt-iron(III) carboxylate solution with water. The magnetic properties of the resulting cobalt ferrite particles are dependent on the particle size, and the magnetization at a magnetic field of 0.8 MA/m and the coercive force were evaluated as 51.7 mWb_m/kg and 4.8 kA/m for the ferrite nanoparticles of 13 nm. So called spinel ferrites ($MeFe_2O_4$) such as Fe(II), Co(II), Ni(II), and Zn(II)), are of importance as magnetic nanomaterials which exhibit the superparamagnetic behavior namely due to their extremely small size.

The superparamagnetic properties of iron (II) oxide particles are used to guide drug microcapsules in target place by external magnetic fields. Another advantage is the ability to heat such particles after their internalization which is known as the hyperthermia effect [215]. For example, ferromagnetic Au-coated cobalt NPs (3 nm in diameter) were incorporated into the polymer walls of microcapsules. Subsequently, application of external alternating magnetic fields of 100–300 Hz and 1,200 Oe strength disturbed the capsule wall structures and dramatically increased their permeability to macromolecules. The benefits of superparamagnetic NPs over classical therapies are minimal invasiveness and minimal side effects. Conventional heating of a tissue by, for example, microwaves or laser light results in the destruction of healthy tissue surrounding the tumor. However, targeted paramagnetic particles provide a powerful strategy for localized heating of target cells. Encapsulation of a drug onto polymer- coated magnetic NPs may be realized in a variety of fashions. One of the most common methods for SPION delivery is to functionalize the SPION with antibodies that have a specific affinity for its intended target. Aside from antibodies, SPIONs may also be labelled with ligands that are attracted to receptors on cell surfaces; these targets have included folate, biphosphate and integrin receptors [216-218]. Many studies have shown optimal Enhanced Permeability and Retention (EPR) effect to occur using particles with diameters between 100 nm and 200 nm [219-221], which can typically be achieved by controlling the SPION diameter and/or coating thickness. Besides drug delivery systems superparamagnetic iron oxide nanoparticles are also promising candidates for bioimaging such as magnetic resonance imaging [7].

Composite Nanomaterials

To exclude composite from polymeric nanocarriers, nanomaterials composed of more than two nanomaterials without polymer encapsulation are considered as composite nanomaterials. Hybrid nanomaterials, consisting of different nanomaterials, have recently been investigated as promising platforms. By combining the specific function of each material, new hybrid nanocomposite materials can also be fabricated. Polymer layered silicate nanocomposites are relatively new class of nanoscale materials [222-229]. Owing to nanometer thick platelets in layered silicates, incorporation of such fillers strongly influences the properties of the composites at very low volume fractions because of much smaller interparticle distances and the conversion of a large fraction of the polymer matrix near their surfaces into an interphase of synergistically improved properties. As a result, the desired properties are usually reached at low filler volume fraction, which allows the nanocomposites to retain the macroscopic homogeneity and low density of the polymer. Polar polymers have been generally reported to achieve better filler dispersion owing to better match of the surface polarities of filler and polymers. On the other hand, the dispersion of filler in the non-polar polymers like polyetylene or polypropylene is challenging owing to the absence of any positive interactions between the organic and inorganic phases. However, either low molecular weight compatibilizers can be added to the system (or filler surface) to specifically modify additional chemical or physical processes. Liposomes and polymeric NPs are the two most widely studied drug delivery platforms, and attempts have been made to combine the advantages of both systems. Another example is the nanocomposite formulation, which consists of an oleic acid-coated magnetic nanocrystal core and a cationic lipid shell, which can be magnetically guided to deliver and silence genes [230].

Nanosimilars and Nanoparticle Application Pitfalls

A notable number of different nanotherapeutics have yet gained marketing authorization by the European Medicines Agency. Liposomal formulations, iron-based preparations, and drug nanocrystals in oral dosage forms are first generation nanotherapeutics, whose effectiveness and safety have been substantiated for long- term clinical use [139]. In the process of their approval, the applications were assessed under a conventional regulatory framework using established principles of benefit/risk analysis. At this point, it is particularly challenging to develop and evaluate "follow-on" nanotherapeutic products that could be authorized after the innovator product patent expiration. This specifically applies to nanotherapeutics that reach the systemic circulation and thereby determine the pharmacokinetics, biodistribution, and therapeutic performance. Any variations in the manufacturing process and the formulation may result in a generic product with different physicochemical properties (*e.g.* size, size distribution, surface properties, drug loading and release profile, aggregation status, and stability), which could lead to a different biopharmaceutical profile with a significant impact on patient safety and drug efficacy. In particular, different physicochemical properties can result in different ratios of free to nanoparticle-incorporated/associated drug, pharmacological effects, specific cell nanotherapeutic interactions, distributions, target organ uptake, immunological effects, or toxicities. To assess these differences, approaches more complex than the simple plasma concentration measurement are required. It is generally considered that the regulatory approach established for similar biological medicinal products (biosimilars) should be adopted for nanosimilars because such an approach includes the stepwise comparison of their quality, safety, and efficacy. The overall experience with the development of nanosimilars has been limited. Interestingly, although being the oldest nanotherapeutic platform, no liposomal nanosimilars have yet been approved in the EU. In addition, significant pitfalls have surfaced during the development of nanosimilar colloidal iron-based formulations with respect to innovator products [5]. For instance, significant differences in tissue distribution and toxicological profiles have been found among nanoirons that have different carbohydrate coatings. When similar differences in the toxicological profiles were observed for nanoparticle iron formulations with the same coating, they were previously ascribed to the differences in the manufacturing process [231]. However, recent advances in nanoscience will undoubtedly lead to the development of new, more complex nanotherapeutics, and therefore also to the need of appropriate regulatory approaches. Despite the assessment of existing nanotherapeutics which has yet provided valuable experience in certain evaluation aspects of emerging (next generation) nanotherapeutics, further scientific research is required to provide an exact evaluation of their quality, safety, and efficacy [126]. A general concern of nanoiron application is their potential for release of free, unbound iron, which may produce both short-term and chronic toxic effects due to oxidative stress [232]. Product characterization is required with respect to the labile iron present at the time of administration and the stability of the formulation in plasma. The differences in the production process or composition with reference to an innovator product may lead to diverse product characteristics that will impact on safety and therapeutic performance [233]. Assessing nanosimilars designed for a treatment and developed with reference to a nanosized colloidal innovator product is of great importance [233]. The overarching principle is that development of a specific product should take into account the critical product attributes of the innovator (reference medicinal product), and the evidence that supports its use when designing the quality, nonclinical and clinical development program. A particular challenge for these products is the scale of the clinical data required, which depends on how accurately the physicochemical and nonclinical characterization can be used to predict differences that could influence the efficacy and safety of the product [139]. The toxicity is an important issue which must be dealt with before nanosimilars enter in a widespread use in the drug delivery. No drug is free from side effects, and these side effects usually arise from nonspecificity in drug action. Furthermore, biocompatible and biodegradable polymeric nanoparticles have higher stability, tunable physiochemical properties and controllable drug profiles. They are self-assembled nanoparticles consisting of amphiphilic diblock copolymers and hence, drug encapsulation may be achieved by mixing the drug in the nanoparticle preparation mixture. Multiple drugs may be directly co-encapsulated into the hydrophobic polymeric core or by incorporating an additional media compartment into the nanoparticle for drug delivery or by covalent conjugation. All these approaches aid in temporal control of drug release associated with improved therapeutic efficacy, offering future promises for multidrug resistance reversal [234, 235]. There are also oxidized carbon nanotubes which are easily taken up by the much resistant leukemic K562 cells [236]. Various studies on the toxicological effects of nanoparticles have indicated that nanoparticles may act with the similar oxidative toxicological mechanisms and these effects are dependent on their physicochemical properties [237]. Solid lipid based nanoparticles are used for pH dependent drug release exploiting the acidic pH of multidrug resistance cells, while polymeric nanoparticles possess the unique property of versatile acid responsive drug release kinetics [238, 239]. Hydrogels are easily synthesized to enable stable platforms for protein conjugation and complement targeted delivery of drugs to resistant cells, such as conjugating with anti-Pgp antibody [240]. The importance of understanding nanobioconjugates such as iron nanobioconjugates at cellular level is needed because functionalized iron oxide particles using different covalently bound linkers exhibited different particle internalization rates and trafficking behavior [241, 242]. A common strategy to achieve targeting at cellular level is to functionalize the nanoparticle surface with biomolecules, whose receptors are over-expressed in the cells to be targeted. The recognition of specific receptors, even if achieved, does not necessarily imply that the targeted particle is trafficked in the same way as the targeting protein. It means that the functionalizing nanoparticle surface with ligands for receptors over-expressed in target cells does not necessarily result in higher accumulation of the nanotherapeutic inside the targeted organ. The promise of nanotechnology lies in the ability to engineer customizable nanoscale multi payload constructs loads [243]. These rationally designed nanovehicles may be equipped with an active-targeting element for enhanced selectivity [113]. As example, an area of intense research currently focuses on stepwise delivery into special intracellular compartments, primarily the nucleus [244-246], which contains the target of many small molecule anticancer drugs. When designing a drug delivery system with a selective target-activated release mechanism, one should consider the drug administration route, and the path of the drug(s) to the target cells, with respect to the physic-chemical and biological conditions (*e.g.* pH, ionic strength, enzymes present, and serum protein entrapment) and the barriers encountered (*e.g.* blood vessel endothelium such as blood-brain barrier, cell membrane, nuclear membrane and nuclear pores) [113, 247]. In the case of extracellular release, the drug delivery system is designed to liberate the drug(s) under the extracellular environment conditions in its target or microenvironment, and the drug has to be able to penetrate the cell membrane either passively by diffusion, or actively by specific transporters, or *via* receptor mediated endocytosis. In the latter case, the drug delivery vehicle carrying the drug payload has to enter the cell in an intact form, yet this drug payload must be successfully liberated in its active form under the lysosomal conditions (*e.g.* acidic pH and various hydrolytic enzymes presence) and then must diffuse or be transported into the cytosol.

Nanoparticle Biosensors

Nanoparticles have also numerous possible applications in biosensors. Functional nanoparticles (electronic, optical and magnetic) bound to biological molecules have been developed as biosensors to detect and amplify various signals. Some of the nanoparticle-based sensors include the acoustic wave, optical, magnetic and electrochemical control systems. When excited with an electromagnetic wave, such as light, most noble metal NPs produce an intense absorption and scattering due to the collective oscillation of the conduction electrons located at the NPs" surface. In the particular case of gold and silver NPs, the LSPR yields exceptionally high absorption coefficients and scattering properties within the UV/visible wavelength range that allows them to have a higher sensitivity in optical detection methods than conventional organic dyes, making them the perfect candidates for colorimetric biosensing applications [248, 249]. Moreover, their LSPR properties can be easily modulated according to their size, shape and composition [250, 251]. Typically, colloidal solutions of spherical gold NPs (<40 nm) present a red color with their LSPR band centered approximately at 520 nm, while spherical silver NPs present a yellow color with their LSPR band centered approximately at 420 nm [248]. Both metals can also be combined in an alloy or core-shell conformation, presenting a LSPR band that can vary within the wavelength limits of pure metal NPs LSPR bands. In the case of the core-shell conformation, a dual LSPR peak characteristic of each pure metal can be observed, depending on the thickness of the metallic shell [251]. These LSPR bands are usually weakly dependent on the size of the NPs and the refractive index of the surrounding media, but strongly change with inter-particle distance, for example aggregation of NPs leads to a pronounced color change as a consequence of the plasmon coupling between NPs and a concomitant red-shift of the LSPR absorption band peak [252]. Some of the colorimetric biosensors are based on the unspecific adsorption of biomolecules to non-functionalized noble metal NPs, while others are based on functionalized noble metal NPs for increased specificity. In other systems such as quartz crystal biosensors, the bioreaction generates a change in the mass, reordering the charges in the surface of the piezoelectric material and giving rise to a change in the resonant frequency of the microbalance. Its detection range and sensitivity are not quite as good as some other label-free sensor technologies (*e.g.* LSPR). The use of NP labels in sandwich assays can increase both the surface stress and the mass of the immune complex, allowing the increase in sensitivity of these electromechanical assays.

Typically, biosensors consist of a macromolecule that is immobilized on the surface of a signal transducer. As the macromolecule binds specifically to the ligand being detected, the signal transducer can measure a change due to the binding event. The transducer usually detects a change in resistance, pH, heat, light, or mass and then converts that data to an signal to be collected and processed. Recently, there has been an interest in biosensors with aptamers as bio-recognition elements. In comparison to antibodies, aptamer receptors have a number of advantages. The aptamer select ion process can be manipulated to obtain aptamers that bind to a specific region of the target and with specific binding properties under different binding conditions [253-257]. Due to the broad range of options in synthesis of magnetic nanoparticles there are possibilities to use external and inhomogeneous magnetic fields [258] or magnetoresistive sensors [259] which allow for magneto based monitoring of magnetically labelled biomolecules. For example, recent advances in synthesis and magnetotransport properties of magnetic Co nanoparticles have shown that magnetic Co nanoparticles self-assembled in nanoparticular monolayers revealing giant magnetoresistance but with additional features resulting from dipolar interactions between small domains. A spin-valve with one magnetic Co nanoparticular electrode was employed as a model to demonstrate that individual magnetic moments of Co nanoparticles can be coupled to a magnetic Co layer which in turn offers tailoring of the resulting giant magnetoresistance characteristics.

Nanotherapeutics in the Treatment of Malignancies

Novel nanotherapeutics are rapidly evolving and are gradually being implemented to overcome lack of water solubility, poor oral bioavailability, low therapeutic indices or dose-limiting toxicity of conventional drugs to healthy tissues. To improve the drug biodistribution, NPs have been designed for optimal size and surface characteristics in order to increase their circulation time in the bloodstream. They are able to carry and deliver active drug payloads to target cells by both passive and active targeting mechanisms, using ligands directed against selected determinants which are differentially over-expressed on the surface of target cells. NPs endowed with the ability to accumulate in malignant cells while for instance evading expulsion by efflux pumps are able to modulate the drug resistance and to enhance the drug efficacy. In addition to a malignant selective targeting moiety, NPs could be loaded with a diagnostic-aid, enabling the accurate localization. This will facilitate the harnessing of complementary therapeutic strategies, like radiotherapy, photodynamic therapy and surgery [113].

Multiple nanoparticle (NP)-based therapeutic systems have recently been proposed to overcome drug resistance by neutralizing, evading or exploiting various drug efflux pumps. A payload of therapeutic drug combinations for the selective targeting as well as the simultaneous overcoming of mechanisms of drug resistance are a subject of intense research efforts, some of which are expected to enter clinical trials in the near future or are currently undergoing preclinical *in vivo* studies and advanced stages of clinical evaluation with promising results [113]. Chemotherapy is a frequent treatment of cancer or hematological malignancies and remains the most *via*ble mode both for the heterogeneous group of hematological diseases and cancers. Unfortunately, a failure of the conventional treatment may be caused by many factors. Multidrug resistance in leukemia, like any other cancer, is a multifactorial effect governed by one or more resistance mechanisms, and therefore targeting a single mechanism is not rational. The use of nanotherapeutics emphasizes novel approaches in the treatment of neoplasma diseases by introducing multifunctional nanoplatforms and circumventing the limitation of the conventional treatment of leukemia and cancers [260]. Conventional chemotherapy failures that have commonly been met are not only due to insufficient transport or delivery of pharmacologically active component but also due to its insufficient concentration in targeted cells. This is because the drug must often be used in its suboptimal and/or intermittent dosing to allow others (not targeted cells) to rest or survive [261]. Moreover, many traditional chemotherapeutics have poor stability and aqueous solubility. Due to this limitation and despite their excellent biological activities, drugs are disregarded at early stages of the treatment and cleared by the monocytes and macrophages of the reticuloendothelial systems. [262]. The therapeutical quantity and the active state of the drug is absent, either due to the insufficient drug penetration through the interstitial space or the damage of healthy cells [143, 144, 263]. Moreover, the cell resistance to chemotherapy may be either innate (intrinsic), as a result of inherent genotypic characteristics or acquired (extrinsic), which develops on course of chemotherapeutic exposure [264, 265]. Therapeutic nanoparticles emerge as tumor-targeted drug delivery systems enhancing potency of chemotherapeutic drugs due to their unique characteristics but without causing life-threatening side effects. By virtue of their unique physicochemical and biological behavior, nanotherapeutics offer myriads of escape mechanisms for circumventing multidrug resistance. The overall merit of deploying nanoparticles is the drastic reduction of the half inhibitory concentration for most of the chemotherapy agents; accomplished by the marked intracellular accumulation and pharmacodynamics of drugs. Nanoconjugates optimize therapeutic doses of the conventional cytotoxic agents, reversing multidrug resistance while providing ample treatment response in patients undergoing chemotherapy. Various nanoconstructs have yet been developed which have revolutionized targeted therapy.

It is still necessary keep in mind that nanoconjugates injected intravenously would be exposed to blood plasma, containing in excess of 3,700 proteins, and many other complex biomolecules, which may bind competitively with the surface of the nanoparticles [56]. The nanoparticle surface can therefore be modified by the adsorption of biomolecules that lower the bare surface energy by a combination of particle surface charge compensation, water displacement, screening of hydrophobic patches and other mechanisms. Typical (unmodified) nanomaterial surfaces are having the surface free energies many times the thermal energy. A tightly bound immobile layer is formed by the proteins with higher affinities for the particle surface (the hard corona) and a weakly associated mobile layer (the soft corona) [266-268]. However in practice, even though biological fluids contain a large variety of proteins, typical final coronas contain much limited numbers and types of biomolecules [269-272]. Early binders (which are typically the more abundant proteins) are quickly displaced by the proteins with higher affinities for the particle surface, leading to a layer of strongly bound proteins which constitutes the final hard corona [196].

There are several nanoplatforms constructed either naturally or synthetically [273]. There are several nanotherapeutic approaches utilized, but the majority include liposomes for the multidrug resistance reversal in leukemia. Liposomal drug delivery system is an eloquent multifunctional design that overcomes the drug efflux mechanism of multidrug resistance tumors [274]. Following are a few successful nanoplatforms employed in combination therapy. Co-encapsulated liposomal nanocarriers are engulfed by non-specific endocytosis which enables crossing the cellular membrane, preventing the drugs from being expelled by efflux pumps. Due to these specifics, the liposomal nanocarriers have been widely studied for their potential in reversing multidrug resistance [275, 276]. A few examples were reported where NPs or liposomes were designed to harness the EPR effect to deliver anti-angiogenic drugs and to combat drug-resistant tumors [277, 278]. EPR is centered on the nano-properties of these vehicles, leading to their accumulation in the tumor microenvironment, where they can release therapeutic payload to interfere with the supportive environmental cross-talk. Various approaches are being developed that aim at targeting the microenvironment of or the cross-talk between malignant cells and their supporting stroma and/or vasculature. For instance, it is known that most mature B-cell malignancies still remain incurable; compelling evidence suggests that cross-talk with accessory stromal cells in specialized tissue microenvironments such as bone marrow and secondary lymphoid organs favors disease progression by promoting malignant B- cell growth, cell proliferation and drug resistance. Therefore, disruption of the cross- talk between malignant B-cells and their stroma is an attractive strategy for treating selected mature hematological malignancies of B-cell origin [279-281]. An example of this approach is the disruption of the interaction between bone marrow stromal cell-secreted, stromal cell-derived factor-1 (SDF-1/CXCL12) and its receptor CXCR4 [280]. Activation of CXCR4 induces leukemia cell trafficking and homing to the bone marrow microenvironment, where CXCL12 retains leukemia cells in close contact with marrow stromal cells that provide cell growth and drug resistance signals. CXCR4 antagonists can disrupt adhesive tumor-stroma interactions and mobilize leukemia cells from their protective stromal microenvironment, thereby rendering them more accessible or vulnerable to cytotoxic drugs. Therefore, targeting the CXCR4-CXCL12 axis is an emerging attractive therapeutic approach that is even explored in on-going clinical trials [280].

Stem Cells, Nanomedicines and Multidrug Resistance

Chemoresistance, a major cause of cancer treatment failure, is closely linked to so called cancer stem cells (cancer stem-like cells (CSCs) or tumor-initiating cells (TICs)). These cells possess unique properties, such as quiescence, mesenchymal morphology, increased DNA repair ability, overexpression of antiapoptotic proteins, drug efflux transporters and detoxifying enzymes [70]. Such properties, together with the favorable tumor microenvironment and hypoxic stability enable them to often escape the conventional pharmaceutical intervention. Having survived through the chemotherapy, they can give rise to metastases and recurrent tumors which increase their malignancy power and drug resistance. Unfortunately, alternative pathways are also activated and resistant mutations enable tumor cells to survive [90, 263, 280]. Consequently multidrug resistance may develop through a process of cross-resistance in which malignant cells mutate and acquire resistance to multiple structurally-related drugs either *via* the over-expression of multidrug transporters or through altered apoptosis [282, 283]. The concept of cancer stem cells has tremendous implications for the management of malignancy [284, 285]. Malignant stem cells have been identified in both hematological and solid malignancies, suggesting that the existence of these stem cells is a common feature of most malignancies. These cells can self-renew and thus regenerate more stem malignant cells. To cure cancer, malignant stem cells must entirely be eradicated; however, they are more chemoresistant compared to their progeny cells [286, 287]. Although the chemoresistance of the malignant stem cells can be overcome with high-dose chemotherapy followed by bone marrow transplantation, the high-dose chemotherapy may be associated with severe toxicity and high therapy-related mortality. Moreover, many elderly patients are not eligible for this treatment because of co-morbidities. Allogeneic hematopoietic stem cell transplantation may also result in severe graft-*versus*-host disease, or malignant stem cell contamination, thus commonly not offered to elderly patients. The development of therapeutic modalities eradicating the malignant stem cells is a critical current medical need yet unmet [288]. Several strategies for the treatment of CSCs/TICs are currently under investigation. A molecular targeting of deregulated signaling pathways, which may contribute to the uncontrolled growth, survival, invasion and resistance of these cancer-regenerating cells, to current cancer therapies is of great therapeutic interest [289]. Potential CSC drug targets include telomerase, anti-apoptotic factors, DNA repair enzymes and proteins, detoxifying enzymes and efflux transporters of the ABC superfamily, as well as distinct oncogenic cascades [290, 291]. Therapeutic innovations may also emerge from a better molecular understanding of the biology of CSCs/TICs and their microenvironment, which can create a niche favoring their survival and proliferation and protect these cells from chemotherapy-induced apoptosis [292]. CSCs/TICs constitute a small subpopulation of malignant cells that have the properties of tumor-initiating ability, self-renewal and differentiation, suggesting that they are responsible for tumor maintenance, recurrence and distant metastasis. An increasing body of evidence establishes that these CSC/TIC populations are more resistant to conventional cancer therapies, including chemotherapy and radiotherapy, than the bulk of non-CSC/TICs. Thus, elimination of these CSCs/TICs appears to be essential in order to cure malignant diseases [293, 294]. Cell surface markers expressed by CSCs/TICs are generally shared by normal somatic stem cells. However, it is hoped that subtle surface antigen differences as well as signaling pathway and metabolic alterations that distinguish between CSCs/TICs and normal somatic stem cells, may be exploited for the selective targeted delivery of NPs harboring therapeutic payloads. The latter can simultaneously overcome the inherent drug resistance of these CSCs/TICs as well as eliminate this crucial subpopulation, hence interfering with the self-renewal properties of the tumor [112]. To demonstrate the feasibility of using drug-loaded nanoparticles in hematological malignancy disease, engineered NPs with the abilities to bind with high affinity and specificity to acute myeloid leukemia stem cells were administrated [295]. Such targeting nanomicelles significantly improved the treatment outcomes. Three mechanisms were considered in this respect: (1) targeting stem cells through direct drug delivery into the interior of stem cells, (2) killing of leukemia cells throughout the body with chemotherapeutic drugs released from nanomicelles into the blood circulation, and (3) formulation of chemotherapeutic drugs inside the nanoparticles allowing administration of high-dose chemotherapy without increasing the toxicity [288].

Multidrug resistance can be surpassed by therapeutic synergism attained by combining two or more drugs administered as combination therapy. Conventional combination strategies are substituted by varying pharmacokinetic profiles of different drugs, which may result in varying rates of drug uptake and inadequate dosages at the target site. Nanoparticles carrying multiple drugs can overcome these inadequacies, ensuring spatial release of ratiometric drug doses aiming at multiple targets including metastasized sites in cancer [247]. The major modalities of malignancy drug resistance may be grouped into at least five categories: decreased drug influx, increased drug efflux predominantly *via* ATP-driven extrusion pumps frequently of the ATP-binding cassette (ABC) superfamily, activation of DNA repair, metabolic modification and detoxification or inactivation of apoptosis pathways with parallel activation of anti- apoptotic cellular defense modalities. Members of the ABC superfamily including P- glycoprotein (P-gp/ABCB1), multidrug resistance proteins (MRPs/ABCC) and breast cancer resistance protein (BCRP/ABCG2) function as ATP-driven drug efflux transporters, which form a unique defense against chemotherapeutics and numerous endo- and exotoxins. These pumps significantly decrease the intracellular concentration of a multitude of endogenous and exogenous cytotoxic agents which are structurally and mechanistically distinct [296-299]. Among the mechanisms of drug resistance, that are independent of drug efflux pumps, a prominent role is played by the activation of anti-apoptotic cellular defense modalities, including the overexpression of BCL2, a pro-survival, anti-apoptosis regulator and nuclear factor kappa B, a master transcription factor which controls the expression of various genes including those involved in suppression of the apoptotic response [300]. Nuclear factor kappa B is ubiquitously expressed in almost all animal cell types and is involved in cellular responses to stimuli such as stress, cytokines, free radicals, ultraviolet irradiation, oxidized LDL, and bacterial or viral antigens. Conversely, impaired regulation of nuclear factor kappa B (*i.e.* activation) and hence chronic inflammation have been recently shown to result in malignant transformation, autoimmune diseases, septic shock, viral infection, and improper immune development [301, 302].

Nanotherapeutics and Metal Nanoparticles in the Management of Microbial Infection

The successful translation of antimicrobial nanotechnologies can be facilitated by developing more clinically relevant animal models, identifying the mechanisms of microbial pathogenesis and new biomarkers, understanding the microenvironment of bacterial infection sites, and reducing the regulatory barriers. Nanotechnology presents a great opportunity for the development of fast, sensitive, specific, and cost-effective techniques for the diagnosis of microbial infection [303, 304] and using nanomaterials as vaccine adjuvants and/or delivery vehicles may evoke more efficient immune responses against microbial infection. Combination antibiotic therapy appears to hold a great deal of potential not only in tackling the existing mechanisms of drug resistance but in preventing its development in the first place [305]. Combining multiple drugs may result in higher potency and higher antimicrobial efficacy by additive or synergistic effects. Development of resistance to multiple agents, each of which has different mechanisms of action, requires multiple simultaneous gene mutations in the same bacterial cell, the chances of which are considered slim. Nanoparticles could facilitate the co-delivery of multiple antibiotics as well as the combination of antibiotics with antimicrobial nanomaterials, while avoiding synergistic/additive off-target toxicities from these combinations [306, 307]. The small size of nanoparticles facilitates their entering into host cells through endocytic/phagocytic pathways and subsequent release of the antibiotic payload to the localities of infection. A beneficial fact is that many different types of intracellular bacteria reside in the mononuclear phagocyte system composed primarily of monocytes and macrophages, which is also responsible for the clearance of nanoparticles administered in the body [308]. The potential impact of nanotechnology on microbial infectious diseases has already been demonstrated by the clinical approval of many nanotechnology-based products for the detection of bacterial infection, the delivery of antibiotics, and the development of medical devices with antimicrobial coatings. In addition, the coating of medical devices with antimicrobial nanomaterials has drastically reduced device-associated bacterial infection and biofilm formation, and enhanced wound healing when used in dressings. Vascular permeability at infection sites is another important issue that has not been well exploited but could be significant in the development of systemic nanoparticle drug delivery.

Interestingly, several features of infection-induced inflammation resemble the tumor environment in terms of pathological processes that result in the EPR effect [309]. It has also been demonstrated that a clinically significant EPR effect is present during infection by major pathogenic bacterial species [310]. Therefore, taking full advantage of the EPR effect in infection sites may lead to new nanotherapeutic approaches for the management of infectious diseases. Antibiotics and vaccines have unequivocally been one of the important medical innovations and their introduction in medical practice has significantly reduced the morbidity and mortality from infectious diseases [311]. However, antimicrobial resistance has also invalidated many routinely used antibiotics and still represents a serious public health concern [312]. Some bacterial strains are capable of developing or acquiring resistance to multiple antimicrobial agents [313]. Another an important issue in antimicrobial therapy is the treatment of chronic infections, which are often caused by the formation of microbial biofilms and/or by intracellular microbes [314]. Biofilm is a matrix consisting of extracellular polymeric substance that accumulates and surrounds bacterial cells [315] acting as a barrier of diffusion by trapping and degrading antibiotic molecules. Bacteria in a biofilm can exhibit up to 1,000 times more resistance to multiple antibiotics than planktonic bacteria [316]. As the effect of some antibiotics depends on their interaction with surface components of the bacterium stronger antibacterial effects could be obtained through the polyvalent effect achieved by conjugating multiple antibiotic copies on nanomaterial surface [317-319]. Nanoparticles exert immunostimulatory effects that could be attributed to several different mechanisms such as better tissue penetration, preferential uptake, depot effect, repetitive antigen and adjuvant display on the particle surface, nanoparticle- mediated escape of antigens into the cytosol [320, 321]. Some nanoparticle systems also show adjuvant properties by themselves. The inherent antibacterial properties of some metals and metal oxides have been known for centuries, causing them to be utilized extensively as bactericidal substances in infection control [322, 323]. Silver (Ag) nanoparticles have been the most intensely studied in this respect. They are capable of killing both Gram-positive and Gram-negative bacteria and are effective against many drug-resistant microbes, such as *Pseudomonas aeruginosa* ampicillin-resistant *Escherichia coli O157:H7*, and erythromycin-resistant *Streptococcus pyogenes* [324].

The bactericidal activity of nanoparticles can be related to several mechanisms. The silver nanoparticles may directly interact with microbial cells. Silver ions are associated with respiratory electron transport from oxidative phosphorylation, which inhibits respiratory chain enzymes or interferes through covering permeability to phosphate and protons, *e.g.*, interrupting transmembrane electron transfer, oxidizing cell components, disrupting, penetrating the cell covering or reactive oxygen species (ROS), or dissolving heavy metal ions that cause damage. Besides Ag, other metal nanomaterials have also been considered for antimicrobial treatment, including tellurium (Te) and bismuth (Bi). Interestingly, Te nanoparticles were reported to exhibit higher antibacterial activity and lower toxicity than Ag nanoparticles [325]. Carbon-based nanomaterials such as single-walled carbon nanotubes, multi-walled carbon nanotubes, and fullerene have also been utilized in antimicrobial applications [326]. These nanomaterials may exert antibacterial activity through cell membrane damage upon direct contact or through their photothermal/photodynamic properties upon irradiation [327, 328]. However, their limited application as antimicrobial therapeutic agents may be partially due to safety concerns [329, 330]. An important advantage of using metal and metal oxide nanoparticles as antimicrobial agents is that it is difficult for microbes to develop resistance to them. The primary reason is that metals/metal oxides have multiple modes of action, which significantly reduces the chance for microbes to gain resistance unless multiple mutations occur simultaneously [331-333]. Cationic antimicrobial peptides are short amphipathic peptides present in virtually every life form as nature's antibiotics, and are potent against a broad spectrum of microbes and multidrug resistant bacteria [334]. Cationic antimicrobial peptides are also considered an integral part of the ancestral system of defense against microbial infection in higher multicellular organisms. The antimicrobial properties are generally based on the cationic and hydrophobic nature of cationic antimicrobial peptides, which can physically damage negatively charged microbial membranes. Although hundreds of cationic antimicrobial peptides sequences have been identified, their antimicrobial application is limited by the inherent drawbacks of cationic peptides, including cytotoxicity (*e.g.*, hemolysis), enzymatic instability, and immune surveillance [335]. Despite the enormous potential of targeted nanoparticles, their translation into clinical development has faced considerable challenges, such as the difficulty of identifying highly selective targeting ligands, the development or adaptation of simple, robust, and reproducible processes that can facilitate scale-up and manufacturing, or the rapid optimization of the biophysicochemical properties of nanoparticles for maximal efficacy [336, 337]. Thus, many efforts have been focused on developing nanoparticles through self-assembly and high-throughput processes to facilitate their screening and optimization as well as subsequent scale-up and manufacturing [338, 339].

Nanomedicine, Public Health Policy and Nanomaterials in the Environment

Around 100 nanomedicine products have already been approved for clinical use ranging from drug delivery and imaging to implantable biomaterials and medical devices [340]. More than 10 nanoparticle-based products have been recently marketed for bacterial diagnosis, antibiotic delivery, and medical devices [303]. Nanotherapeutic regulations and nanomedicine safety Specific safety issues relating to novel nanomaterials have been widely reviewed [341, 342]. Definition of nanomedicine safety is distinct from the broader issues relating to nanomaterial toxicology as the product is designed for use at a specific dose, with a specific route and frequency of administration, and in the context of a specific clinical setting/patient population. Safety and efficacy must be established under these conditions as well as the environmental impact of all medicinal products [343]. The biointerface is very important in determining the safety profile of nanomedicines and may influence pharmacokinetics and tissue/cell distribution. Nonclinical studies aimed to define the biodistribution and metabolic fate of nanomedicines/nanomaterials with respect to their acute and long-term safety and immunological properties represent also a challenging and important issue [139]. Particular challenges are becoming apparent in two distinct areas. First, in relation to the evaluation of "follow-on" nanomedicine products now beginning to arise as first generation products come off patent. Such products are described as nanosimilars and they have recently become the subject of discussion [344]. These are new nanomedicines that are claimed to be "similar" to a reference (originator/innovator) nanomedicine that has been granted a marketing authorization (licensing). In order to demonstrate similarity there is a need for stepwise comparability studies to generate evidence substantiating the similar nature, in terms of quality, safety and efficacy of the nanosimilar and the chosen reference (originator/innovator) nanomedicine. As nanomedicines differ significantly in their complexity and nature, a case-by-case or product class-specific approach for their evaluation might be necessary. It is important to stress that any drug developed to be comparable to the reference patented drug (brand, originator or innovator) product must demonstrate equivalence in terms of quality, safety and efficacy before a market authorization can be granted. Given the degree of complexity of many nanomedicine products, special scientific considerations is needed to ensure this equivalence of performance. Second, it is becoming clear that recent advances in nanoscience are bringing novel opportunities to bring off material at nanoscale sizes and this is leading to the creation of even more complex, hybrid structures by both new top-down fabrication and bottom-up manufacturing techniques. With regard to novels pharmaceuticals imaging agents and especially to combination products termed usually "next-generation" nanomedicines accountable regulatory consideration is required prior to approval of market authorization [345]. Nanotherapeutics should be considered as a novel entity in disease treatment and as such should extensively be studied for their pharmacokinetic and pharmacodynamic characteristics, despite the first generation nanomedicines, including liposomal formulations, iron-based preparations and drug nanocrystal technologies in oral dosage forms, have been established as safe and effective for many years [118, 346-348].

A better understanding of various novel concepts of their behavior in the biological microenvironment, their influence on cell membrane, pH variation in the intracellular and extracellular niche, transport dynamics, active/passive delivery mechanisms *etc.* will help towards improving the application and efficiency of nanotherapeutics [260]. Currently, the European Medicines Agency has evaluated 11 marketing authorization applications for nanomedicines, out of which eight have been authorized and three have been withdrawn. The assessment reports are publicly available [349]. During the development of 12 nanomedicines scientific advice was requested from the European Medicines Agency to enhance the chances for a positive development outcome. In addition, the European Commission has granted orphan status for 10 nanomedicines under development. Approximately 48 nanomedicines and nanoimaging agents are under clinical development (Phase I–III) in Europe, with others progressing through earlier stages of drug discovery and nonclinical development. In the USA, approximately 70 cancer clinical trials are on-going involving nanomedicines [350]. Many nanomedicines approved as products and/or undergoing development include, as an integral component of their design, either a noncovalent or covalent coating. Such coatings have been typically used to minimize aggregation and improve stability, or to minimize rapid clearance by the reticuloendothelial system after intravenous administration. They have also been used to improve hematocompatibility and limit antigenicity. Both are phenomena that can arise due to the inherent physicochemical nature of the product or the surface adsorption of biomolecules from the physiological environment to which they are exposed. It has been noted that the interface between a nanomedicine coating and the biological environment depends on the proposed clinical application and the route of delivery. Current research suggests that a much wider range of approaches to control the surface of nanomedicines will be employed in the future and more sophisticated surface modifications designed to facilitate cell- specific targeting will be in clinical development [139]. Some of the current issues that require consideration during the development and lifecycle of coated nanomedicines designed for parenteral administration are under discussion. It is evident that the coating material is a critical determinant of the biological behavior of the product. The physicochemical nature of the coating, the uniformity of surface coverage, and the coating stability can all govern pharmacokinetics and biodistribution. In some clinical studies involving certain coated nanomedicines, infusion-related reactions have been observed. It has been shown that this may be mediated by biomolecular interactions, complement activation, the presence of circulating antibodies against the coating/product and/or cell surface receptor activation/binding. Even in some cases, the coating material itself may elicit unexpected biological responses not observed for either the coating material or the unmodified nanocarrier alone. These observations underline the need for careful consideration of the impact of the coating material and the nature of its linkage (covalent or noncovalent). These factors will be critical in the determination of final physicochemical properties and the biological behavior of the product, with the potential to impact product safety and efficacy. Many first-generation nanomedicines have already been clinically established as successful medicines. The current regulatory framework is sufficiently robust to evaluate and authorize pharmaceuticals as specific scientific expertise is involved and experts are consulted. By establishing expert groups, organizing international collaboration and convening stakeholders in public workshops, the European Medicines Agency continues to adapt and prepare the regulatory system for the development, evaluation and successful market entry of nano-similars and second- generation nanomedicines. In order to assess potential needs for regulatory requirements for the evaluation and assessment of nanomedicines, in 2006, the European Medicines Agency created a cross-agency Nanomedicine Expert Group. In 2009, this was further expanded *via* establishment of the International Regulators Subgroup on Nanomedicines, an initiative jointly launched by the medical regulatory agencies of the EU (European Medicines Agency), USA (US FDA), Japan (Ministry of Health, Labor and Welfare) and Canada (Health Canada) and the First International Workshop on Nanomedicines took place in London in September 2010 [351]. An on-going goal is the early identification of gaps in scientific knowledge as nanotechnologies bring not only opportunities to improve current treatments but also the potential to change the way in healthcare and disease treatment [352]. Early and interactive communication between regulators and stakeholders is required to establish the trust that avoids risks misperception in this new field. European Medicines Agency initiatives aim at enabling dialog and guidance regarding the safe development and clinical use of nanomedicine products. In order to bring together regulatory experience with first-generation nanomedicines the European Medicines Agency's main Scientific Committee for Human Medicinal Products established a multidisciplinary expert group on nanomedicines in 2011. The remit of the expert group is to provide scientific input for well-founded Scientific Advice, collate the current regulatory reflection for the safe approval of nanosimilar nanomedicines, and to monitor the uptake of technical advances in the development and evaluation of upcoming new nanomedicines, for example, innovative block copolymer micelle products that are being developed as nanomedicines to assist targeted drug delivery and control drug release [137]. A series of recommendation protocols has been drafted on principles for the development and evaluation of nanosimilars developed with reference to first generation nanomedicines on principles to be considered when generating supporting evidence to changes made to the manufacture and control of these products [145, 233] and on principles for the development and evaluation of emerging nanomedicines (second-generation nanomedicines) progressing towards first-in-man studies [353].

Many biological processes typically operate at the nanoscale and many biological entities, from proteins/peptides, DNA/RNA, ATP, to viruses, are nanosized. Nanoparticles may be released into the environment directly from the organism (*e.g.*, mucoprotein exudates, dispersion of virus particles) or during the degradation of biological matter (*e.g.*, humic and fulvic acids) and may occur freely in natural waters and soils [354, 355]. Despite the fact that living organisms have clearly been exposed to nano-sized natural materials throughout evolution and evolved to deal with natural nanomaterials and their fluctuations over millennia, it is not known how organisms will currently cope with high discharges of myriads of novel anthropogenic nanomaterials into the environment [9, 356]. It is worth to note that under certain circumstances, naturally occurring NPs can be toxic to life forms. In this respect serious concerns should be considered about manufactured NPs. Many natural NPs are transient in the environment, often disappearing through dissolution, or becoming larger through particle growth or aggregation. On the contrary, manufactured NPs may persist a long time because they can be stabilized by capping or fixing agents [357]. They may contain chemically toxic components in concentrations or structural forms that do not occur naturally and might also inadvertently additively incorporate toxic material from the environment [354]. Therefore, nanotechnology is not exclusively a positive concept, and like any other emerging field it is not without risk. There are no proven toxicity screening methods to evaluate a long term nanoparticle behavior in the living systems. Research of nanoparticle actions is still at the beginning to declare how these may impact public health safety, human or animal life and environment in the future [260]. Robust methodology is essential to ensure long-term safety/risk management. Specific tools may be needed to ensure adequate product characterization and manufacturing control (on a product-by-product basis), and to define those critical product attributes that determine pharmacokinetics, body distribution, nonclinical safety and enable pharmacological proof of concept. Determination of stability *in vitro* and *in vivo* should be studied as well as an in depth understanding of physicochemical functionality of the nanosystem in other systems [139].Recent research has called for food and ecosystem-scale research in the areas of nanomaterials [358, 359]. It has been declared that some important differences exist with regard to the application or release of nanomaterials in the environment. Widely used herbicides and insecticides are applied in support of crop production and are applied only at specific times during the growing season, or in response to observed pest outbreaks. Conversely, nanomaterials commonly tend to enter the environment continuously [360]. On the contrary to conventional agrochemicals, nano-encapsulated agrochemicals can be designed in such a way that they possess all necessary properties such as effective concentration with high solubility, stability and effectiveness, time controlled release in response to certain stimuli, enhanced targeted activity and less ecotoxicity with safe and easy mode of delivery, avoiding repeated applications [361-364]. Properly functionalized nanocapsules provide better penetration through cuticle and allow slow and controlled release of active ingredients on reaching the target weed.

Current agrochemicals are conventionally applied to crops by spraying and/or broadcasting [365]. Usually only a very low concentration of chemicals which is much below the minimum effective concentration required has reached the target site of crops due to problems such as leaching of chemicals, degradation by photolysis, hydrolysis and by microbial degradation. Hence, a repeated application is necessary to have an effective control which might cause some unfavorable effects such as soil or water pollution. On the other hand, the application of nanoparticle technology in plant pathology may target specific agricultural problems in plant–pathogen interactions and may provide new ways for plant protection. The agro-nanotechnology holds the promise of controlled release of nanochemicals and site targeted delivery of various macromolecules needed for improved plant disease resistance, efficient nutrient utilization and enhanced plant growth. Despite processes such as nanoencapsulation and safer handling of agrochemicals with less exposure to the environment that guarantees ecoprotection the uptake efficiency and effects of various nanoparticles on the growth and metabolic functions varied differently among plants, leading to specific impacts. Chromosomal aberrations and other undesirable outcomes were, however also described [366-369]. Similarly, while the risk assessment of conventional substances is based on the notion that their chemical identity governs the biological effects of a substance, there is general agreement that the toxicity of NM must be determined by a set of nanoparticle characteristics. Given the substantial diversity within each group of nanomaterials and the complexity of nanosystems (*e.g.* stability of dispersions under different conditions), a large number of property combinations need to be considered in order to assess the overall hazard of a single material type. Currently, it is widely accepted that the hazard/risk assessment of NM can sufficiently be addressed on a case-by-case basis [36]. There is also a need to consider physicochemical properties of NM that could generate unexpected hazardous biological outcomes. In order to assess engineered nanomaterial hazard, reliable and reproducible screening approaches are needed to test the basic materials as well as nano-enabled products [370]. Due to the diversity of NM in regard to type of material, shape, size, surface modification, crystalline structure, *etc.*, for efficient eco-toxicological testing it is of special importance to define specific groups which may comprise NM with comparable fate, behavior and effects in the environment [371]. To obtain comprehensive information on the properties and effects of a given group, a sufficient number of members should be given. In the context of REACH, the choice of test systems and test organisms to be used for eco-toxicological testing of conventional chemicals depends on the production volume of the respective substance. In the case of NM, however, it seems more appropriate to base the testing scheme on expected concerns based on routes of exposure, findings on physical-chemical properties and on fate studies, rather than production volume. Such an adaptation of the testing scheme seems permissible since the REACH guidance documents [372] allow adapting the testing strategy in specific cases. A higher exposure does not always equate with a higher risk. It remains to be determined whether nanotriggers comparable with those used for the testing of conventional chemicals can be applied for ecotoxicity testing of NM or whether other modifications are required [36]. For NM, expression of the environmental concentration on mass basis may be less suitable, and particle number concentration instead of mass concentration may be an appropriate alternative. To address these important issues, further knowledge on NM exposure and fate is necessary. Especially one group of compounds and their fate in the environment (*i.e.* pharmaceutical and personal care products) has captured an interest of both scientific community and the general public. There are approximately 4.000

pharmaceuticals on the market [373] and most of these compounds were primarily developed for human or veterinary pharmaceutical uses and, as such, are designed to be biologically active. Toxicology of some pharmaceutical and personal care products is basically understood, but their indirect secondary effect, including their metabolites on ecological structure and function of ecosystems is largely unknown. Similarly, the number of anthropogenic compounds that occur in ecosystems today is in the thousands, many at trace concentrations [360]. Currently, research on personal care products covers two major topics, (1) describing the occurrence and concentration of various personal care products, or (2) single- species examinations of the effect of personal care products on mortality, growth, or reproduction. Notably absent from the literature is an ecosystem-based approach for assessing the effect of personal care products in the special environment. As a result, it is known very little about how these compounds, alone or in combination, might affect ecosystem functions [374]. A research on the interactions between personal care products and functional properties of special ecosystems has the potential to provide information on a contemporary environmental problem, as well as advance understanding of how various ecosystem functions respond to anthropogenic stresses. Most of the nanocompounds are not regulated as pollutants and especially novel personal care products are supposed to be continuously developed in the future [360]. The personal care products and pharmaceuticals are introduced to the environment through municipal and other waste water [375-377]. The personal care products and/or their metabolites and transformation products are transported to surface waters where they contribute to the contaminant load from agricultural and industrial activities. The emission of nanopharmaceuticals and nanomaterials from human activities to the environment can be expected to increase, either due to an increase in life expectancy or increase in living standard and affordability of pharmaceuticals [378]. Several personal care products have even been shown to be acutely toxic [14, 379, 380].

A relatively large proportion of investigated pharmaceuticals (approx. 30%) are predicted to be intrinsically toxic. For instance, one-fifth of antibiotics studied were predicted to be very toxic to algae. Sixteen percent of the antibiotics were found to be extremely toxic ($EC_{50} < 0.1$ mg L^{-1}) and 44% were predicted to be very toxic ($EC_{50} < 1$ mg L^{-1}) to daphnids. Almost 1/3 were predicted to be very toxic to fish, and more than half are toxic ($EC_{50} < 10$ mg L^{-1}) to fish [381]. Although colloidal silver (or nanosilver) has been used as an form of antibiotics since ancient times, it has recently found many more applications in medicine, optics, sensing, painting and cosmetics [382-384]. According to the Project on

Emerging Nanotechnologies more than 400 consumer products have been claimed to contain nanosilver [385]. Taking into consideration such increasing use in commercial products, the potential for the release of nanosilver into the environment and its effects on environmental health are of serious concern. Nanosilver undergoes a variety of transformations in environmental and biological media A number of studies focused on the exposure, environmental fate, and *in vivo* and *in vitro* nanosilver toxicities. [386-395]. One of the most widely known lesions caused by nanosilver is argyria, although the mechanism causing the lesion is still unknown. Other studies have widely investigated nanosilver toxicities supposed to be associated with oxidative stress and damage of cellular systems by free radicals [396, 397]. However, the environmental fate, state of agglomeration or aggregation, and dissolution in environmental and biological media are dependent on how nanosilver is prepared, what types of surface coating are used, and the conditions under which they are used. As a result, environmental fate is highly variable within a range of surface functionality that can make the same material biocompatible or biohazardous.

Many chemicals enter the environment through sewage treatment plants. Two major groups of chemicals in sewage waste water are pharmaceuticals and surfactants [398]. In spite of the extensive research on surfactants there is still serious concern for their environmental impact in the forthcoming era. One reason is their extensive use. Another reason is that there is a continuous development of new surfactants for various applications, however the new ones receive less scientific interest than the older ones previously. The effects of individual personal care products have been widely studied [376, 398, 399], however, mixtures of personal care products have been studied to a limited extent [14, 400]. Pharmaceuticals are bioactive compounds but commonly resistant to biodegradation which can make them problematic when they are released in the environment. A majority of toxic compounds are believed to act through a baseline toxicity which is assumed being caused by hydrophobicity-dependent and nonspecific interaction with biological membranes and membrane associated proteins [401, 402]. In this respect, the mode of action of some nanoparticles might be expected similar as chemicals that are sufficiently lipophilic to accumulate in the lipid or the lipid-aqueous interface of biological membranes [402]. This effect leads to disruption of membrane functions. The baseline toxicity of several biocides and their interaction with other organic contaminants are not fully known. It has been observed that compounds causing the same type of effect or having a similar baseline toxicity can be additive [14]. The combined effects of chemicals have been studied by application of the two widely used prediction models for additive effects, the concentration addition and independent action prediction models. Combined effects of pharmaceuticals or biocides have been shown to be mostly additive [14, 403-405]. Similar effects of nanotherapeutics or nanobiocides are still supposed to be evaluated. Current toxicological approaches to assess hazards of NM are either based on methods adopted from classical toxicology or on alternative methods. These approaches do not fully consider the unique aspects of NM, as mentioned above: 1) different forms of a given NM in different biological media; 2) uptake/absorption, distribution, corona formation and elimination/deposition (absorption, distribution, metabolism, excretion; since, for most NM, metabolism, unlike corona formation, does not play a major role); 3) functional impacts at the organ and cellular levels. Slow elimination and persistence are main drivers for bioaccumulation. Similar to conventional hydrophobic persistent chemicals that are resistant to environmental and biological degradation, NM have the potential to accumulate in humans and biota (food chain) when exposure takes place on a daily or other time basis. Presently, namely the structure and dynamics of protein corona are considered to be key to the nanoparticle's rate of uptake and transport into cells and final subcellular localization [406, 407]. A number of proteomics methods to identify the nature, composition and dynamics of the biomolecules associated with NM has been developed [408]. Without suitable information on the potential for NM to bioaccumulate, it is not possible to carry out higher-tier human health or environmental risk assessment as well as to derive environmental quality standards [409].

Risk reduction in the exposed human or environmental populations should focus on limiting or avoiding exposures that trigger toxicological responses as well as implementing safer design of potentially hazardous engineered nanomaterials. There is considerable debate about how to proceed with engineered nanomaterial toxicity testing, with the major discussion points focusing on toxicological endpoints to assess [406, 410, 411]. It is of paramount importance for adequate testing to ensure that the testing conditions applied (including NM characteristics and exposure conditions) are appropriate to assess the risk under relevant real-life exposure situations. One aspect is that the physical-chemical properties of the nanomaterial during testing are known, either by analytical techniques or standardized techniques when suspending or dispersing NM for toxicity testing. Since a multitude of different NM in different exposure scenarios is expected, it will not be possible to perform comprehensive testing of all NM in all relevant scenarios. Instead, the testing must be targeted to the actual concerns for a given NM making use of realistic exposure scenarios. Moreover as mentioned above, a testing strategy should include possibilities for the grouping of NM (*e.g.* by applying a "read-across" methodology some tests could be waived based on a categorization of NM), and should also aid the grouping concept itself (*e.g.* the testing strategy should provide information that is relevant for grouping) [36].

It is the responsibility of the manufacturer to establish the toxicity profile for the engineered nanoparticles to reassure the regulators, workers, and consumers that the engineered nanoparticles can be used safely. Similarly, scientists and researchers should generate the toxicity profile for the engineered nanoparticles, specifically prior to be established the beneficial outcomes [412]. To date, no new type of toxic effect has been described for NM (*i.e.* no effects which have not been observed with any other substance or particle before). A concern- driven NM toxicity testing scheme is thus consisting of three main tiers. Tier 1 includes a concern assessment based on the physic-chemical characteristics of the specific NM and relevant scenarios for potential exposure, depending on its envisaged use. Such relevant exposure scenarios should take into account realistic dose levels that for example workers or consumers might be exposed to and whether nanomaterial exposure is likely to be in the aggregated or agglomerated states. For NM of concern identified in tier 1, tier 2 focuses on identifying their basic toxicological concerns, while tier 3 provides options to study specific endpoints of concern in more detail. The need for the performance of specific tests in the last tier is determined by the combined results of the first two tiers [413, 414]. Tiers 2 and 3 each consist of three parts, the so-called toxicity domain, the decision making process and the description of options for testing within each domain. Each toxicity domain reflects a specific toxicity end point or type of testing (*e.g.* repeated dose, biokinetics) to be addressed and contains a number of options for testing or non-testing (*e.g.* grouping/waiving) that can be selected in the decision-making process based on the concerns identified in the preceding tier(s) [36]. The European Union (EU) Commission in its second regulatory review on nanomaterials, has reinforced its continuous commitment to promoting research and development in this area [37]. Warranting the safety of nanotechnological products is seen as a crucial element in ensuring that the benefits of the new technology can be fully exploited. In parallel to on-going basic scientific research, a number of different international initiatives have also been launched in this respect, such as the OECD Working Party on Manufactured Nanomaterials [415-418]. Of particular importance for the above discussions are the OECD WPMN Steering Groups (SG) 6 and 7. SG 6 deals with risk assessment approaches, such as "read-across" methodologies, and SG 7 addresses the use of alternative test methods and integrated testing strategies for NM hazard assessment. This SG has proposed a new short term inhalation study (STIS) for NM testing [420] and has compiled a list of *in vitro* methods that might be used for NM human hazard identification. Furthermore, it has initiated a similar discussion for environmental impacts. Comparable structures of the human and environmental hazard identification frameworks aim at allowing a better integration of human health and ecological risk assessment into one coherent strategy and at facilitating its implementation for regulatory purposes [36]. Up until the end of 2012, the Commission has funded a total of 46 nanosafety projects representing a total EU investment of 130 million EUR [37]. To facilitate the formation of consensus on nanotoxicology, the Commission has requested EU-funded nanosafety projects to join forces through the so-called NanoSafety Cluster [37]. The members of this initiative have assigned themselves the goal to identify key areas of nanosafety research which are likely to be of special significance in the coming years. In delineating a timeframe for meeting such challenges, the NanoSafety Cluster has taken on the deadline of 2020, a year that has been spelled out by the Commission both in the Europe 2020 strategy for smart, sustainable and inclusive growth and for the upcoming Research Framework Program Horizon 2020 [36, 37, 417]. The NanoSafety Cluster vision 2020 foresees the development of a concern-driven guidance for investigating potential risks of NM. This will enable focused research on NM that may be of particular concern based on (expected) exposure levels, exposure routes and material properties. By advocating that animals are only used for crucial and focused studies minimizing the numbers of animals used and the distress inflicted upon them. This approach further meets the provisions of EU Directive 2010/63/EU [417]. This vision 2020 of the NanoSafety Cluster should be realized in the context of existing international and national chemical regulatory frameworks taking into account existing OECD Test Guidelines (TG) and adapted as appropriate to take into account the specific properties of NM, as recommended by the OECD Chemicals Committee [418]. Considering the large number of existing and emerging novel nanoformulations in the future, it is particularly relevant to speed up their risk assessment process on one side but also, on the other side, to reduce testing costs and animal use. This seems a difficult task because of the 3Rs principle to replace, reduce and refine animal testing [419] implemented in EU Directive 2010/63/EU on the protection of animals used for scientific purposes [417]. However, such integrated processes are highly demanded because it should consolidate all available information (including both exposure and toxicity information) and identify the involvement of 3Rs principles that best addresses these concerns [420, 421]. A variety of methods and approaches for assessing the eco-toxicity and bioaccumulation of NM is already available, and further ones are under development. Underlining the fundamental difference between human health hazard and environmental hazard assessment, (*i.e.* different types of test methods and purposes of testing) novel laboratory tests and environmental simulation studies are proposed. For assessing the impact of NM on the environment, both eco-toxicity and bioaccumulation should be determined making use of standardized laboratory test systems (similar to the standardized procedures established for conventional chemicals (*e.g.*, according to OECD test guidelines). In these test systems, standardized and frequently artificial test media are employed. In this sense a detailed characterization of NM effects and verification of results from such tests (taking into consideration environmentally relevant groups of organisms and realistic environmental exposure scenarios) should take place [422]. Naturally, the complex interactions between NM, test media and organisms require amendment of existing test guidelines. Some NM will not raise specific concerns, but will readily dissolve or form other, larger, particles (*e.g.* granular, biopersistent NM with low surface area and reactivity). Many NM are likely to possess properties raising concerns for direct toxic effects. These concerns may be general or linked to specific uses or exposures. For instance, NM solely used in non-spray cosmetic sunscreen lotions may be of low concern for inhalation toxicity for consumers. The output of tier 1 might be a prioritization list for the testing of an individual NM in tier 2 (basic testing) and tier 3 (specific testing). The testing strategy is largely based on modified OECD standard methods extended with selected new methods, which partly were specifically developed for NM, such as short- term nanotoxicity studies [423, 424]. Further development of toxicological methods and increasing knowledge on NM adverse effects, their mechanism of action and the relevance of test results for the situation in humans will alter the choice of test methods and improve NM toxicity testing strategies. Furthermore, the information obtained in standardized test systems can be limited with regard to determining modes of action of the NM. Such information may, however, be required for classifying NM, and it is likely to be a prerequisite for reduced, efficient testing aimed at a practical assessment of NM environmental hazard. Suitable and novel test systems beyond standardized testing, however, remain to be defined with regard to NM safety or toxicity [425, 426]. Further research is required to define which test systems are applicable for screening or confirmatory testing of the compliance of standardized laboratory tests using more specific tests with environmentally relevant organisms. Depending on the potential environmental concerns identified in basic laboratory testing, further test species, endpoints and food webs (interspecies transfer) might be studied in environmental simulation studies applying more realistic environmental exposure scenarios. Since NM bioavailability is highly dependent on the chosen test conditions, care has to be taken when simulating environmentally relevant exposure [426, 427].

CONCLUSIONS

Nanotherapeutics have already yielded significant breakthroughs in the detection, diagnosis and treatment, and appear to have the potential to yield many more, with extensive and focused routes of research planned for the future and the possibility of nanotechnology-based disease prevention. However, nanotechnology must be thoroughly understood and its risks assessed if it is to be developed safely.

Advanced therapy can only be achieved through the rational design of nanotherapeutics that lead to the development of nanoplatforms of particular size, shape, and surface properties. It is highly important with regard to safety and efficacy of nanomedicines that can be influenced by minor variations in multiple parameters and need to be carefully examined, particularly in context of the biodistribution, targeting to intended sites, and potential immune toxicities. Nanomedicines may present additional development and regulatory considerations compared with conventional medicines. While there is currently a lack of regulatory standards in the examination of nanoparticle-based agents, appropriate efforts should be made in this respect. When assessing the risk to patients, it is important to take into consideration that preclinical trials of nanodrugs may be less indicative of human risks than trials of standard medicines. Nanotherapeutics can be less invasive than conventional diagnosis and treatment methods. This leads to shorter recovery times and a decreased risk of infection, and these advantages in turn should lead to a reduction in cost and improved life expectancy and quality. Despite the potential of nanotherapeutics only a relatively small number of nanoparticle-based medicines have been approved for clinical use. Nanomaterials has a great potential for their application in many fields. However, it is not without risk and nanomaterials may present greater risks than bulk materials because of their greater relative surface and unique properties. Furthermore, their potential to cause harm is harder to predict. The lack of information in this respect about how nanomaterials may impact safety, health and the environment has thus raised serious concerns.

CONFLICT OF INTEREST

The author confirms that he has no conflict of interest to declare for this publication.

ACKNOWLEDGEMENTS

Supported by IGA, Ministry of Health, Czech Republic, NT 14060-3/2013.

REFERENCES

[1]   Zhang L, Gu FX, Chan JM, Wang AZ, Langer RS, Farokhzad OC. Nanoparticles in medicine: therapeutic applications and developments. Clin Pharmacol Ther 2008; 83: 761-9.

[2]   Venditto VJ, Szoka FC, Jr. Cancer Nanomedicines: So Many Papers and So Few Drugs! Adv Drug Deliv Rev 2013; 65(1): 80-8.

[3]   Wang S, Su R, Nie S, Sun M, Zhang J, Wu D, Moustaid-Moussa N. Application of nanotechnology in improving bioavailability and bioactivity of diet-derived phytochemicals. J Nutr Biochem. 2014; 25(4): 363-76.

[4]   Jahn MR, Andreasen HB, Futterer S, *et al.* A comparative study of the physicochemical properties of iron isomaltoside 1000 (Monofer), a new intravenous iron preparation and its clinical implications. Eur J Pharm Biopharm 2011; 78(3): 480-91.

[5]   Borchard G, Fluhmann B, Muhlebach S. Nanoparticle iron medicinal products - Requirements for approval of intended copies of non-biological complex drugs (NBCD) and the importance of clinical comparative studies. Regul Toxicol Pharmacol 2012; 64(2): 324-28.

[6]   Ferrari M. Cancer Nanotechnology: Opportunities and Challenges. Nature Reviews Cancer 2005; 5: 161-71.

[7]   Talekar M, Kendall J, Denny W, Garg S. Targeting of Nanoparticles in Cancer: Drug Delivery and Diagnostics. Anticancer Drugs 2011; 22(10): 949-62.

[8]   Kim KY. Nanotechnology Platforms and Physiological Challenges for Cancer Therapeutics. Nanomedicine NBM 2007; 3(2): 103-10.

[9]   Baker TJ, Tyler CR, Galloway TS. Impacts of metal and metal oxide nanoparticles on marine organisms. Environ Pollut 2014; 186:257-71.

[10]   Kolpin DW, Furlong ET, Meyer MT, *et al.* Pharmaceuticals, hormones, and other organic wastewater contaminants in US streams, 1999- 2000: a national reconnaissance. Environ Sci Technol 2002; 36:1202-11.

[11]   Kim Y, Choi K, Jung JY, Park S, Kim PG, Park J. Aquatic toxicity of acetaminophen, carbamazepine, cimetidine, diltiazem and six major sulfonamides, and their potential ecological risks in Korea. Environ Int 2007; 33: 370-5.

[12]   Focazio MJ, Kolpin DW, Barnes KK, *et al.* A national reconnaissance for pharmaceuticals and other organic wastewater contaminants in the United States—II) untreated drinking water sources. Sci Total Environ 2008; 402:201-16.

[13]   Fick J, Soderstrom H, Lindberg RH, Phan C, Tysklind M, Larsson DGJ. Contamination of surface, ground, and drinking water from pharmaceutical production. Environ Toxicol Chem 2009; 28(12): 2522-7.

[14]   Backhaus T, Porsbring T, Arrhenius A, Brosche S, Johansson P, Blanck H. Single-substance and mixture toxicity of five pharmaceuticals and personal care products to marine periphyton communities. Environ Toxicol Chem 2011; 30:2030-40.

[15]   Kortenkamp A. Low dose mixture effects of endocrine disrupters: impli-cations for risk assessment and epidemiology. Int J Androl 2008; 31(2): 233-7.

[16]   Munaron D, Tapie N, Budzinski H, Andral B, Gonzalez JL. Pharmaceuticals, alkylphenols and pesticides in Mediterranean coastal waters: results froma pilot survey using passive samplers. Estuar Coast Shelf Sci 2012; 114: 82-92.

[17]   Petersen K, Heiaas HH, Tollefsen KE. Combined effects of pharmaceuticals, personal care products, biocides and organic contaminants on the growth of Skeletonema pseudocostatum. Aquat Toxicol. 2014; 150: 45-54.

[18]   Fu PP. Introduction to the Special Issue: Nanomaterials— Toxicology and medical applications. J Food Drug Anal 2014; 22 (1): 1-2.

[19]   Maiorano G, Sabella S, Barbara Sorce S, *et al.* Effects of cell culture media on the dynamic formation of protein–nanoparticle complexes and influence on the cellular response. ACS Nano 2010; 4:7481-91.

[20]   Lesniak A, Fenaroli F, Monopoli MP, Åberg Ch, Dawson KA, Salvati A. Effects of the presence or absence of a protein corona on silica nanoparticle uptake and impact on cells. ACS Nano 2012;6: 5845-57.

[21]　Deng ZJ, Liang M, Monteiro M, Toth I, Minchin RF. Nanoparticle-induced unfolding of fibrinogen promotes Mac-1 receptor activation and inflammation. Nat. Nanotechnol., 2011; 6: 39-44.

[22]　Gea C, Dua J, Zhaoa L, *et al*. Binding of blood proteins to carbon nanotubes reduces cytotoxicity. Proc Natl Acad Sci USA. 2011; 108: 16968-73.

[23]　Wang F, Yu L, Monopoli MP, *et al*. The biomolecular corona is retained during nanoparticle uptake and protects the cells from the damage induced by cationic nanoparticles until degraded in the lysosomes. Nanomedicine 2013; 9: 1159-68.

[24]　Hu W, Peng Ch, Lv M, *et al*. Protein corona-mediated mitigation of cytotoxicity of graphene oxide. ACS Nano 2011;5: 3693-700.

[25]　Oberdörster G. Safety assessment for nanotechnology and nanomedicine: concepts of nanotoxicology. J Int Med 2009; 267(1): 89-105.

[26]　Landsiedel R, Ma-Hock L, Kroll A, *et al*. Testing metal-oxide nanomaterials for human safety. Adv Mat 2010; 22:2601-27.

[27]　Lankveld DP, Oomen AG, Krystek P, *et al*. The kinetics of the tissue distribution of silver nanoparticles of different sizes. Biomaterials 2010; 31:8350-61.

[28]　Dan M, Wu P, Grulke EA, Graham UM, Unrine JM, Yokel RA. Ceria engineered nanomaterial distribution in and clearance from blood: size matters. Nanomed 2012; 7(1): 95-110.

[29]　Taniguchi N. On the basic concept of nanotechnology. Proc Intl Conf Prod Eng Tokyo. Part II, Japan Soc. Precis. Eng. 1974.

[30]　Ganguly A, George JJ, Kar S, Bandyopadhyay A, Bhowmick AK. Rubber Nanocomposites based on miscellaneous nanofillers. in Current Topics in Elastomers Research, CRC Press, 2008.

[31]　Mamalis A, Vogtländer L O, Markopoulos A. Nanotechnology and nanostructured materials: trends in carbon nanotubes. Precis Eng 2004;28(1):16-30.

[32]　Li J, Ma PC, Chow WS, To CK, Tang BZ, Kim JK. Correlations between percolation threshold, dispersion state and aspect ratio of carbon nanotubes. Adv Funct Mater 2007;17(16):3207-15

[33]　Kostoff RN, Koytcheff, Raymond G. Lau, Clifford G.Y. Global Nanotechnology Research Literature Overview. Technol Forecast Soc Change 2007; 74(9): 1733-47.

[34]　Ishiyama N. Nanomedicine: nanocarriers shape up for long life Nat Nanotechnol 2007; 2: 203-4.

[35]　Commission Recommendation of 18 October 2011 on the definition of nanomaterial (Text with EEA relevance) (2011/696/EU)L 275/38 Official Journal of the European Union 20.10.2011

[36]　Oomen AG, Bos PMJ, Fernandes TF, *et al*. Concern-driven integrated approaches to nanomaterial testing and assessment - report of the NanoSafety Cluster Working Group 10. Nanotoxicology 2014; 8(3):334-48.

[37]　Anon. 2012. Communication from the Commission to the European Parliament, the Council and the European Economic and Social Committee. Second regulatory review on nanomaterials. COM (2012)572 final. Brussels, 3 October 2012. http://eur-lex.europa. eu/LexUriServ/LexUriServ.do?uri=COM:2012:0572:FIN:en:PDF.

[38]　Albanese A, Tang PS, Chan WC. The effect of nanoparticle size, shape, and surface chemistry on biological systems. Annu Rev Biomed Eng 2012; 14: 1-16.

[39]　Murthy SK. Nanoparticles in modern medicine: state of the art and future challenges. Int J Nanomedicine 2007; 2(2): 129-141

[40]　Mura S, Couvreur P. Nanotheranostics for personalized medicine. Adv Drug Deliv Rev 2012; 64(13): 1394-1416.

[41]　Rizzo LY, Theek B, Storm G, Kiessling F, Lammers T. Recent progress in nanomedicine: therapeutic, diagnostic and theranostic applications. Curr Opin Biotechnol 2013; 24(6): 1159-66.

[42]　Wagner W, Huüsing B, Gaisser S, Bock A. Nanomedicine: drivers for development and possible impacts. Centre ECJR, Studies IfPT, trans: European Commission Joint Research Centre, Institute for Prospective Technological Studies; 2006.

[43]　Webster TJ. Projections for nanomedicine into the next decade: but is it all about pharmaceuticals? Int J Nanomedicine 2008; 3(1): i.

[44]　Huber FX, McArthur N, Heimann L, *et al*. Evaluation of a novel nanocrystalline hydroxyapatite paste Ostim in comparison to Alpha-BSM - more bone ingrowth inside the implanted material with Ostim compared to Alpha BSM. BMC Musculoskelet Disord. 2009; 10:164.

[45]　Fleischer CC, Payne CK. Nanoparticle-cell interactions: molecular structure of the protein corona and cellular outcomes. Acc Chem Res 2014; 47(8): 2651-9.

[46] Handbook of immunological properties of engineered nanomaterials: Chapter 11, b1429:Salvador-Morales C, Sinn RB. Complement Activation. Available from: http://www.complementsystem.se/assets/upload/files/SalvadorMorales%20Rev%20Method%2013%2 0O%20ED.pdf

[47] Zolnik BS, Gonzalez-Fernandez A, Sadrieh N, Dobrovolskaia MA. Nanoparticles and the immune system. Endocrinology 2010; 151(2): 458-65.

[48] Dobrovolskaia MA, McNeil SE. Immunological properties of engineered nanomaterials Nat Nanotechnol 2007; 2(8): 469-78.

[49] Dobrovolskaia MA, Aggarwal P, Hall JB, McNeil SE. Preclinical studies to understand nanoparticle interaction with the immune system and its potential effects on nanoparticle biodistribution Mol Pharm 2008; 5(4): 487-95.

[50] Bosi S, Feruglio L, Da Ros T, *et al.* Hemolytic effects of water-soluble fullerene derivatives. J Med Chem 2004; 47(27): 6711-15.

[51] Domanski DM, Klajnert B, Bryszewska M. Influence of PAMAM dendrimers on human red blood cells. Bioelectrochemistry. 2004; 63(1-2):189-91.

[52] Malik N, Wiwattanapatapee R, Klopsch R, *et al.* Dendrimers: relationship between structure and biocompatibility *in vitro*, and preliminary studies on the biodistribution of 125I-labelled polyamidoamine dendrimers *in vivo*. J Control Release 2000; 65(1-2):133-48.

[53] Cohen JJ. Apoptosis: the physiologic pathway of cell death. Hosp Pract (Off Ed) 1993; 28(12): 35-43.

[54] Schroit AJ, Madsen JW, Tanaka Y. *In vivo* recognition and clearance of red blood cells containing phosphatidylserine in their plasma membranes. J Biol Chem 1985; 260(8): 5131-38.

[55] Greish K, Thiagarajan G, Herd H, *et al.* Size and surface charge significantly influence the toxicity of silica and dendritic nanoparticles. Nanotoxicology 2012; 6(7): 713-23.

[56] Mahon E, Salvati A, Baldelli Bombelli F, Lynch I, Dawson KA. Designing the nanoparticle-biomolecule interface for "targeting and therapeutic delivery". J Control Release 2012; 161(2): 164-74.

[57] Li SD, Huang L. Stealth nanoparticles: High density but sheddable PEG is a key for tumor targeting. J Control Release 2010; 145: 178-81.

[58] Zheng M, Li ZG, Huang XY. Ethylene glycol monolayer protected nanoparticles: synthesis, characterization, and interactions with biological molecules. Langmuir 2004; 20: 4226-35.

[59] Yang A, Liu W, Li Z, Jiang L, Xu H, Yang X. Influence of polyethyleneglycol modification on phagocytic uptake of polymeric nanoparticles mediated by immunoglobulin G and complement activation. J Nanosci Nanotechnol 2010; 10: 622-28.

[60] Owens DE, Peppas NA. Opsonization, biodistribution and pharmacokinetics of polymeric nanoparticles. Int J Pharm 2006; 307: 93-102.

[61] Vonarbourg A, Passirani C, Saulnier P, Benoit JP. Parameters influencing the stealthiness of colloidal drug delivery systems. Biomaterials 2006; 27: 4356-73.

[62] Furumotoa K, Ogawaraa K, Nagayamaa S, *et al.* Important role of serum proteins associated on the surface of particles in their hepatic disposition. J Control Release 2002; 83: 89-96.

[63] Hamad I, Al-Hanbali O, Hunter AC, Rutt KJ, Andresen TL, Moghimi SM. Distinct polymer architecture mediates switching of complement activation pathways at the nanosphere–serum interface: implications for stealth nanoparticle engineering. ACS Nano 2010; 4: 6629-38.

[64] Kim HR, Andrieux K, Delomenie C, *et al.* Analysis of plasma protein adsorption onto PEGylated nanoparticles by complementary methods: 2-DE, CE and Protein Labon- chip® system, Electrophoresis. 2007; 28: 2252-61.

[65] Harris L, Batist G, Belt R, *et al.* Liposome-encapsulated doxorubicin compared with conventional doxorubicin in a randomized multicenter trial as first-line therapy of metastatic breast carcinoma. Cancer 2002; 94: 25-36.

[66] Müller RH, Rühl D, Runge SA. Biodegradation of solid lipid nanoparticles as a function of lipase incubation time. Int J Pharm 1996; 122: 115-21.

[67] Olbrich C, Muller RH. Enzymatic degradation of SLN—effect of surfactant and surfactant mixtures Int J Pharm 1999; 180: 31-9.

[68] Heurtault B, Saulnier P, Pech B, Proust JE, Benoit JP. A novel phase inversion-based process for the preparation of lipid nanocarriers. Pharm Res 2002, 19. 875-80.

[69] Peer D, Karp JM, Hong S, Farokhzad OC, Margalit R, Langer R. Nanocarriers as an emerging platform for cancer therapy. Nat Nanotechnol 2007; 2: 751-60.

[70] Vinogradov S, Wei X. Cancer stem cells and drug resistance: the potential of nanomedicine. Nanomedicine 2012; 7(4): 597-615.

[71] McNeil SE. Nanoparticle therapeutics: a personal perspective. Wiley Interdiscip Rev Nanomed Nanobiotechnol 2009; 1(3): 264-71.

[72] Tiwari M. Nano cancer therapy strategies. J Cancer Res Ther 2012; 8(1): 19-22.

[73] Buse J, El-Aneed A. Properties, engineering and applications of lipid-based nanoparticle drug-delivery systems: current research and advances. Nanomedicine (Lond) 2010; 5:1237-60.

[74] Huang HC, Barua S, Sharma G, Dey SK, Rege K. Inorganic nanoparticles for cancer imaging and therapy. J Control Release 2011;155:344-57.

[75] Maham A, Tang Z, Wu H, Wang J, Lin Y. Protein-based nanomedicine platforms for drug delivery. Small 2009;5:1706-21.

[76] Poon Z, Lee JA, Huang S, Prevost RJ, Hammond PT. Highly stable, ligand-clustered "patchy" micelle nanocarriers for systemic tumor targeting. Nanomedicine 2011;7:201-9.

[77] Choi HS, Liu W, Misra P, Tanaka E, Zimmer JP, Itty Ipe B. Renal clearance of quantum dots. Nat Biotechnol 2007;25:1165-70.

[78] Jiang J, Tong X, Morris D, Zhao Y. Toward photocontrolled release using light-dissociable block copolymer micelles. Macromolecules 2006; 39:4633-40.

[79] Fomina N, McFearin C, Sermsakdi M, Edigin O, Almutairi A. UV and near-IR triggered release from polymeric nanoparticles. J Am Chem Soc 2010;132:9540-2.

[80] Shen Y, Tang H, Radosz M, Van Kirk E, Murdoch WJ. pH-responsive nanoparticles for cancer drug delivery. Methods Mol Biol 2008; 437:183-216.

[81] Gao W, Chan JM, Farokhzad OC. pH-Responsive nanoparticles for drug delivery. Mol Pharm 2010; 7:1913-20.

[82] You JO, Auguste DT. Nanocarrier cross-linking density and pH sensitivity regulate intracellular gene transfer. Nano Lett 2009; 9:4467-73.

[83] Lee ES, Na K, Bae YH. Doxorubicin loaded pH-sensitive polymeric micelles for reversal of resistant MCF-7 tumor. J Control Release 2005;103:405-18.

[84] Cohen JL, Almutairi A, Cohen JA, Bernstein M, Brody SL, Schuster DP. Enhanced cell penetration of acid-degradable particles functionalized with cell-penetrating peptides. Bioconjug Chem 2008;19:876-81.

[85] Matsumura Y, Maeda H. A new concept for macromolecular therapeutics in cancer chemotherapy: mechanism of tumoritropic accumulation of proteins and the antitumor agent SMANCS. Cancer Res 1986; 46: 6387-92.

[86] Wang X, Li J, Wang Y. *et al.* HFT-T, a targeting nanoparticle, enhances specific delivery of paclitaxel to folate receptor-positive tumors. ACS Nano 2009; 3: 3165-74.

[87] Choi CHJ, Alabi CA, Webster P, Davis ME. Mechanism of active targeting in solid tumors with transferrin-containing gold nanoparticles. Proc Natl Acad Sci USA 2010; 107: 1235-40.

[88] Ruoslahti E, Bhatia SN, Sailor MJ. Targeting of drugs and nanoparticles to tumors. J Cell Biol 2010; 188: 759-68.

[89] Uskokovic V. Challenges for the Modern Science in its Descent Towards Nano Scale. Curr Nanosci 2009; 5: 372-89.

[90] Hu CM, Zhang L. Nanoparticle-based Combination Therapy Toward Overcoming Drug Resistance in Cancer. Biochem Pharmacol 2012; 83(8): 1104-11.

[91] Nel AE, Xia T, Meng H, *et al.* Nanomaterial toxicity testing in the 21st century: use of a predictive toxicological approach and high-throughput screening. Acc Chem Res 2013; 46 (3): 607-21.

[92] Dekkers S, Bouwmeester H, Bos PMJ, Peters RJB, Rietveld AG, Oomen AG. 2012. Knowledge gaps in risk assessment of nanosilica in food: evaluation of the dissolution and toxicity of different forms of silica. Nanotoxicology 2013; 7(4): 367-77.

[93] Dekkers S, Krystek P, Peters RJB, *et al.* Presence and risks of nanosilica in food products. Nanotoxicology 2011; 5(3): 393-405.

[94] EFSA. European Food Safety Authority - Guidance on the risk assessment of the application of nanoscience and nanotechnologies in the food and feed chain. EFSA J 2011; 9(5): 2140.

[95]   Barenholz Y. Doxil—the first FSA approved nano-drug: lessons learned. J Control Release 2012;160(2):117-34.

[96]   Gabizon A. Pegylated liposomal doxorubicin: metamorphosis of an old drug into a new form of chemotherapy. Cancer Invest 2001;19(4): 424-36.

[97]   Gabizon A, Shmeeda H, Barenholz Y. Pharmacokinetics of pegylated liposomal doxorubicin: review of animal and human studies. Clin Pharmacokinet 2003;. 42(5): 419-36.

[98]   Gabizon A, Catane R, Uziely B *et al.* Prolonged circulation time and enhanced accumulation in malignant exudates of doxorubicin encapsulated in polyethylene-glycol coated liposomes. Cancer Res 1994; 54(4): 987-92.

[99]   Malam Y, Loizidou M, Seifalian AM. Liposomes and nanoparticles: nanosized vehicles for drug delivery in cancer. Trends Pharmacol Sci 2009; 30(11):592-9.

[100]   Gabizon A, Goren D, Horowitz AT, Tzemach D, Lossos A, Siegal T. Long-circulating liposomes for drug delivery in cancer therapy: a review of biodistribution studies in tumor-bearing animals. Adv Drug Deliv Rev 1997; 24(2-3): 337-44.

[101]   Waterhouse DN, Tardi PG, Mayer LD, Bally MB. A comparison of liposomal formulations of doxorubicin with drug administered in free form: changing toxicity profiles. Drug Saf 2001; 24(12):. 903-20.

[102]   Dromi S, Frenkel V, Luk A, *et al.* Pulsed-high intensity focused ultrasound and low temperature— sensitive liposomes for enhanced targeted drug delivery and antitumor effect. Clin Cancer Res 2007;13(9): 2722-7.

[103]   Hawkins MJ, Soon-Shiong P, Desai N. Protein nanoparticles as drug carriers in clinical medicine. Adv Drug Deliv Rev 2008; 60(8):876-85.

[104]   Funkhouser J. Reinventing pharma: the theranostic revolution. Curr Drug Discov 2002; 2:17-19.

[105]   Wang LS, Chuang MC, Ho JA. Nanotheranostics - a review of recent publications. Int J Nanomedicine 2012; 7:4679-95.

[106]   Liu Z, Cai W, He L, *et al. In vivo* biodistribution and highly efficient tumour targeting of carbon nanotubes in mice. Nat Nano 2007; 2:47-52.

[107]   Sefah K, Shangguan D, Xiong X, O'Donoghue MB, Tan W. Development of DNA aptamers using Cell-SELEX. Nat Protoc 2010; 5:1169-85.

[108]   Zhu G, Zhao Z, Chen Z, Zhang X, Tan W. Noncanonical self-assembly of multifunctional DNA nanoflowers for biomedical applications, J Am Chem Soc 2013; 135:16438-45.

[109]   Stephen W. Morton SW, Lee MJ, Deng ZJ, *et al.* A nanoparticle-based combination chemotherapy delivery system for enhanced tumor killing by dynamic rewiring of signaling pathways. Sci Signal, 2014: 7: ra44.

[110]   Zhu G, Mei L, Tan W. Nanomedicine. From bioimaging to drug delivery and therapeutics, nanotechnology is poised to change the way doctors practice medicine. The Scientist. August 1, 2014. www.the-scientist.com

[111]   Bharali DJ, Mousa SA. Emerging Nanomedicines for Early Cancer Detection and Improved Treatment: Current Perspective and Future Promise. Pharmacol Ther 2010; 128(2):324-35.

[112]   Jabir NR, Tabrez S, Ashraf GM, Shakil S, Damanhouri GA, Kamal MA. Nanotechnology- based Approaches in Anticancer Research. Int J Nanomedicine 2012; 7: 4391-408.

[113]   Shapira A, Livney YD, Broxterman 85. Shapira A, Livney YD, Broxtermanc HJ, Yehuda G. Assaraf GY. Nanomedicine for targeted cancer therapy: Towards the overcoming of drug resistance. Drug Resistance Updates 2011; 14: 150-63.

[114]   Wheeler HE, Maitland ML, Dolan ME, Cox NJ, Ratain MJ. Cancer Pharmacogenomics: Strategies and Challenges. Nat Rev Genet. 2013; 14(1): 23-34.

[115]   Perche F, Torchilin VP. Recent trends in Multifunctional Liposomal Nanocarriers for Enhanced Tumor Targeting. J Drug Deliv 2013; 2013: 705265.

[116]   Heller M, Heller MJ. Nanotechnology for Cancer Diagnostics and Therapeutics. Nanomedicine NBM 2006; 2(4): 301.

[117]   Pepić I, Hafner A, Lovrić J, Pirkić B, Filipović-Grčić J. A nonionic surfactant/chitosan micelle system in an innovative eye drop formulation. J Pharm Sci 2010; 99(10): 4317-25.

[118]   Junghanns JU, Müller RH. Nanocrystal technology, drug delivery and clinical applications. Int J Nanomedicine 2008; 3(3): 295-309.

[119]   Zhang XX, Eden HS, Chen X. Peptides in cancer nanomedicine: drug carriers, targeting ligands and protease substrates. J Control Release 2012; 159(1): 2-13.

[120]   Crielaard BJ, Lammers T, Schiffelers RM, Storm G. Drug targeting systems for inflammatory disease: one for all, all for one. J Control Release 2012; 161(2): 225-34.

[121]   Hafner A, Lovrić J, Voinovich D, Filipović-Grčić J. Melatonin-loaded lecithin/chitosan nanoparticles: physicochemical characterisation and permeability through Caco-2 cell monolayers. Int J Pharm 2009; 381(2): 205-13.

[122]   Pavelić Z, Škalko-Basnet N, Filipović-Grčić J, Martinac A, Jalšenjak I. Development and *in vitro* evaluation of a liposomal vaginal delivery system for acyclovir. J Control Release 2005; 106(1-2): 34-43.

[123]   Pepić I, Lovrić J, Filipović-Grčić J. How do polymeric micelles cross epithelial barriers? Eur J Pharm Sci 2013; 50(1): 42-55.

[124]   Hafner A, Lovrić J, Pepić I, Filipović-Grčić J. Lecithin/chitosan nanoparticles for transdermal delivery of melatonin. J Microencapsul 2011; 28(8): 807-15.

[125]   Lammers T, Kiessling F, Hennink WE, Storm G. Drug targeting to tumors: principles, pitfalls and (pre-) clinical progress. J Control Release 2012; 161(2): 175-87.

[126]   Hafner A, Lovrić J, Lakoš GP, Pepić I. Nanotherapeutics in the EU: an overview on current state and future directions. Int J Nanomedicine 2014; 9: 1005-23.

[127]   Filipović-Grčić J, Mrhar A, Junginger H. Thematic Issue on Emerging nanopharmaceuticals for nonparenteral application routes. Eur J Pharm Sci 2013; 50(1):1.

[128]   European Commission/ETP Nanomedicine. Roadmaps in Nanomedicine Towards 2020: Joint European Commission/ETP Nanomedicine expert report. European Commission/ETP Nanomedicine; 2009. [Accessed December 17, 2013]. Available from: http://www.etp-nanomedicine.eu/public/press-documents/publications/etpn-publications/091022_ETPN_Report_2009.pdf.

[129]   Chen H, Khemtong C, Yang X, Chang X, Gao J. Nanonization strategies for poorly water-soluble drugs. Drug Discov Today 2011; 16(7-8): 354-60.

[130]   Xia D, Cui F, Piao H, *et al.* Effect of crystal size on the *in vitro* dissolution and oral absorption of nitrendipine in rats. Pharm Res 2010; 27(9): 1965-76.

[131]   Desai PR, Date AA, Patravale VB. Overcoming poor oral bioavailability using nanoparticle formulations - opportunities and limitations. Drug Discov Today Technol 2012; 9(2): e87-e95.

[132]   Lu Y, Park K. Polymeric micelles and alternative nanonized delivery vehicles for poorly soluble drugs. Int J Pharm 2013; 453: 198-214.

[133]   Tziomalos K, Athyros VG. Fenofibrate: a novel formulation (Triglide) in the treatment of lipid disorders: a review. Int J Nanomedicine 2006; 1(2): 129-47.

[134]   Florence AT, Attwood D. Physicochemical Principles of Pharmacy. Fourth edition. Pharmaceutical Press; London, UK: 2006. p. 22.

[135]   Lim EK, Jang E, Lee K, Haam S, Huh YM. Delivery of cancer therapeutics using nanotechnology. Pharmaceutics 2013; 5: 294-317.

[136]   Mattheolabakis G, Rigas B, Constantinides PP. Nanodelivery strategies in cancer chemotherapy: biological rationale and pharmaceutical perspectives. Nanomedicine 2012; 7(10): 1577-90.

[137]   Matsumura Y. Polymeric micellar delivery systems in oncology. Jpn J Clin Oncol 2008; 38(12): 793-802.

[138]   Oerlemans C, Bult W, Bos M, Storm G, Nijsen JF, Hennink WE. Polymeric micelles in anticancer therapy: targeting, imaging and triggered release. Pharm Res 2010; 27(12): 2569-89.

[139]   Ehmann F, Sakai-Kato K, Duncan R, *et al.* Next-generation nanomedicines and nanosimilars: EU regulators' initiatives relating to the development and evaluation of nanomedicines. Nanomedicine (Lond). 2013; 8(5): 849-56.

[140]   Alexis F, Rhee JW, Richie JP, Radovic-Moreno AF, Langer R, Farokhzad OC. New Frontiers in Nanotechnology for Cancer Treatment. Urol Oncol 2008; 26(1): 74-85.

[141]   Misra R, Acharya S, Sahoo SK. Cancer nanotechnology: application of nanotechnology in cancer therapy. Drug Discov Today 2010; 15(19-20): 842-50.

[142]   European Medicines Agency/Committee for Medicinal Products for Human Use. Joint MHLW/EMA reflection Paper on the development of Block Copolymer Micelle Medicinal Products. London: European Medicines   Agency;   2013.   Available   from: http://www.ema.europa.eu/docs/en_GB/document_library/Scientific_guideline/2013/02/WC 00138390.pdf.

[143] Jain RK, Stylianopoulos T. Delivering Nanomedicine to Solid Tumors. Nat Rev Clin Oncol 2010; 7(11): 653-64.

[144] Brannon-Peppas L, Blanchette JO. Nanoparticle and Targeted Systems for Cancer Therapy. Adv Drug Deliv Rev 2004; 56(11): 1649-59.

[145] European Medicines Agency. Reflection paper on the data requirements for intravenous liposomal products developed with reference to an innovator liposomal product.www.ema.europa.eu/docs/en_GB/document_library/Scientific_guideline/2013/03/WC50014 0351.pdf

[146] European Medicines Agency. Guideline on similar biological medicinal products containing biotechnology-derived proteins as active substance: non-clinical and clinical issues. www.ema.europa.eu/docs/en_GB/document_library/Scientific_ guideline/2009/09/WC500003920.pdf

[147] Langer R. New methods of drug delivery. Science 1990; 249: 1527-33.

[148] Mozafari MR, Pardakhty A, Azarmi S, Jazayeri JA, Nokhodchi A, Omri A. Role of nanocarrier systems in cancer nanotherapy. J Liposome Res 2009; 19: 310-21.

[149] Abreu AS, Castanheira EM, Queiroz MJ, Ferreira PM, Vale-Silva LA, Pinto E. Nanoliposomes for encapsulation and delivery of the potential antitumoral methyl 6-methoxy-3-(4-methoxyphenyl)-1H-indole-2-carboxylate. Nanoscale Res Lett 2011; 6: 482.

[150] Mozafari MR, Johnson C, Hatziantoniou S, Demetzos C. Nanoliposomes and their applications in food nanotechnology. J Liposome Res 2008; 18: 309-27.

[151] Liu F, Liu D. Long-circulating emulsions (oil-in-water) as carriers for lipophilic drugs. Pharm Res 1995; 12:1060-4.

[152] Mun S, Decker EA, McClements DJ. Influence of droplet characteristics on the formation of oil-in-water emulsions stabilized by surfactant-chitosan layers. Langmuir 2005; 21:6228-34.

[153] McClements DJ, Decker EA, Weiss J. Emulsion-based delivery systems for lipophilic bioactive components. J Food Sci 2007; 72: R109-24.

[154] Gursoy RN, Benita S. Self-emulsifying drug delivery systems (SEDDS) for improved oral delivery of lipophilic drugs. Biomed Pharmacother 2004; 58: 173-82.

[155] Khan AW, Kotta S, Ansari SH, Sharma RK, Ali J. Potentials and challenges in selfnanoemulsifying drug delivery systems. Expert Opin Drug Deliv 2012; 9: 1305-17.

[156] Singh B, Bandopadhyay S, Kapil R, Singh R, Katare O. Self-emulsifying drug delivery systems (SEDDS): formulation development, characterization, and applications. Crit Rev Ther Drug Carrier Syst 2009; 26: 427-521.

[157] Narang AS, Delmarre D, Gao D. Stable drug encapsulation in micelles and microemulsions. Int J Pharm 2007; 345(1-2): 9-25.

[158] Puri A, Loomis K, Smith B, Lee JH, Yavlovich A, Heldman E, *et al.* Lipid-based nanoparticles as pharmaceutical drug carriers: from concepts to clinic. Crit Rev Ther Drug Carrier Syst 2009; 26: 523-80.

[159] Muller RH, Maassen S, Weyhers H, Mehnert W. Phagocytic uptake and cytotoxicity of solid lipid nanoparticles (SLN) sterically stabilized with poloxamine 908 and poloxamer 407. J Drug Target 1996; 4:161-70.

[160] Uner M. Preparation, characterization and physico-chemical properties of solid lipid nanoparticles (SLN) and nanostructured lipid carriers (NLC): their benefits as colloidal drug carrier systems. Pharmazie 2006; 61:375-86.

[161] Das S, Chaudhury A. Recent advances in lipid nanoparticle formulations with solid matrix for oral drug delivery. AAPS PharmSciTech 2011; 12: 62-76.

[162] Yoo JW, Irvine DJ, Discher DE, Mitragotri S. Bio-inspired, bioengineered and biomimetic drug delivery carriers. Nat Rev Drug Discov 2011; 10(7): 521-35.

[163] Tong R, Hemmati HD, Langer R, Kohane DS. Photoswitchable nanoparticles for triggered tissue penetration and drug delivery. J Am Chem Soc 2012; 134: 8848-55.

[164] Shao N, Jin J, Wang H, Zheng J and al. Design of bis-spiropyran ligands as dipolar molecule receptors and application to *in vivo* glutathione fluorescent probes. J Am Chem Soc 2010; 132:725-36.

[165] Jonsson F, Beke-Somfai T, Andréasson J, Nordén B. Interactions of a photochromic spiropyran with liposome model membranes. Langmuir 2013;29: 2099-103.

[166]   Han G, You CC, Kim BJ *et al.* Light-regulated release of DNA and its delivery to nuclei by means of photolabile gold nanoparticles. Angew Chem Int Ed 2006; 45: 3165-9.

[167]   Subramani C, Yu X, Agasti SS *et al.* Direct photopatterning of light-activated gold nanoparticles. J Mater Chem 2011; 21: 14156-8.

[168]   Zhu MQ, Wang LQ, Exarhos G J, Li ADQ. Thermosensitive gold nanoparticles. J Am Chem Soc 2004; 126:2656-7.

[169]   Liu Y, Han X, He L, Yin Y. Thermoresponsive assembly of charged gold nanoparticles and their reversible tuning of plasmon coupling. 2012; Angew Chem Int Ed 51: 6373-7.

[170]   Cheng ZL, Zaki AA, Hui JZ, Muzykantov VR, Tsourkas A. Multifunctional nanoparticles: cost *versus* benefit of adding targeting and imaging capabilities. Science 2012;338: 903-10.

[171]   Discher DE, Eisenberg A. Polymer vesicles. Science 2002;297: 967-73.

[172]   Davis ME, Chen ZG., Shin DM. Nanoparticle therapeutics: an emerging treatment modality for cancer. Nat Rev Drug Discov 2008; 7: 771-82.

[173]   Ejima H, Richardson JJ, Liang K *et al.* One-step assembly of coordination complexes for versatile film and particle engineering. Science 2013;341:154-7.

[174]   Zhang Z, Jia J, Lai Y, Ma Y, Weng, J, Sun, L. Conjugating folic acid to gold nanoparticles through glutathione for targeting and detecting cancer cells. Bioorg Med Chem 2010; 18(15):5528-34.

[175]   Orive G, Hernández RM, Gascón AR, Pedraz JL. Micro and nano drug delivery systems in cancer therapy. Cancer Ther. 2005;3:131-8.

[176]   Medina C, Santos-Martinez MJ, Radomski A, Corrigan OI, Radomski MW. Nanoparticles: pharmacological and toxicological significance. Br. J. Pharmacol. 2007;150:552-8.

[177]   Rawat M, Singh D, Saraf S, Saraf S. Nanocarriers: promising vehicle for bioactive drugs. Biol. Pharm. Bull 2006;.29(9):1790-8.

[178]   Adili A, Crowe S, Beaux M *et al.* Differential cytotoxicity exhibited by silica nanowires and nanoparticles. Nanotoxicology 2008;2:1-8.

[179]   Ohulchanskyy TY, Roy I, Goswami LN *et al.* Organically modified silica nanoparticles with covalently incorporated photosensitizer for photodynamic therapy of cancer. 2007;Nano Lett.7(9):2835-42.

[180]   Vallet-Regí M. Nanostructured mesoporous silica matrices in nanomedicine. J Intern Med 2010;267:22-43.

[181]   Vallet-Regí M, Balas F, Arcos D. Mesoporous materials for drug delivery. Angew Chem Int Ed Engl 2007;46:7548-58.

[182]   Santos HA, Bimbo LM, Lehto VP, Airaksinen AJ, Salonen J, Hirvonen J. Multifunctional porous silicon for therapeutic drug delivery and imaging. Curr Drug Discov Technol 2011;8:228-49.

[183]   Salonen J, Kaukonen AM, Hirvonen J, Lehto VP. Mesoporous silicon in drug delivery applications. J Pharm Sci 2008;97:632-53.

[184]   Salonen J, Lehto V-P. Fabrication and chemical surface modification of mesoporous silicon for biomedical applications. Chem Eng J 2008;137:162-72.

[185]   Kinnari P, Mäkilä E, Heikkilä T, Salonen J, Hirvonen J, Santos HA. Comparison of mesoporous silicon and non-ordered mesoporous silica materials as drug carriers for itraconazole. Int J Pharm 2011;414:148-56.

[186]   Santos HA, Salonen J, Bimbo LM, Lehto VP, Peltonen L, Hirvonen J. Mesoporous materials as controlled drug delivery formulations. J Drug Deliv Sci Tech 2011;21:139-55.

[187]   Bimbo LM, Mäkilä E, Laaksonen T, *et al.* Drug permeation across intestinal epithelial cells using porous silicon nanoparticles. Biomaterials 2011;32:2625-33.

[188]   Bimbo LM, Mäkilä E, Raula J, *et al.* Functional hydrophobin-coating of thermally hydrocarbonized porous silicon microparticles. Biomaterials 2011;32:9089-99.

[189]   Bimbo LM, Sarparanta M, Santos HA *et al.* Biocompatibility of thermally hydrocarbonized porous silicon nanoparticles and their biodistribution in rats. ACS Nano 2010;4:3023-32.

[190]   Sarparanta M, Mäkilä E, Heikkilä T, *et al.* ^{18}F-labeled modified porous silicon particlesfor investigation of drug delivery carrier distribution *in vivo* with positron emission tomography. Mol Pharm 2011;8:1799-806.

[191]   Sarparanta MP, Bimbo LM, Mäkilä EM, *et al.* The mucoadhesive and gastroretentive properties of hydrophobin-coated porous silicon nanoparticle oral drug delivery systems. Biomaterials 2012;33:3353-62.

[192]   Bimbo LM, Sarparanta M, Mäkilä E, *et al.* Cellular interactions of surface modified nanoporous silicon particles. Nanoscale 2012;4:3184-92.

[193]   Limnell T, Santos HA, Mäkilä E, *et al.* Drug delivery formulations of ordered and nonordered mesoporous silica: comparison of three drug loading methods. J Pharm Sci 2011;100:3294-306.

[194]   Limnell T, Heikkilä T, Santos HA, *et al.* Physicochemical stability of high indomethacin payload ordered mesoporous silica MCM-41 and SBA-15 microparticles. Int J Pharm 2011;416:242-51.

[195]   Hon NK, Shaposhnik Z, Diebold ED, Tamanoi F, Jalali B. Tailoring the biodegradability of porous silicon nanoparticles. J Biomed Mater Res A. 2012;100(12):3416-21.

[196]   Park JH, Gu L, von Maltzahn G, Ruoslahti E, Bhatia SN, Sailor MJ. Biodegradable luminescent porous silicon nanoparticles for *in vivo* applications. Nat Mater 2009;8:331-6.

[197]   Tang H, Guo J, Sun Y, Chang B, Ren Q, Yang W. Facile synthesis of pH sensitive polymer-coated mesoporous silica nanoparticles and their application in drug delivery. Int J Pharm 2011;421:388-96.

[198]   Wu J, Sailor MJ. Chitosan hydrogel-capped porous SiO2 as a pH responsive nano-valve for triggered release of insulin. Adv Funct Mater 2009;19:733-41.

[199]   Yang X, Liu X, Liu Z, Pu F, Ren J, Qu X. Near-infrared light-triggered, targeted drug delivery to cancer cells by aptamer gated nanovehicles. Adv Mater 2012;24:2890-5.

[200]   Wang LS, Wu LC, Lu SY, *et al.* Biofunctionalized phospholipid-capped mesoporous silica nanoshuttles for targeted drug delivery: improved water suspensibility and decreased nonspecific protein binding. ACS Nano 2010;4:4371-9.

[201]   Wu EC, Park JH, Park J, Segal E, Cunin F, Sailor MJ. Oxidation-triggered release of fluorescent molecules or drugs from mesoporous Si microparticles. ACS Nano 2008;2:2401-9.

[202]   Descalzo AB, Martinez-Manez R, Sancenon F, Hoffmann K, Rurack K. The supramolecular chemistry of organic-inorganic hybrid materials. Angew. Chem. Int. Ed. Engl. 2006;45:5924-48.

[203]   Trewyn BG, Slowing II, Giri S, Chen HT, Lin VSY. Synthesis and functionalization of a mesoporous silica nanoparticle based on the sol-gel process and applications in controlled release. Acc. Chem. Res 2007; 40,846-53.

[204]   Angelos S, Johansson E, Stoddart JF, Zink JI. Mesostructured silica supports for functional materials and molecular machines. Adv. Funct. Mater. 2007;17:2261-71.

[205]   Slowing II, Vivero-Escoto JL, Wu CW, Lin VSY. Mesoporous silica nanoparticles as controlled release drug delivery and gene transfection carriers. Adv. Drug Deliv. Rev. 2008;60:1278-88.

[206]   Klumpp C, Kostarelos K, Prato M, Bianco A. Functionalized carbon nanotubes as emerging nanovectors for the delivery of therapeutics. Biochim Biophys Acta 2006;1758:404-12.

[207]   Bekyarova E, Ni Y, Malarkey EB, *et al.* Applications of carbon nanotubes in biotechnology and biomedicine. J Biomed Nanotechnol 2005;1:3-17.

[208]   Ebbesen TW, Ajayan PM. Large-scale synthesis of carbon nanotubes. Nature 1992;358:220-2.

[209]   Feazell RP, Nakayama-Ratchford N, Dai H, Lippard SJ. Soluble single-walled carbon nanotubes as longboat delivery systems for platinum (IV) anticancer drug design. J Am Chem Soc 2007;129:8438-9.

[210]   Hirsch LR, Stafford RJ, Bankson JA, *et al.* Nanoshell-mediated near-infrared thermal therapy of tumors under magnetic resonance guidance. Proc Natl Acad Sci USA. 2003;100:13549-54.

[211]   Dumortier H, Lacotte S, Pastorin G *et al.* Functionalized carbon nanotubes are non-cytotoxic and preserve the functionality of primary immune cells. Nano Lett. 2006;6(7):1522-8.

[212]   Magres A, Kasas S, Salicio V *et al.* Cellular toxicity of carbon-based nanomaterials. Nano Lett. 2006;6:1121-5.

[213]   Dave SR, Gao X. Monodispersed magnetic nanoparticles for biodetection, imaging, and drug delivery: a versatile and evolving technology. Wiley Interdiscip Rev Nanomed Nanobiotechnol 2009; 1(6):583-609.

[214]   Kodama RH. Magnetic nanoparticles. J Magn Magn Mater 1999; 200(1-3):359-372.

[215]   Lu Z, Prouty MD, Guo Z *et al.* Magnetic Switch of permeability for polyelectrolyte microcapsules embedded with Co@Au nanoparticles. Langmuir 2005;21:2042-50.

[216]   Lee CS, Chang HH, Bae PK, Jung J, Chung BH. Bifunctional nanoparticles constructed using one-pot encapsulation of a fluorescent polymer and magnetic (Fe3O4) nanoparticles in a silica shell. Macromol Biosci 2013; 13: 321-31.

[217] Panahifar A, Mahmoudi M, Doschak MR. Synthesis and *in vitro* evaluation of bone-seeking superparamagnetic iron oxide nanoparticles as contrast agents for imaging bone metabolic activity. ACS Appl Mater Interfaces 2013; 5: 5219-26.

[218] Liu C, Liu DB, Long GX, *et al.* Specific targeting of angiogenesis in lung cancer with RGD-conjugated ultrasmall superparamagnetic iron oxide particles using a 4.7T magnetic resonance scanner. Chin Med J (Engl) 2013; 126 (12), 2242-7.

[219] Charrois GJR, Allen TM. Multiple injections of pegylated liposomal Doxorubicin: pharmacokinetics and therapeutic activity. J Pharmacol Exp Ther 2003; 306: 1058-67.

[220] Hashizume H, Baluk P, Morikawa S, *et al.* Openings between defective endothelial cells explain tumor vessel leakiness. Am J Pathol 2000; 156 (4): 1363-80.

[221] Hobbs SK, Monsky WL, Yuan F, *et al.* Regulation of transport pathways in tumor vessels: role of tumor type and microenvironment. Proc Natl Acad Sci USA 1998; 95 (8), 4607-12.

[222] Chin, IJ, Thurn - Albrecht T, Kim HC, Russell TP, Wang J. On exfoliation of montmorillonite in epoxy. Polymer 2001;42: 5947-52.

[223] Lim SK, Kim JW, Chin IJ, Kwon YK, Choi HJ. Preparation and interaction characteristics of organically modified montmorillonite nanocomposite with miscible polymer blend of poly(ethylene oxide) and poly(methyl methacrylate). Chem Mater 2002; 14 (5): 1989-94.

[224] Wang Z, Pinnavaia TJ. Nanolayer reinforcement of elastomeric polyurethane. Chem Mater 1998; 10(12): 3769-71.

[225] Messersmith PB, Giannelis EP. Synthesis and barrier properties of poly(ε-caprolactone)-layered silicate nanocomposites. J Polym Sci 1995; 33(7): 1047 -57.

[226] Yano K, Usuki A, Okada A. Synthesis and properties of polyimide-clay hybrid films. J Polym Sci 1997; 35(11): 2289 -94.

[227] Shi H, Lan T, Pinnavaia TJ. Interfacial effects on the reinforcement properties of polymer–organoclay nanocomposites. Chem Mater 1996; 8(8): 1584-87.

[228] Giannelis EP. Polymer Layered Silicate Nanocomposites. Adv Mater 1996; 8(1):29-35.

[229] LeBaron PC, Wang Z, Pinnavaia TJ. Polymer-layered silicate nanocomposites: an overview. Appl Clay Sci 1999; 15: 11-29.

[230] Namiki Y, Namiki T, Yoshida H *et al.* A novel magnetic crystal-lipid nanostructure for magnetically guided *in vivo* gene delivery. Nat. Nanotechnol. 2009;4(9):598-606.

[231] European Medicines Agency. Reflection Paper on Non-Clinical Studies for Generic Nanoparticle Iron Medicinal Product Applications. London: European Medicines Agency; 2011. Available from: http://www.ema.europa.eu/docs/en_GB/document_library/Scientific_guideline/2011/04/WC5001050 48.pdf

[232] Garneata L. Intravenous iron, inflammation, and oxidative stress: is iron a friend or an enemy of uremic patients? J Ren Nutr 2008; 18(1): 40-45.

[233] European Medicines Agency. Reflection paper on non clinical studies for generic nanoparticle iron medicinal product applications. www.ema.europa.eu/docs/en_GB/document_library/Scientific_guideline/2011/04/WC500105048.pdf

[234] Soma CE, Dubernet C, Bentolila D, Benita S, Couvreur P. Reversion of multidrug resistance by co-encapsulation of doxorubicin and cyclosporin A in polyalkylcyanoacrylate nanoparticles. Biomaterials 2000; 21: 1-7.

[235] Pramanik D, Campbell NR, Das S *et al.* A composite polymer nanoparticle overcomes multidrug resistance and ameliorates doxorubicin-associated cardiomyopathy. Oncotarget 2012; 3: 640- 650.

[236] Li R, Wu R, Wu M *et al.* MEKC-LIF analysis of rhodamine123 delivered by carbon nanotubes in K562 cells. Electrophoresis 2009; 30: 1906-12.

[237] Elsayed HH, Al-Sherbini ASAM. M., Abd-Elhady EE, Ahmed KAEA. Treatment of Anemia Progression *via* Magnetite and Folate Nanoparticles *In Vivo*. ISRN Nanotechnology 2014; 2014: 287575.

[238] Li B, Xu H, Li Z *et al.* Bypassing multidrug resistance in human breast cancer cells with lipid/polymer particle assemblies. Int J Nanomedicine 2012; 7: 187-97.

[239] Kang KW, Chun MK, Kim O *et al.* Doxorubicin-loaded solid lipid nanoparticles to overcome multidrug resistance in cancer therapy. Nanomedicine. 2010; 6: 210-13.

[240] St'astny M, Plocova D, Etrych T, Kovar M, Ulbrich K, Rihova B. HPMA-hydrogels containing cytostatic drugs. Kinetics of the drug release and *in vivo* efficacy. J Control Release 2002; 81: 101-11.

[241] Högemann-Savellano D, Bos E, Blondet C, *et al*. The transferrin receptor: a potential molecular imaging marker for human cancer. Neoplasia 2003; 5: 495-506.

[242] Hogemann D, Josephson L, Weissleder R, Basilion JP. Improvement of MRI probes to allow efficient detection of gene expression. Bioconjug Chem 2000; 11: 941-6.

[243] Basu S, Chaudhuri P, Sengupta S. Targeting oncogenic signaling pathways by exploiting nanotechnology. Cell Cycle 2009; 8: 3480-7.

[244] Akita H, Kudo A, Minoura A, *et al*. Multi-layered nanoparticles for penetrating the endosome and nuclear membrane *via* a step-wise membrane fusion process. Biomaterials 2009; 30: 2940-9.

[245] Schroder T, Niemeier N, Afonin S, Ulrich AS, Krug HF, Brase S. Peptoidic amino- and guanidinium-carrier systems: targeted drug delivery into the cell cytosol or the nucleus. J Med Chem 2008; 51: 376-9.

[246] Xu ZP, Niebert M, Porazik K, *et al*. Subcellular compartment targeting of layered double hydroxide nanoparticles. J Control Release 2008; 130: 86-94.

[247] Lee JH, Nan A. Combination drug delivery approaches in metastatic breast cancer. J Drug Deliv 2012; 2012: ID 915375.

[248] Wilcoxon J. Optical absorption properties of dispersed gold and silver alloy nanoparticles. J Phys Chem B 2009; 113: 2647-56.

[249] Jain PK, Lee KS, El-Sayed IH, El-Sayed MA. Calculated absorption and scattering properties of gold nanoparticles of different size, shape, and composition: Applications in biological imaging and biomedicine. J Phys Chem B 2006; 110: 7238-48.

[250] Murphy CJ, Sau TK, Gole AM, *et al*. Anisotropic metal nanoparticles: Synthesis, assembly, and optical applications. J Phys Chem B 2005;109:13857-70.

[251] Liz-Marzán LM. Tailoring surface plasmons through the morphology and assembly of metal nanoparticles. Langmuir 2006; 22: 32-41.

[252] Jain PK, El-Sayed MA. Universal scaling of plasmon coupling in metal nanostructures: Extension from particle pairs to nanoshells. Nano Lett 2007;7:2854-8.

[253] Kim YS, Niazi JH, Gu MB. Specific detection of oxytetracycline using DNA aptamer-immobilized interdigitated array electrode chip. Anal Chim Acta 2009; 634(2): 250-4.

[254] Proske D, Blank M, Buhmann R, Resch A. Aptamers—basic research, drug development and clinical applications. Appl Microbiol Biotechnol 2005; 69(4): 367-74.

[255] Xie S, Walton SP. Application and analysis of structure-switching aptamers for small molecule quantification. Anal Chim Acta 2009; 638(2):213-9.

[256] Fang L, Lu Z, Wei H, Wang E. A electrochemiluminescence aptasensor for detection of trombin incorporating the capture aptamer labeled with gold nanoparticles immobilized onto the thio-silanized ITO electrode. Anal Chim Acta 2008; 628: 80-6.

[257] Farokhzad OC, Jon S, Khademhosseini A, Tran TN, Lavan DA, Langer, R. Nanoparticle-aptamer bioconjugates: A new approach for targeting prostate cancer cells. Cancer Res 2004; 64(21): 7668-72.

[258] Pamme N. Magnetism and microfluidics. Lab Chip 2006;.6(1):24-38.

[259] Loureiro J, Ferreira R, Cardoso S, *et al*. Toward a magnetoresistive chip cytometer: Integrated detection of magnetic beads flowing at cm/s velocities in microfluidic channels. Appl Phys Lett 2009;95:034104.

[260] Rose-James A, Sreelekha TT, George SK. Nanostrategies in the war against multidrug resistance in leukemia. OncoDrugs 2013; 1(1): 3e-9e.

[261] Kane B. Cancer Chemotherapy: Teaching Old Drugs New Tricks. Ann Intern Med 2001; 135(12): 1107-10.

[262] Blanco E, Hsiao A, Mann AP, Landry MG, Meric-Bernstam F, Ferrari M. Nanomedicine in cancer therapy: innovative trends and prospects. Cancer Sci 2011; 102(7): 1247-52.

[263] Cao Y, Wang B, Lou D, Wang Y, Hao S, Zhang L. Nanoscale Delivery Systems for Multiple Drug Combinations in Cancer. Future Oncol 2011; 7(11): 1347-57.

[264] Kerbel RS, Kobayashi H, Graham CH. Intrinsic or acquired drug resistance and metastasis: are they linked phenotypes ? J Cell Biochem 1994; 56: 37-47.

[265] Gottesman MM, Fojo T, Bates SE. Multidrug resistance in cancer: role of ATP-dependent transporters. Nat Rev Cancer 2002; 2: 48-58.

[266] Monopoll MP, Walczyk D, Campbell A, *et al*. Physical-chemical aspects of protein corona: relevance to *in vitro* and *in vivo* biological impacts of nanoparticles. J Am Chem Soc 2011; 133: 2525-34.

[267]   Dobrovolskaia MA, Patri AK, Zheng JW, *et al.* Interaction of colloidal gold nanoparticles with human blood: effects on particle size and analysis of plasma protein binding profiles. Nanomedicine 2009; 5: 106-17.

[268]   Casals E, Pfaller T, Duschl A, Oostingh GJ, Puntes V. Time Evolution of the Nanoparticle Protein Corona. ACS Nano 2010; 4: 3623-32.

[269]   Lundqvist M, Stigler J, Elia G, Lynch I, Cedervall T, Dawson KA. Nanoparticle size and surface properties determine the protein corona with possible implications for biological impacts. Proc Natl Acad Sci USA 2008; 105:14265-70.

[270]   Cedervall T, Lynch I, Lindman S, *et al.* Understanding the nanoparticle-protein corona using methods to quantify exchange rates and affinities of proteins for nanoparticles. Proc Natl Acad Sci USA 2007; 104: 2050-55.

[271]   Tenzer S, Docter D, Rosfa S, *et al.* Nanoparticle size is a critical physicochemical determinant of the human blood plasma corona: a comprehensive quantitative proteomic analysis. ACS Nano 2011; 5: 7155-67.

[272]   Martel J, Young D, Young A, *et al.* Comprehensive proteomic analysis of mineral nanoparticles derived from human body fluids and analyzed by liquid chromatography-tandem mass spectrometry. Anal Biochem 2011; 418: 111-25.

[273]   Gao Z, Zhang L, Sun Y. Nanotechnology applied to overcome tumor drug resistance. J Control Release 2012; 162: 45-55.

[274]   Immordino ML, Dosio F, Cattel L. Stealth liposomes: review of the basic science, rationale, and clinical applications, existing and potential. Int J Nanomedicine 2006; 1: 297-315.

[275]   Allen TM, Mehra T, Hansen C, Chin YC. Stealth liposomes: an improved sustained release system for 1-beta-Darabinofuranosylcytosine. Cancer Res 1992; 52: 2431-9.

[276]   Wang J, Goh B, Lu W, *et al. In vitro* cytotoxicity of Stealth liposomes co-encapsulating doxorubicin and verapamil on doxorubicin resistant tumor cells. Biol Pharm Bull 2005; 28: 822-8.

[277]   Koziara JM, Whisman TR, Tseng MT, Mumper RJ. In-vivo efficacy of novel paclitaxel nanoparticles in paclitaxel-resistant human colorectal tumors. J Control Release 2006; 112: 312-9.

[278]   Rouf MA, Vural I, Renoir JM, Hincal AA. Development and characterization of liposomal formulations for rapamycin delivery and investigation of their antiproliferative effect on MCF7 cells. J Liposome Res 2009; 19: 322-31.

[279]   Morin PJ. Drug resistance and the microenvironment: nature and nurture. Drug Resist. Update 2003; 6: 169-72.

[280]   Burger JA, Peled A. CXCR4 antagonists: targeting the microenvironment in leukemia and other cancers. Leukemia 2009; 23: 43-52.

[281]   Konopleva M, Tabe Y, Zeng Z, Andreeff M. Therapeutic targeting of microenvironmental interactions in leukemia: mechanisms and approaches. Drug Resist Update 2009; 12: 103-113.

[282]   Ayers D, Nasti A. Utilisation of Nanoparticle Technology in Cancer Chemoresistance. J Drug Deliv 2012; 2012: 265691.

[283]   Chen ZG. Small-molecule Delivery by Nanoparticles for Anticancer Therapy. Trends Mol Med 2010; 16(12): 594-602.

[284]   Pan CX, Zhu W, Cheng L. Implications of cancer stem cells in the treatment of cancer. Future Oncol 2006; 2: 723-31.

[285]   Misaghian N; Ligresti G, Steelman LS, *et al.* Targeting the leukemic stem cell: the Holy Grail of leukemia therapy. Leukemia 2009; 23: 25-42.

[286]   Terpstra W, Ploemacher RE, Prins A, *et al.* Fluorouracil selectively spares acute myeloid leukemia cells with long-term growth abilities in immunodeficient mice and in culture. Blood 1996; 88: 1944-50.

[287]   Copland M, Hamilton A, Elrick LJ, *et al.* Dasatinib (BMS-354825) targets an earlier progenitor population than imatinib in primary CML but does not eliminate the quiescent fraction. Blood 2006; 107: 4532-9.

[288]   Zhang H, Luo J, Li Y, *et al.* Characterization of high-affinity peptides and their feasibility for use in nanotherapeutics targeting leukemia stem cells. Nanomedicine: Nanotechnology, Biology, and Medicine 2012; 8: 1116-24.

[289]   Wang Z, Li Y, Ahmad A, Azmi, *et al.* Targeting miRNAs involved in cancer stem cell and EMT regulation: an emerging concept in overcoming drug resistance. Drug Resist Update 2010; 13:109-18.

[290] Liu JH, Kopeckova P, Buhler P, *et al.* Biorecognition and subcellular trafficking of HPMA copolymer-anti-PSMA antibody conjugates by prostate cancer cells. Mol Pharm 2009; 6: 959-70.

[291] Mimeault M, Batra SK. Recent advances in the development of novel anticancer drugs targeting cancer stem/progenitor cells. Drug Dev Res 2008; 69: 415-30.

[292] Besancon R, Valsesia-Wittmann S, Puisieux A, de Fromentel CC, Maguer-Satta V. Cancer stem cells: the emerging challenge of drug targeting. Curr Med Chem 2009; 16: 394-416.

[293] LaBarge MA. The difficulty of targeting cancer stem cell niches. Clin Cancer Res 2010; 16: 3121-9.

[294] Lacerda L, Pusztai L, Woodward WA. The role of tumor initiating cells in drug resistance of breast cancer: implications for future therapeutic approaches. Drug Resist Update 2010; 13: 99-108.

[295] Bakker AB, van den Oudenrijn S, Bakker AQ, *et al.* C-type lectin-like molecule-1: a novel myeloid cell surface marker associated with acute myeloid leukemia. Cancer Res 2004; 64: 8443-50.

[296] Szakacs G, Paterson JK, Ludwig JA, Booth-Genthe C, Gottesman MM. Targeting multidrug resistance in cancer. Nat Rev Drug Discov 2006; 5: 219-34.

[297] Assaraf YG. Molecular basis of antifolate resistance. Cancer Metast Rev 2007; 26: 153-81.

[298] Broxterman HJ, Gotink,KJ, Verheul HM. Understanding the causes of multidrug resistance in cancer: a comparison of doxorubicin and sunitinib. Drug Resist Update 2009; 12: 114-26.

[299] Qiao L, Wong BC. Targeting apoptosis as an approach for gastrointestinal cancer therapy. Drug Resist Update 2009; 12: 55-64.

[300] Chen AM, Zhang M, Wei DG, *et al.* Co-delivery of doxorubicin and Bcl-2 siRNA by mesoporous silica nanoparticles enhances the efficacy of chemotherapy in multidrug-resistant cancer cells. Small 2009; 5: 2673-7.

[301] Grivennikov SI, Greten FR, Karin M. Immunity, inflammation and cancer. Cell 2010; 140: 883-99.

[302] Gyrd-Hansen M, Meier P. IAPs: from caspase inhibitors to modulators of NF-_B, inflammation and cancer. Nat Rev Cancer 2010; 10: 561-74.

[303] Zhu X, Radovic-Moreno AF, Wu J, Langer R2, Shi J. Nanomedicine in the Management of Microbial Infection - Overview and Perspectives. Nano Today 2014; 9(4): 478-98.

[304] Kaittanis C, Santra S, Perez JM. Emerging nanotechnology-based strategies for the identification of microbial pathogenesis. Adv Drug Deliv Rev 2010; 62 (4-5): 408-423.

[305] Chow JW, YuVL. Combination antibiotic therapy *versus* monotherapy for gram-negative bacteraemia: A commentary. Int J Antimicrob Agents 1999; 11(1): 7-12.

[306] Toti US, Guru BR, Hali M, *et al.* Targeted delivery of antibiotics to intracellular chlamydial infections using PLGA nanoparticles. Biomaterials 2011; 32(27): 6606-13.

[307] Carmona D, Lalueza P, Balas F, Arruebo M, Santamaría J. Mesoporous silica loaded with peracetic acid and silver nanoparticles as a dual-effect, highly efficient bactericidal agent. Microporous Mesoporous Mater 2012; 161(1): 84-90.

[308] Abed N, Couvreur P. Nanocarriers for antibiotics: A promising solution to treat intracellular bacterial infections. Int J Antimicrob Agents 2014; 43(6): 485-496.

[309] Fang J, Nakamura H, Maeda H. The EPR effect: Unique features of tumor blood vessels for drug delivery, factors involved, and limitations and augmentation of the effect. Adv Drug Deliv Rev 2011; 63: 136-51.

[310] Azzopardi EA, Ferguson EL, Thomas DW. The enhanced permeability retention effect: a new paradigm for drug targeting in infection. J Antimicrob Chemother 2013; 68: 257-74.

[311] Cohen ML. Changing patterns of infectious disease. Nature 2000; 406:762-7.

[312] Kåhrström CT. Environmental microbiology: plant bacteria thrive in storm clouds. Nat Rev Microbiol 2013; 11(3):146.

[313] Alekshun MN, Levy SB. Molecular mechanisms of antibacterial multidrug resistance. Cell 2007; 128(6): 1037-50.

[314] Ray K, Marteyn B, Sansonetti PJ, Tang CM. Life on the inside: the intracellular lifestyle of cytosolic bacteria. Nature Reviews Microbiology 2009; 7(5): 333-40.

[315] Mah TF, O'Toole GA. Mechanisms of biofilm resistance to antimicrobial agents. Trends Microbiol. 2001; 9(1): 34-9.

[316] Huang L, Dai T, Xuan Y, Tegos GP, Hamblin MR. Synergistic combination of chitosan acetate with nanoparticle silver as a topical antimicrobial: efficacy against bacterial burn infections. Antimicrob Agents Chemother 2011; 55: 3432-8.

[317]   Pissuwan D, Cortie CH, Valenzuela SM, Cortie MB. Functionalised gold nanoparticles for controlling pathogenic bacteria. Trends Biotechnol 2010; 28 (4): 207-13.

[318]   Gu H, Ho PL, Tong E, Wang L, Xu B. Presenting vancomycin on nanoparticles to enhance antimicrobial activities Nano Lett 2003; 3 (9): 1261-3.

[319]   Zhao Y, Tian Y, Cui Y, Liu W, Ma W, Jiang X. Small molecule-capped gold nanoparticles as potent antibacterial agents that target gram-negative bacteria. J Am Chem Soc 2010; 132 (35):12349-56.

[320]   Smith DM, Simon JK, Baker JR Jr. Applications of nanotechnology for immunology. Nat Rev Immunol 2013; 13(8): 592-605.

[321]   Reddy ST, van der Vlies AJ, Simeoni E, *et al.* Exploiting lymphatic transport and complement activation in nanoparticle vaccines. Nat Biotechnol 2007; 25(10): 1159-64.

[322]   Huh AJ, Kwon YJ. "Nanoantibiotics": A new paradigm for treating infectious diseases using nanomaterials in the antibiotics resistant era. J Controlled Release 2011; 156 (2): 128-45.

[323]   Tran N, Tran PA. Nanomaterial-based treatments for medical device-associated infections. Chem Phys Chem 2012; 13(10): 2481-94.

[324]   Lara HH, Ayala-Núnez NV, Turrent LdCI, Padilla CR, Bactericidal effect of silver nanoparticles against multidrug-resistant bacteria. World J Microbiol Biotechnol 2010; 26 (4): 615-21.

[325]   Zare B, Faramarzi MA, Sepehrizadeh Z, Shakibaie M, Rezaie S, Shahverdi AR. Biosynthesis and recovery of rod-shaped tellurium nanoparticles and their bactericidal activities. Mater Res Bull 2012; 47 (11): 3719-25.

[326]   Veerapandian M, Yun K. Functionalization of biomolecules on nanoparticles: Specialized for antibacterial applications. Appl Microbiol Biotechnol 2011; 90 (5): 1655-67.

[327]   Kang S, Herzberg M, Rodrigues DF, Elimelech M. Antibacterial effects of carbon nanotubes: Size does matter! Langmuir 2008; 24 (13): 6409-13.

[328]   Kotagiri N, Lee JS, Kim JW. Selective pathogen targeting and macrophage evading carbon nanotubes through dextran sulfate coating and PEGylation for photothermal theranostics. J Biomed Nanotechnol 2013; 9 (6): 1008-16.

[329]   Schrand AM, Rahman MF, Hussain SM, Schlager JJ, Smith DA, Syed AF. Metal-based nanoparticles and their toxicity assessment. Wiley Interdiscip Rev Nanomed Nanobiotechnol 2010; 2 (5): 544-68.

[330]   Karlsson HL, Cronholm P, Gustafsson J, Möller L. Copper oxide nanoparticles are highly toxic: A comparison between metal oxide nanoparticles and carbon nanotubes. Chem Res Toxicol 2008; 21 (9): 1726-32.

[331]   Roe D, Karandikar B, Bonn-Savage N, Gibbins B, Roullet JB. Antimicrobial surface functionalization of plastic catheters by silver nanoparticles. J Antimicrob Chemother 2008; 61(4): 869-76.

[332]   Lellouche J, Kahana E, Elias S, Gedanken A, Banin E. Antibiofilm activity of nanosized magnesium fluoride. Biomaterials 2009; 30 (30): 5969-78.

[333]   Kwak SY, Kim SH, Kim SS. Hybrid organic/inorganic reverse osmosis (RO) membrane for bactericidal anti-fouling. Preparation and characterization of TiO_2 nanoparticle self-assembled aromatic polyamide thin-film-composite (TFC) membrane. Environ Sci Technol 2001; 35 (11): 2388-94.

[334]   Hancock RE, Sahl HG. Antimicrobial and host-defense peptides as new anti-infective therapeutic strategies. Nat Biotechnol 2006; 24 (12): 1551-57.

[335]   Peschel A, Sahl HG. The co-evolution of host cationic antimicrobial peptides and microbial resistance. Nat Rev Microbiol 2006; 4 (7): 529-36.

[336]   Kamaly N, Xiao Z, Valencia PM, Radovic-Moreno AF, Farokhzad OC. Targeted polymeric therapeutic nanoparticles: design, development and clinical translation. Chem Soc Rev 2012; 41(7): 2971-3010.

[337]   Bertrand N, Wu J, Xu X, Kamaly N, Farokhzad OC. Cancer nanotechnology: the impact of passive and active targeting in the era of modern cancer biology. Adv Drug Deliv Rev 2014; 66: 2-25.

[338]   Shi J, Xiao Z, Kamaly N, Farokhzad OC. Self-assembled targeted nanoparticles: evolution of technologies and bench to bedside translation. Acc Chem Res 2011; 44(10):1123-34.

[339]   Gu F, Zhang L, Teply BA, *et al.* Precise engineering of targeted nanoparticles by using self-assembled biointegrated block copolymers. Proc Natl Acad Sci USA 2008; 105: 2586-91.

[340] Etheridge ML, Campbell SA, Erdman AG, Haynes CL, Wolf SM, McCullough J. The big picture on nanomedicine: the state of investigational and approved nanomedicine products. Nanomedicine: NBM 2013; 9(1): 1-14.

[341] Nyström AM, Fadeel B. Safety assessment of nanomaterials: implications for nanomedicine. J Control Release 2012; 161(2): 403-8.

[342] Fadeel B, Garcia-Bennett AE. Better safe than sorry: understanding the toxicological properties of inorganic nanoparticles manufactured for biomedical applications. Adv. Drug Deliv Rev 2010; 62:362-74.

[343] Linkov I, Satterstrom K, Corey L. Nanotoxicology and nanomedicine: making hard decisions. Nanomedicine 2008; 4: 167-71.

[344] Schellekens H, Klinger E, Mühlebach S, Brin JF, Storm G, Crommelin DJ. The therapeutic equivalence of complex drugs. Regul Toxicol Pharmacol 2011; 59(1): 176-83.

[345] Duncan R, Gaspar R. Nanomedicine(s) under the microscope. Mol Pharmaceutics 2011; 8(6): 2101-41.

[346] Gaspar R. Regulatory issues surrounding nanomedicines: setting the scene for the next generation of nanopharmaceuticals. Nanomedicine (Lond.) 2007; 2(2): 143-7.

[347] Soloman R, Gabizon AA. Clinical pharmacology of liposomal anthracyclines: focus on pegylated liposomal doxorubicin. Clin Lymphoma Myeloma 2008; 8(1): 21-32.

[348] Auerbach M, Ballard H. Clinical use of intravenous iron: administration, efficacy, and safety. Hematology Am Soc Hematol Educ Program 2010, 2010; 338-347.

[349] European Medicines Agency. The European Medicines Agency publishes a full scientific assessment report called a European public assessment report (EPAR) for every medicine granted a central marketing authorisation by the European Commission. www.ema.europa.eu/ema/index. jsp curl=pages/medicines/landing/epar_search.jsp&mid= WC0b01ac058001d124

[350] Farrell D, Ptak K, Panaro NJ, Grodzinski P. Nanotechnology- based cancer therapeutics promise and challenge-lessons learned through the NCI Alliance for Nanotechnology in Cancer. Pharm Res 2011; 28(2): 273-8.

[351] European Medicines Agency. European Medicines Agency's workshop on nanomedicines. www.ema.europa.eu/ema/index. jsp curl=pages/news_and_events/events/2009/12/event_detail_000095. jspmurl=menus/news_and_events/news_and_events.jsp&mid= WC0b01ac058004d5c3

[352] European Medicines Agency. European Medicines Agency's workshop on nanomedicines. www.ema.europa.eu/ema/index. jsp curl=pages/news_and_events/news/2010/09/news_detail_001108. jsp&mid=WC0b01ac058004d5c1

[353] European Medicines Agency. Joint MHLW/EMA reflection paper on the development of block copolymer micelle medicinal products. www.ema.europa.eu/docs/en_GB/document_library/Scientific_guideline/2013/02/WC500138390.pdf

[354] Handy RD, Owen R, Valsami-Jones E. The ecotoxicology of nanoparticles and nanomaterials: current status, knowledge gaps, challenges, and future needs. Ecotoxicology 2008; 17 (5): 315-25.

[355] Lead JR, Wilkinson KJ. Aquatic colloids and nanoparticles: current knowledge and future trends. Environ Chem 2006; 3:159-71.

[356] Landsiedel R, Fabian E, Ma-Hock L, et al. Toxico-/biokinetics of nanomaterials. Arch Toxicol 2012; 86(7): 1021-60.

[357] Handy RD, von der Kammer F, Lead JR, Hassellöv M, Owen R, Crane M. The ecotoxicology and chemistry of manufactured nanoparticles. Ecotoxicology 2008; 17(4):287-314.

[358] Relyea R, Hoverman J. Assessing the ecology in ecotoxicology: a review and synthesis in freshwater systems. Ecol Lett 2006; 9:1157-71.

[359] Bernhardt ES, Colman BP, Hochella MF, et al. An ecological perspective on nanomaterial impacts in the environment. J Environ Qual 2010; 39:1954-65.

[360] Rosi-Marshall EJ, Royer TV. Pharmaceutical compounds and ecosystem function: an emerging research challenge for aquatic ecologists. Ecosystems 2012; 15: 867-80.

[361] Green JM, Beestman GB. Recently patented and commercialized formulation and adjuvant technology. Crop Prot 2007; 26: 320-27.

[362] Wang L, Li X, Zhang G, Dong J, Eastoe J. Oil-in-water nanoemulsions for pesticide formulations. J Colloid Interface Sci 2007; 314: 230-5.

[363]   Boehm AL, Martinon I, Zerrouk R, Rump E, Fessi H. Nanoprecipitation technique for the encapsulation of agrochemical active ingredients. J Microencapsul 2003; 20: 433-41.

[364]   Tsuji K. Microencapsulation of pesticides and their improved handling safety. J Microencapsul 2001; 18: 137-47.

[365]   Nair R, Varghese SH, Nair BG, Maekawa T, Yoshida Y, Kumar DS. Nanoparticulate material delivery to plants. Plant Science 2010; 179 (3): 154-63.

[366]   Racuciu M, Creanga D. Cytogenetic changes induced by beta-cyclodextrin coated nanoparticles in plant seeds. Romanian J Phys 2009; 54: 125-31.

[367]   Racuciu M, Creanga D. Cytogenetic changes induced by aqueous ferrofluids in agricultural plants. J Magn Magn Mater 2007; 311: 288-90.

[368]   Pavel A, Creanga D. Chromosomal aberrations in plants under magnetic fluid influence. J Magn Magn Mater 2005; 289; 469-72.

[369]   Pavel A, Trifan M, Bara II, Creanga D, Cotae C. Accumulation dynamics and some cytogenetical tests at C. majus and P. somniferum callus under the magnetic liquid effect. J Magn Magn Mater 1999; 201: 443-5.

[370]   Nel A, Xia T, Meng H, Wang X, Lin S, Ji Z, Zhang H. Nanomaterial toxicity testing in the 21st century: use of a predictive toxicological approach and high-throughput screening. Acc Chem Res 2013; 46(3): 607-21.

[371]   Stone V, Nowack B, Baun A, *et al.* Nanomaterials for environmental studies: classification, reference material issues, and strategies for physicochemical characterisation. Sci Total Env 2010; 408(7): 1745-54.

[372]   ECHA. 2012. European Chemicals Agency - Guidance on information requirements and chemical safety assessment. Appendix R7-1 Recommendations for nanomaterials applicable to Chapter R7a Endpoint specific guidance. ECHA; 12-G-03-EN.

[373]   Monteiro SC, Boxall ABA. Occurrence and fate of human pharmaceuticals in the environment. In: Whitacre DM, Ed. Reviews of environmental contamination and toxicology, Vol. 202. New York: Springer. 2010; pp. 153-4.

[374]   Likens GE. Biogeochemistry: some opportunities and challenges for the future. Water Air Soil Pollut Focus 2004; 4: 5-24.

[375]   Daughton CG, Ternes TA. Pharmaceuticals and personal care products in the environment: agents of subtle change? Environ Health Perspect 1999; 107: 907-38.

[376]   Fent K, Weston AA, Caminada D. Ecotoxicology of human pharmaceuticals. Aquat Toxicol 2006; 76 (2): 122-59.

[377]   Kummerer K. The presence of pharmaceuticals in the environment due to human use - present knowledge and future challenges. J Environ Manage 2009; 90(8): 2354-66.

[378]   Kummerer K. Pharmaceuticals in the environment. Annu Rev Environ Resour 2010; 35: 57-75.

[379]   Liu BY, Nie XP, Liu WQ, Snoeijs P, Guan C, Tsui MTK. Toxic effects of erythromycin, ciprofloxacin and sulfamethoxazole on photosyn-thetic apparatus in Selenastrum capricornutum. Ecotoxicol Environ Saf 2011; 74 (4): 1027-35.

[380]   Nunes B, Carvalho F, Guilhermino L. Acute toxicity of widely used pharmaceuticals in aquatic species: Gambusia holbrooki, Artemia parthenogenetica and Tetraselmis chuii. Ecotoxicol Environ Saf 2005; 61 (3): 413-9.

[381]   Sanderson H, Brain RA, Johnson DJ, Wilson CJ, Solomon KR. Toxicity classification and evaluation of four pharmaceuticals classes: antibiotics,antineoplastics, cardiovascular, and sex hormones. Toxicology 2004; 203 (1-3): 27-40.

[382]   Chen X, Schluesener HJ. Nanosilver: a nanoproduct in medical application. Toxicol Lett 2008; 176: 1-12.

[383]   Nowack B, Krug HF, Height M. 120 years of nanosilver history: implications for policy makers. Environ Sci Technol 2011;45:1177-83.

[384]   Schluesener JK, Schluesener HJ. Nanosilver: application and novel aspects of toxicology. Arch Toxicol 2013;87:569-76.

[385]   Consumer products inventory: an inventory of nanotechnology-based consumer products introduced on the market. Washington DC: Woodrow Wilson Center: Project on Nanotechnology; 2013.http://www. nanotechproject.org/cpi [accessed 10.12.13].

[386] Liu J, Wang Z, Liu FD, *et al.* Chemical transformations of nanosilver in biological environments. ACS Nano 2012; 6: 9887-99.

[387] Reidy B, Haase A, Luch A, *et al.* Mechanisms of silver nanoparticle release, transformation and toxicity: a critical review of current knowledge and recommendations for future studies and applications. Materials 2013; 6: 2295-350.

[388] Yu S-J, Yin Y-G, Liu J-f. Silver nanoparticles in the environment. Environ Sci Proc Impacts 2013;15:78-92.

[389] Wijnhoven SWP, Peijnenburg WJGM, Herberts CA, *et al.* Nano-silver: a review of available data and knowledge gaps in human and environmental risk assessment. Nanotoxicity 2009;3:109-38.

[390] Unrine JM, Colman BP, Bone AJ, *et al.* Biotic and abiotic interactions in aquatic microcosms determine fate and toxicity of Ag nanoparticles. Part 1. Aggregation and dissolution. Environ Sci Technol 2012;46:6915-24.

[391] Bone AJ, Colman BP, Gondikas AP, *et al.* Biotic and abiotic interactions in aquatic microcosms determine fate and toxicity of Ag nanoparticles: part 2 e toxicity and Ag speciation. Environ Sci Technol 2012;46:6925-33.

[392] Calder AJ, Dimkpa CO, McLean JE, *et al.* Soil components mitigate the antimicrobial effects of silver nanoparticles towards a beneficial soil bacterium, Pseudomonas chlororaphis O6. Sci Total Environ 2012;429:215-22.

[393] Nair PMG, Choi J. Modulation in the mRNA expression of ecdysone receptor gene in aquatic midge, Chironomus riparius upon exposure to nonylphenol and silver nanoparticles. Environ Toxicol Pharmacol 2012;33:98-106.

[394] Yeo MK, Kang M. Effects of nanometer sized silver materials on biological toxicity during zebrafish embryogenesis. Bull Kor Chem Soc 2008;29:1179-84.

[395] Volker C, Oetken M, Oehlmann J. The biological effects and possible modes of action of nanosilver. Rev Environ Contam Toxicol 2013;223:81-106.

[396] Wu Y, Zhou Q. Silver nanoparticles cause oxidative damage and histological changes in medaka (Oryzias latipes) after 14 days of exposure. Environ Toxicol Chem 2013;32:165-73.

[397] McShan D, Ray PC, Yu H. Molecular toxicity mechanism of nanosilver. J Food Drug Anal 2014; 22: 116-27.

[398] Dave G, Herger G. Determination of detoxification to Daphnia magna of four pharmaceuticals and seven surfactants by activated sludge. Chemosphere 2012; 88 (4): 459 66.

[399] Ellesat KS, Tollefsen KE, Asberg A, Thomas KV, Hylland K. Cytotoxicity of atorvastatin and simvastatin on primary rainbow trout (Oncorhynchus mykiss) hepatocytes. Toxicol *In Vitro* 2010; 24 (6): 1610-18.

[400] DeLorenzo ME, Fleming J. Individual and mixture effects of selected pharmaceuticals and personal care products on the marine phytoplankton species Dunaliella tertiolecta. Arch. Environ Contam Toxicol 2008: 54 (2): 203-10.

[401] Mayer P, Reichenberg F. Can highly hydrophobic organic substances cause aquatic baseline toxicity and can they contribute to mixture toxicity? Environ Toxicol Chem 2006; 25: 2639-44.

[402] van Wezel AP, Opperhuizen A. Narcosis due to environmental pollutants in aquatic organisms: residue-based toxicity, mechanisms, and membrane burdens. Crit Rev Toxicol 1995; 25 (3): 255-79.

[403] Cleuvers M. Aquatic ecotoxicity of pharmaceuticals including the assessment of combination effects. Toxicol Lett 142 2003; 3: 185-94.

[404] Cleuvers M. Mixture toxicity of the anti-inflammatory drugs diclofenac, ibuprofen, naproxen, and acetylsalicylic acid. Ecotoxicol Environ Saf 2004; 59 (3): 309-15.

[405] Faust M, Altenburger R, Backhaus T, *et al.* Joint algal toxicity of 16 dissimilarly acting chemicals is predictable by the concept of independent action. Aquat Toxicol 2003; 63 (1): 43-63.

[406] Nel AE, Mädler L, Velegol D, *et al.* Understanding biophysicochemical interactions at the nanobiointerface. Nat Mater 2009; 8(7): 543-57.

[407] Lundqvist M, Stigler J, Cedervall T, *et al.* The evolution of the protein corona around

[408] Lai ZW, Yan Y, Caruso F, Nice EC. Emerging techniques in proteomics for probing nano-bio interactions. ACS Nano 2012; 6(12):10438-48.

[409] Anon. 2008. Directive 2008/105/EC of the European Parliament and of the Council of 16 December 2008 on environmental quality standards in the field of water policy, amending and subsequently repealing Council Directives 82/176/EEC, 83/513/EEC, 84/156/EEC, 84/491/EEC, 86/280/EEC and

amending Directive 2000/60/EC of the European Parliament and of the Council. OJ L 348/84, 24 December 2008.

[410]　Nel A, Xia T, Madler L, Li N. Toxic potential of materials at the nanolevel. Science 2006; 311: 622-27.

[411]　Xia T, Li N, Nel AE. Potential health impact of nanoparticles. Annual review of public health 2009; 30: 137-50.

[412]　Chidambaram M, Krishnasamy K. Nanotoxicology: Toxicity of engineered nanoparticles and approaches to produce safer nanotherapeutics. Int J Pharm Sci 2012; 2(4): 117-22.

[413]　Zuin S, Micheletti C, Critto A, *et al.* Weight of evidence approach for the relative hazard ranking of nanomaterials. Nanotoxicol 2011; 5(3): 445-58.

[414]　Cockburn A, Bradford R, Buck N, *et al.* Approaches to the safety assessment of engineered nanomaterials (ENM) in food. Food Chem Toxicol 2012; 50(6):2224-42.

[415]　OECD. 2010. Guidance manual for the testing of manufactured nanomaterials: OECD's sponsorship programme; first revision. ENV/JM-MONO(2009)20/REV, OECD; 2 June 2010; Paris, France.

[416]　OECD. 2011. A steering group 7 case-study for hazard identification of inhaled nanomaterials: an integrated approach with short-term inhalation studies. ENV/CHEM/NANO(2011)6/REV1. 10th Meeting of the Working Party on Manufactured Nanomaterials, OECD; 27-29 June 2012; Paris, France.

[417]　OECD. 2012. Proposal for a template, and guidance on developing and assessing the completeness of adverse outcome pathways. 17. http://www.oecd.org/env/ehs/testing/49963554.pdf.

[418]　OECD. 2013. Revised Draft Recommendation of the Council on the safety testing and assessment of manufactured nanomaterials. ENV/CHEM/NANO(2013)3/REV1 Working Party on Manufactured Nanomaterials. OECD; 25 February 2013; Paris, France.

[419]　Russell WMS, Burch RL. 1959. The principles of humane experimental technique. London, UK: Methuen; Reprinted by UFAW, Hamilton Close, South Mimms, Potters Bar, Herts EN6 3QD England 238. 1992; pp: 8.

[420]　van Leeuwen CJ, Patlewicz GY, Worth AP. Intelligent testing strategies. In: van Leeuwen CJ, Vermeire TG, Vermeire T, editors. Risk assessment of chemicals: an introduction. Heidelberg, Germany: Springer. 2007; pp. 467-509.

[421]　Hartung T. Toxicology for the twenty first century. Nature 2009; 460 (7252): 208-212.

[422]　von der Kammer F, Ferguson PL, Holden PA, *et al.* Analysis of engineered nanomaterials in complex matrices (environment and biota): general considerations and conceptual case studies. Environ Toxicol Chem 2012; 31(1): 32-49.

[423]　Ma-Hock L, Burkhardt S, Strauss V, *et al.* Development of a short-term inhalation test in the rat using nano titanium dioxide as a model substance. Inhal Toxicol 2009; 21: 102-118.

[424]　Klein CL, Wiench K, Wiemann M, Ma-Hock L, van Ravenzwaay B, Landsiedel R. Hazard identification of inhaled nanomaterials: making use of short-term inhalation studies. Arch Toxicol 2012; 86 (7): 1137-51.

[425]　Klaine SJ, Alvarez PJ, Batley GE, *et al.* Nanomaterials in the environment: behavior, fate, bioavailability, and effects. Environ Toxicol Chem 2008; 27(9): 1825-51.

[426]　Handy RD, van den Brink N, Chappell M, *et al.* Practical considerations for conducting ecotoxicity test methods with manufactured nanomaterials: what have we learnt so far? Ecotoxicology 2012; 21(4): 933-72.

[427]　Handy RD, Cornelis G, Fernandes TF, *et al.* Ecotoxicity test methods for engineered nanomaterials: Practical experiences and recommendations from the bench. Environ Toxicol Chem 2012; 31:15-31.

CHAPTER 5

Monoclonal Antibodies in Lymphoma

Esmeralda Chi-yuan Teo and Colin Phipps[*]

Department of Hematology, Singapore General Hospital, Singapore

Abstract: Monoclonal antibodies (moAbs) have changed the landscape of lymphoma therapy. The chimeric anti-CD20 antibody, rituximab was the first to show significant activity in B-cell lymphomas. Since its FDA approval in 1997, rituximab has become a standard of care whether in combination with chemotherapy, or as a single-agent in induction and maintenance regimens in B-cell lymphoma. The success with rituximab paved the way for a steady stream of anti-lymphoma moAb therapies. Ofatumumab and obinutuzumab are next generation anti-CD20 antibodies designed to improve the cytotoxic effect of rituximab and enhance tumor cell killing. The breath of receptors amenable to Moab targeting extends far beyond CD20 and this chapter will review the most clinically impactful moAbs that have emerged over the past decade.

Keywords: Monoclonal antibodies, lymphoma, antibody-drug conjugates, rituximab.

INTRODUCTION

Interactions between antibodies and its cognate antigen form a major part of humoral immunity. In a similar way, monoclonal antibodies (moAbs) have been designed to recognize cancer-related antigens, including cell surface molecules, and soluble effectors with the goal of delivering targeted cytotoxicity. In most B-cell lymphoid malignancies, moAb therapy has become a standard of care.

MoAbs may be 'naked' antibodies or antibody conjugates which consists of an antibody (or antibody fragment) covalently linked to cytotoxic compounds, like drugs, immunotoxins or radioisotopes. These are designed as a homing device to carry the cytotoxic straight to the target antigen.

'Bare' antibodies employ various mechanisms of action including complement-dependent cytotoxicity (CDC), antibody-dependent cellular cytotoxicity (ADCC), antibody-dependent cellular phagocytosis (ADCP), and programmed cell death.

---

*Corresponding author Colin Phipps. Department of Haematology, Singapore General Hospital, 20 College Road, Singapore 169856; Tel: +65 6576 7795; Fax: +65 6226 0237; E-mail: phipps_8@msn.com

The degree to which these are employed depends on the antibody construct and degree of glycoengineering. A brief summary of the mechanisms of action of the various moAbs discussed is given in Table **1**.

The most widely used 'bare' antibody in lymphoma is rituximab. CDC may be triggered by the binding of serum C1q to the Fc portion of rituximab which initiates a signaling cascade that ultimately generates a membrane attack complex that disrupts the cell membrane of antibody-bound cells. ADCC is evoked by antigen-bound antibody and affected by cells expressing the FC gamma receptor (FcγR), including NK-cells and macrophages which express FcγRIIIA (CD16a) and neutrophils, dendritic cells, and macrophages which have FcγRIIA (CD32a). The recruitment of phagocytic cells also results in ADCP of antibody-bound cells. Single nucleotide polymorphisms of the FcγR have been variably associated with differential responses to rituximab. Genomic polymorphisms involving the FcγRIIIA result in exchanges of amino acids, valine, or phenylalanine at position 158 (V158F). This residue at position 158 interacts with the lower hinge region of IgG$_1$. It has been shown that NK cells from donors homozygous for FcγRIIIA 158V (V/V) bound more IgG1 resulting in greater ADCC compared with cells from donors who were homozygous for FcγRIIIA 158F (F/F) [1]. Studies using single-agent rituximab in FL have shown a beneficial effect of the FcγRIIIA 158 V/V genotype although the PRIMA study reported that the FcγRIIIA polymorphism did not influence outcomes of FL patients treated with rituximab, whether it was combined with chemotherapy or used as maintenance [2-4]. Another genomic polymorphism that results in an exchange of amino acids, histidine, or arginine at position 131 of FcγRIIA has also been shown to influence outcomes in FL patients treated with rituximab [5]. There is also evidence that anti-CD20 moAbs may induce apoptotic cell death in a manner that is not dependent on a robust immune response. This mechanism seems to be related to functional reorganization of CD20 into lipid rafts and the inhibition of multiple signaling pathways, including p38 mitogen-activated protein kinase, nuclear factor-kappa B, and AKT antiapoptotic survival pathway.

In this chapter, we will review the most clinically relevant moAbs being used in the treatment of lymphoid malignancies.

Table 1. Mechanism of action of various monoclonal antibodies used in lymphoma treatment.

Antibody	Structure	Main Mechanism of Action
Anti-CD20		
Rituximab	Chimeric IgG$_1$	Mainly CDC and ADCC
Ofatumumab	Human IgG$_1$	Strong CDC
Obinutuzumab	Humanized IgG$_1$	Strong ADCC and ADCP
Ibritumomab tiuxetan	Murine (Y90)	Cytotoxicity through yttrium-90
Anti-CD30		
Brentuximab vedotin	ADC with MMAE	Cytotoxicity through MMAE
Anti-CD52		
Alemtuzumab	Human IgG$_1$	Mainly CDC and ADCC
Anti-CD22		
Inotuzumab ozogamicin	ADC with calicheamicin	Cytotoxicity through calicheamicin
Moxetumomab pasudotox	Immunotoxin conjugated	Cytotoxicity through PE38
Others		
Otlertuzumab (anti-CD37)	SMIP	Direct apoptosis and ADCC
Polatuzumab vedotin (anti-CD79)	ADC with MMAE	Cytotoxicity through MMAE

CDC = complement-dependent cytoxicity; ADCC = antibody-dependent cellular cytotoxicity; ADCP = antibody-dependent cellular phagocytosis; ADC = antibody-drug conjugate; MMAE = monomethyl auristatin E; PE38 = Pseudomonas endotoxin 38; SMIP = small modular immunopharmaceutical

ANTI CD20 ANTIBODIES

The CD20 antigen was first characterized in 1980, and is present on the normal B cell from its development at the stage of cytoplasmic heavy chain expression to differentiation into antibody-producing plasma cells. Monocytes, T cell, non-lymphoid cells and stem cells are CD20-negative, making CD20 antigen a reliable B cell marker [6].

The CD20 molecule is a transmembrane protein that consists of a large loop and smaller loop. Monoclonal antibodies to this molecule are classified into type I or type II. Type I antibodies are able to translocate CD20 into detergent-insoluble fractions, or 'lipid rafts' which function as platforms for cell signaling and receptor trafficking [7, 8]. They are most effective in the activation of complement directed cytotoxicity (CDC). Type II antibodies, however, do not induce lipid rafts but are effective in antibody dependent cellular toxicity (ADCC) *via* the recruitment of cells displaying FcγR. They also play an important role in the induction of direct cell death *via* apoptotic or non-apoptotic mechanisms (Fig. **1**).

The CD20 antigen forms part of a B lymphocyte signal transduction complex and serves as a calcium channel initiating intracellular signals. This regulates B-cell growth and differentiation following activation [9]. The antigen is expressed at high density (90,000 molecules/cell) on 90% of all non-Hodgkin lymphomas (NHL), although in CLL, the antigen density is lower at approximately 8,000 – 15,000 molecules/cell [10]. CD20 is not internalized or down-modulated following antibody binding, thereby rendering it an excellent therapeutic target for most B-cell malignancies [11]. A summary of important trials conducted with the anti-CD20 moAbs is given in Table **2**.

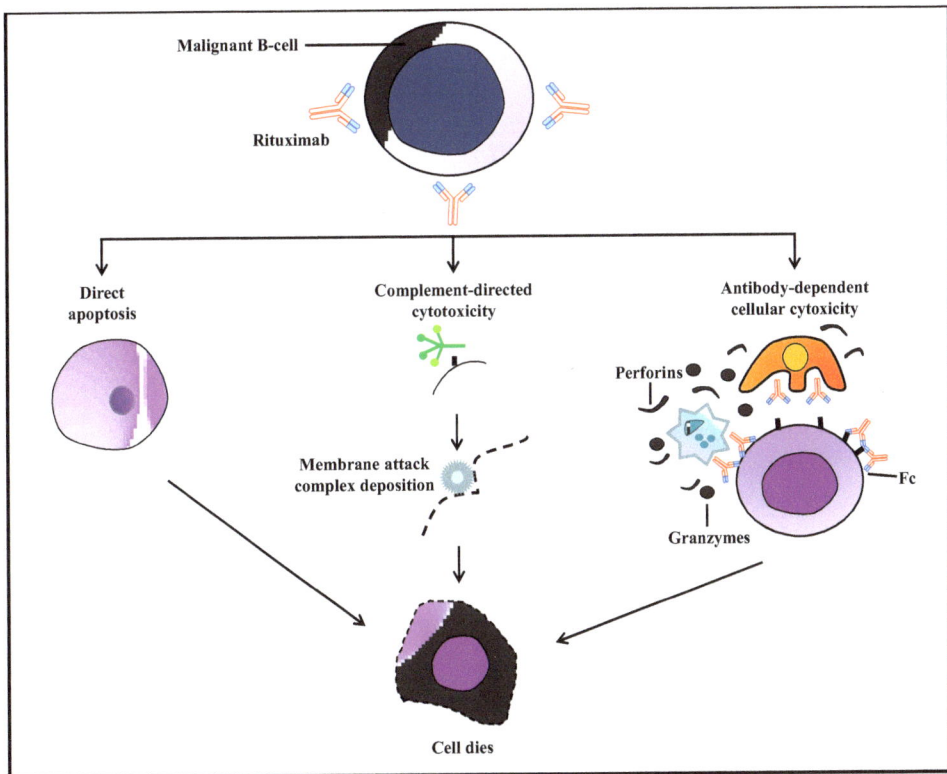

Figure 1: Mechanism of action of rituximab (courtesy of Roche pharmaceuticals).

Rituximab

Rituximab (Mabthera®) is a type I monoclonal antibody, which binds to the large loop of CD20 – an antigen expressed on B cells [12]. *In vitro* studies demonstrated that the binding of anti-CD20 moAbs inhibited proliferation of malignant B-cells and caused apoptosis by altering intracellular calcium concentration. Clinical trials initially investigating the role of radioimmunotherapy against lymphoma discovered that an iodine-labeled moAb called 'anti-B1' exerted a marked anti-tumor effect *via* ADCC [13]. The mechanism of destruction was later visualized in 2013 using high-resolution microscopy, in which Davis *et al.* described how rituximab induced the formation of a 'cap' comprising of CD20, intercellular adhesion molecule 1 (ICAM-1), moesin and the microtubule organizing center (MTOC) on the cell surface; this polarized the surface of the B cells and enhanced its interaction with NK cells, preferentially killing them [14].

The first experimental treatment of a patient with monoclonal murine 'anti-idiotype antibody' was reported in 1982. Levy *et al.* reported a dramatic anti-tumour response in a patient with lymphoma which had previously failed conventional chemotherapy [15]. Rituximab was subsequently constructed as a chimeric antibody with a murine variable region derived from monoclonal anti-CD20 antibody IDEC-2B8, and a human IgG_1-kappa constant region. This construct ensured high affinity and strong ADCC, and was thousand times more cytotoxic compared to its wholly murine equivalent [11]. *In vivo* studies demonstrated this chimeric antibody to selectively deplete CD20-positive B-cells in the peripheral lymph nodes and bone marrow of cynomolgus monkeys when administered weekly [16]. Therefore, based on these pre-clinical studies, rituximab was approved for clinical trial under the name IDEC-C2B8.

Early Trials and Single-Agent Rituximab

The first phase I clinical trial was conducted in 1994 in which 15 patients with relapsed low grade B-cell NHL were enrolled. Single intravenous infusions of IDEC-C2B8 were administered at escalating doses and resulted in tumor regression in half of the patients, with the most notable outcome occurring at a dosage of 500 mg/m^2. $CD20^+$ B cells were monitored and remained depleted for up to 3 months [17]. The mean half-life was 209 hours and the clearance rate was 9.2 mL/hour [18]. Minimal short term side-effects were observed, therefore, trials involving multiple administrations of rituximab as four weekly infusions were initiated, which demonstrated durable remission in half of the patients with indolent but chemotherapy-resistant lymphomas [19].

Following these successes, Coiffeur *et al.* conducted a multicenter phase II trial to assess the tolerability of single-agent rituximab in more aggressive lymphomas like DLBCL, MCL and other intermediate to high grade lymphomas. They enrolled 54 patients to receive 8 weekly infusions of 375 mg/m^2 or 500 mg/m^2

dosage. There was no significant difference in the overall survival (OS) or overall response rate (ORR) between the two dosage groups, although slightly more infusional-related reactions manifesting as anaphylaxis, chills, fevers, bronchospasms and hypotension were seen with the higher dose. These were mitigated by premedication with paracetamol, prednisolone, diphenhydramine and intravenous fluids. Other common adverse events were rash, arthralgia, fever, and cytopenias [20]. Following this study, weekly rituximab was deemed most tolerable and optimally effective at 375mg/m^2.

At present, rituximab monotherapy may be offered as first-line therapy for follicular lymphoma (FL) with low tumour burden as well as for maintenance therapy following completion of chemotherapy or high-dose therapy. Colombat *et al.* demonstrated sustainable clinical effectiveness in 50 untreated FL patients, of whom 62% achieved bcl-2 PCR negativity up to 12 months later [21]. The PRIMA study was a multicenter study in previously untreated FL patients requiring systemic therapy that had 2 randomization steps. The first was between 3 different immunochemotherapy regimens, and the second between rituximab maintenance and observation in patients whom achieved at least a partial remission (PR) after induction. Rituximab maintenance of 375 mg/m^2 was given ($N = 505$) every 2 months for up to 2 years and resulted in a sustained CR (at 2 years) of 71.5% in the maintenance arm compared to 52.2% in the observation arm ($P =.0001$) with no difference in OS observed [22]. In the setting of relapsed/refractory FL, a meta-analysis of 5 randomized controlled trials using rituximab maintenance showed an improvement in OS (pooled HR of death =.72) [23]. A randomized trial from the European Group for Blood and Marrow Transplantation also found a PFS benefit for rituximab maintenance post-autologous transplantation [24]. In the setting of mantle cell lymphoma (MCL), maintenance rituximab was shown to decrease the risk of progression compared to interferon-alpha. The latter was shown in a trial involving patients $\geq$ 60 years old who were randomized to induction therapy with either FCR ($N = 246$) or R-CHOP ($N = 239$) with a second randomization in responding patients to maintenance with rituximab or interferon alfa. FCR resulted in more early progressions and more toxic events. A longer remission duration was seen with rituximab maintenance with a 45% risk reduction in progression or death (HR 0.55, $P = 0.01$) [25].

Combination with Chemotherapy and the Advent of 'R-CHOP'

For more than twenty five years, the standard treatment of DLBCL has traditionally involved cyclophosphamide, doxorubicin, vincristine and prednisolone (CHOP).

The 'landmark' study was reported in 1993 by Fisher *et al.* that involved 1198 patients with advanced stage, intermediate- or high- grade NHL randomized to receive one out of four different chemotherapy regimens. In the CHOP arm, the ORR was 80% (estimated using clinical examination and chest or abdominal radiographs), with a complete response (CR) of 44%, and a 3 year OS of 54%. The CHOP regimen was associated with the least toxicity, was more cost effective but had similar efficacy compared to the other newer generation regimens [26]. Although CHOP proved to be the balance between efficacy and toxicity among other chemotherapeutic regimens, combination with non-cross resistant agents to further enhance treatment responses would be ideal. Later studies with CHOP alone in DLBCL employing more stringent response assessment techniques like computed tomography scans and gallium scans highlighted the significant proportion of patients (up to 22%) who have disease progression while on therapy [27]. Studies in advanced FL showed even lower rates of CR compared to DLBCL and persistence of b-cell lymphoma 2 (bcl-2) positivity despite the application of induction and salvage chemotherapy [28, 29]. This brought light to the need for more efficacious agents to eradicate minimal residual disease.

Rituximab has a synergistic or additive effect with chemotherapy. This was demonstrated in *in vitro* studies where the combination of rituximab with dexamethasone resulted in supra-additive anti-proliferative and apoptotic effects in NHL cell lines [30]. Studies demonstrating that pre-treatment of NHL cell lines with rituximab, augmented tumor cell susceptibility and allowed them to overcome their previous resistance to chemotherapy [31].

Follicular Lymphoma and Indolent NHL

The first combination study with CHOP reported in 1999, showed in 40 patients with low-grade B cell lymphoma, an ORR of 95%, CR of 55%, and a PR rate of 40%. In this study, patients who had newly diagnosed or relapsed/refractory low-grade, including FL were given rituximab with infusions 1 and 2 on days 1 and 6 before first cycle CHOP, then infusions 3 and 4 before third and fifth CHOP cycles, and infusions 5 and 6 after sixth CHOP cycle. With the addition of rituximab, they were able to achieve bcl-2 negativity in previously PCR positive patients [32].

Diffuse Large B-Cell Lymphoma (DLBCL)

In aggressive lymphomas, a phase II multicenter study ($N = 33$) using R-CHOP showed an ORR of 89 100% with disease IPI < 2 responding better. Twenty-nine of the 31 patients who responded remained in remission after a median observation of 26 months. Out of these, 16 patients had an IPI, more than 2 [33]. These studies confirmed that rituximab with CHOP had good safety profiles, higher CR rates and did not antagonize one another.

A European GELA group subsequently conducted a phase III randomized prospective international multicenter trial which further established superiority of R-CHOP over CHOP. Three hundred ninety nine untreated elderly patients with DLBCL were enrolled and given either eight cycles of CHOP or R-CHOP. CR was achieved in 76% of those given R-CHOP and 63% given CHOP alone (P=0.005) [27]. The 5 year OS was significantly longer in the R-CHOP group at 47.5% *versus* 29 % in the CHOP group [34]. A US Intergroup study revealed similar CR rates, whether rituximab was given upfront on day 1 with CHOP or given as maintenance after 6 to 8 cycles of CHOP [35]. The MInT study involving 824 untreated DLBCL patients showed that rituximab was most beneficial in young patients, as nearly twice as many young patients failed chemotherapy compared to chemotherapy plus rituximab (41% *vs.* 21%) [36]. These studies proved that the increased CR rate, decreased treatment failures and increased OS was undoubtedly attributable to the addition of rituximab.

Chronic Lymphocytic Leukemia (CLL)

In patients with CLL, rituximab demonstrated single-agent activity in 44 newly diagnosed patients who received 375 mg/m^2 for 4 consecutive weeks, then repeated the course every 6 months for 4 courses if there were objective responses. They achieved ORR rates of 58% with CR of 9% and a 2-year progression free survival (PFS) of 49% [37]. Previous trials had demonstrated high response and PFS with fludarabine based treatment, but none had demonstrated statistically significant survival. In a retrospective phase III trial comparing previous outcomes with fludarabine based therapy with the addition of rituximab, the US intergroup and CALG B group showed that the rituximab/fludarabine group had increased CR (0.38 *versus* 0.20, $P=.002$) and ORR (0.84 *versus* 0.63, $P=.0003$) rates, and had significantly improved OS ($P=.003$) [38]. A large phase III randomized study compared fludarabine and chlorambucil (FC) with the addition of rituximab (FCR) and showed improvement in PFS (20.6 months *vs.* 30.6 months), and ORR rates, but no improvement in OS [39]. However, the study did clearly established FCR as the standard of care for fit young patients. In relapsed CLL, chemoimmunotherapy with bendamustine and rituximab has shown to be effective and safe, with an ORR rate of 59% (CR 9%, PR 47%) [40]. Rituximab combinations have significantly improved the treatment of CLL.

The robust activity that rituximab demonstrated in these early trials, has led to the widespread use of this agent in the treatment of most B-cell lymphomas, whether in the upfront or relapsed setting. The rituximab era has borne endless possibilities in further improving cancer cure rates as more combinations will prove efficacious, especially as evolving novel agents are discovered. The effect of rituximab has extended beyond malignant diseases, especially autoimmune driven conditions. However the downside, as with all chemotherapy agents, is the development of resistance. Indeed, Fc receptor polymorphisms and down-regulation of CD20 following repeated rituximab exposure have proved to be more challenging, highlighting the need for advancement in alternative treatment options. In addition, the presence of circulating CD20 antigen was shown to diminish binding of rituximab to CLL cells and also correlated with poorer PFS and OS when detected on post-treatment testing. Nonetheless, the advent of rituximab as the first therapeutic monoclonal antibody has etched its place in altering the history of lymphoma treatment.

Ofatumumab

Ofatumumab (Arzerra®) is a fully humanized second generation CD20 type I antibody which mediates cytotoxicity through CDC and ADCC mechanisms. It was chosen for the development from a panel of novel CD20 moAbs derived from human immunoglobulin transgenic mice as named 2F2 or HuMax-CD20 was found to be the most active in lysis of lymphoma B-cell lines (Su-DHL-4, Daudi and Raji cells) and rituximab-resistant fresh tumor isolates. *In vitro* studies have shown it to be more potent in inducing CDC activity compared to rituximab, even in cells with low CD20 expression [41]. Ofatumumab binds to a novel region of the CD20 molecule (encompassing the small and large loops), which was seen through epitope-mapping studies. It was postulated that this binding was more stable compared to rituximab and therefore was more effective in recruiting C1q [42]. This, together with slow off-rates, appears to be two independent factors that increase the potency of this moAb in activating CDC [12].

CLL

Studies have demonstrated that the mechanism of rituximab's induction of cytotoxicity is dependent on CD20 expression [43]. As CLL has low CD20 expression [44], monotherapy with rituximab has shown only limited activity [45]. Two trials involving high risk CLL patients treated with rituximab and methylprednisolone showed CR rates of only 22% to 36% [46, 47]. In patients who are refractory to fludarabine and alemtuzumab or have bulky lymphadenopathy, salvage chemotherapy is associated with a low ORR of 23% and a median OS of 8 months [48]. The Hx-CD20-406 trial studied this very population of refractory CLL patients and based on their findings, ofatumumab received accelerated United States Food and Drug Administration (FDA) approval, in 2009. Coiffier *et al.* reported this open-label phase I/II trial of ofatumumab in 2008, which enrolled 33 patients with relapsed/ refractory CLL who had previously received rituximab, alemtuzumab or fludarabine. They received weekly infusions at 100, 300 or 500 mg on week 1, then 3 weekly infusions of 500 mg (cohort A; n = 3), 1000 mg (cohort B; n = 3) or 2000 mg (cohort C; n = 27). Of note, 51% of the patients reported infections (mostly nasopharyngitis), one was fatal from infectious interstitial lung disease and hematologic toxicity was reported in 15%. In this study, PR or near PR was seen in half of the patients and there was a clear relationship between the dose and the efficacy. The median PFS was 15 weeks and the median duration of response was 3.7 months. It is noteworthy that only 7 of 33 patients had previously received rituximab and none received it within the preceding 6 months of trial entry [49]. Concurrently, an international multicenter trial was conducted in which 154 patients received 8 weekly intravenous ofatumumab infusions, followed by 4 monthly infusions. At the interim analysis, the ORR was similar to the study by Coiffier *et al.* at 58% (99% CI, 40 – 74%) in the fludarabine-resistant group and 47% (99% CI, 32% to 62%) in the bulky disease group. Although not powered to statistically analyze subgroups, it was noted that some cytogenetics particularly 17p deletions had poorer response. Fortunately, patients previously considered high risk due to advanced age (more than 70 years), poor performance status and 11q deletion did respond well [50]. Wierda *et al.* also investigated 206 patients, of whom 117 were heavily pretreated with rituximab. The ORR was 43% and 53% in the rituximab and rituximab-naive groups, respectively [51]. The fact that ofatumumab bound to a distinct epitope was vital in overcoming previous rituximab refractoriness.

An international phase II combination trial was reported in 2011, in which 67 previously untreated CLL patients were randomly assigned to receive six courses of 500 mg ofatumumab with fludarabine and cyclophosphamide (FC) ($N = 31$) or six courses of 1000 mg ofatumumab and FC, every four weeks. All patients were pre-medicated with steroids, antihistamines and acetaminophen prior to infusions. Sixty four percent completed all 6 courses, for which the main cause for premature withdrawal was cytopenia. The CR rate was 32% in the 500 mg cohort and 50%in the 1000 mg group, which was not statistically different. The ORR was 77% and 73%, respectively. Two patients died of febrile neutropenia and one of dyspnoea. Similar rates of infusional reactions during ofatumumab were experienced [52]. Another phase II trial combining ofatumumab with bendamustine demonstrated an ORR of 72.3% and CR of 17% in relapse/refractory CLL patients. This study included high risk patients and demonstrated that the ORR was not affected by immunoglobulin heavy chain variable (IGHV) mutation status, deletion 11q, age over 70 or prior exposure to fludarabine and rituximab. However, it was significantly affected by bulky disease, advanced Binet's stage, TP53 disruptions, deletion 17p (del (17p)) and splicing factor 3B subunit 1 (SF3B1) mutations [53]. Phase III trials of comparing ofatumumab and chlorambucil with chlorambucil alone, and ofatumumab maintenance, are currently ongoing. So far, combination studies with ofatumumab have shown to have good effect in previously treated or untreated CLL.

FL

Ofatumumab was first clinically evaluated in a phase I/II trial in 2008 by Hagenbeek *et al.* for patients with relapsed/refractory FL with rituximab resistance. Forty patients received 4 doses of weekly intravenous infusions at escalating doses of 300, 500, 700 or 1000mg. Adverse events were infusion related and commonly reported as chills, rash, pruritus, pyrexia, vomiting and urticaria, and the frequency of symptoms decreased following subsequent infusions. Treatment led to immediate and profound B-cell depletion for 6 to 10 months. This was longer and deeper than with rituximab, which the authors proposed was attributable to the increasing concentration of the drug from the first and fourth dose. After the infusions, ofatumumab had a longer terminal half-life ($t_{1/2}$) (410 hours) and drug clearance rate was 9.5 mL/hour, so cellular exposure to the antibody likely extended beyond 4 weeks. Clinical response rates ranged from 20-63% but were not dose-dependent (unlike experience with CLL patients) and previous rituximab-treated patients had an ORR across dose groups of 64%. Sixty five percent of patients reverted to a bcl-2 negative state and median time-to-progression (TTP) was 32 months [54]. Ofatumumab monotherapy demonstrated activity in 27 heavily treated rituximab –resistant FL patients in another phase II trial reported in 2012. Czuzcman *et al.* demonstrated an ORR of 13% in the 500 mg cohort and 10% in the 1000 mg group [55]. In refractory DLBCL patients ineligible for ASCT, the 81 patients enrolled achieved an ORR of 11%, with a median duration of response of 9.5 months with single-agent ofatumumab [56]. Ofatumumab was also recently investigated in refractory MCL, but results were poor with an ORR of 8.3% and median OS of 11.2 months [57]. These phase II trials of ofatumumab monotherapy proved that this antibody demonstrated modest activity alone in different subtypes of rituximab-resistant lymphoma, but brought light to the need for studies exploring combination therapy.

An international trial combining ofatumumab with CHOP (O-CHOP) was published in 2012, which looked at the efficacy and safety of two dose levels of ofatumumab in previously untreated FL. Fifty nine patients were randomized to receive six cycles of either 500 mg or 1000 mg ofatumumab. The combination proved efficacious; the ORR was 90% in the 500 mg group with 24% CR and 21% PR. In the 1000 mg cohort, the ORR was 100% with 38% CR and 45% PR. When stratified by FLIPI scores; 59% of the low risk (n=17), 50% of intermediate risk (n=20) and 76% of high risk (n=21), attained CR. O-CHOP was tolerable, although 90% of patients developed neutropenia and required granulocyte-colony stimulating factor (GCSF) support [58]. These rates are higher than historical R-CHOP related cytopenias [59]. The trial concluded that O-CHOP was a highly active regimen and there was no difference between both doses, although the investigators chose the 1000 mg dose for future trials.

DLBCL

In transplant eligible patients with relapsed DLBCL and grade 3b or transformed FL failing R-CHOP, 61 patients were enrolled to either receive ifosfamide, carboplatin and etoposide (ICE) or dexamethasone, cytarabine and cisplatin (DHAP) in combination with ofatumumab followed by consolidation with high-dose therapy. With this regimen, the ORR was 61% with a CR rate of 37%, and it was encouraging to note that similar CR and PR rates were also attained in the subgroups who had early-relapsing or primary refractory disease (ORR 55% and CR 30%) and with patients with sa-International Prognostic Index (IPI) scores of 2 or 3 (ORR 59%, CR 31%). Febrile neutropenia was more common in the O-DHAP group (31% *vs.* 3%), despite all patients being routinely given GCSF. Importantly, these salvage regimens did not interfere with stem cell mobilization and 74% patients went on to ASCT [60]. However, a recently reported phase III trial which randomized relapsed/refractory DLBCL patients between O-DHAP (N = 222) and R-DHAP (N = 225) showed no difference between treatment arms in terms of 2-year PFS of 21% and 26% (P =.27) and OS 41% and 36% (P =.25), respectively [61].

The advantage of ofatumumab over rituximab in terms of an enhanced CDC has translated into a modest clinical benefit when used as a single-agent therapy in rituximab-resistant B-NHL. Nonetheless, the accelerated FDA approval in CLL that was awarded, testifies to a meaningful response that some patients with prior fludarabine, alemtuzumab and rituximab exposure may achieve. However, the small molecules like the Bruton's tyrosine kinase inhibitors have already started to displace single-agent ofatumumab use in CLL. The recent phase III RESONATE trial demonstrated superiority of ibrutinib over ofatumumab monotherapy with respect to PFS (9.4 *vs.* 8.1 months), 12-month OS (90% *vs.* 81%) and ORR (43% *vs.* 4%), with these benefits extending to the high-risk chromosome 17p13 deletion subgroup [62]. This notwithstanding, phase II trials suggest that ofatumumab might find its place in lymphoma treatment in combination with chemotherapy and phase III comparisons with rituximab and other regimens are currently ongoing.

Obinutuzumab

Obinutuzumab (Gazyva®) is a next-generation humanized type II anti-CD20 antibody. This antibody binds to CD20 and induces intracellular signaling transduction cascades that mediate homotypic cell adhesions which rearrange the actin cytoskeleton causing the lysosome membrane to increase in permeability [63, 64]. This proposed mechanism culminates in the generation of reactive oxygen species (ROS) scavengers, which compromises the plasma membrane integrity and causes non-apoptotic cytoplasmic cell death [65]. Unlike type I antibodies, type II antibodies do not stabilize CD20 into lipid rafts, and therefore exhibit lower CDC. During fine epitope mapping studies, obinutuzumab was seen to bind to CD20 in a completely different orientation than type I antibodies with a wider elbow angle, forming different molecular assemblies to elicit different cellular responses [66].

During the humanization process, the variant at the elbow hinge region that had superior binding to the CD20 type II epitope was chosen. Additionally, obinutuzumab was also glycoengineered to have an enhanced afucosylated Fc segment which more effectively engaged effector cells displaying CD16 receptors (*e.g.* FcγRIIIA in NK cells). The resultant design increased its affinity by up to 50-fold [67]. Other cells like polymorphic neutrophils (PMN) expressing FcγRIIIB (CD16B), and monocytes and macrophages (*via* FcγRIIIA) are also activated by obinutuzumab and its glycoengineered structure consequently results in enhanced antibody-dependent cellular phagocytosis (ADCP).

Obinutuzumab demonstrated superior ADCC activity, direct cell death and B-cell depleting activities compared to rituximab and ofatumumab in Raji lymphoma cell lines. In *ex vivo* CLL cells from whole blood samples and the CLL cell line Mec1, obinutuzumab demonstrated increased B-cell depletion compared to rituximab and ofatumumab, when assessed with ADCC or direct cytotoxicity assays [68, 69].

Indolent NHL

Obinutuzumab was first evaluated in the phase I GAUDIN trial, reported in 2012, in 21 patients with heavily pre-treated, relapsed/refractory CD20-positive indolent NHL. Obinutuzumab was administered at a dose-escalation regimen for eight 21-days cycles. The most common side effects were infusion related. Pharmacokinetic studies indicated that the targets were fully saturated at doses above 400 mg, and the serum concentrations reached a steady state after cycle 4. Hence, subsequent trials gave three times the dose in cycle one to reach the steady state earlier. No clear dose-response relationship was identified. This antibody had good tolerability up to 2000 mg and was effective as monotherapy with an ORR of 43% in B-NHL, achieving 5 CRs and 4 PRs [70]. In a similar GAUSS trial involving relapsed NHL and CLL, obinutuzumab was given weekly for 4 weeks followed by 3-monthly maintenance for up to 2 years. At the end of induction, 23% achieved PR, 54% had stable disease. After maintenance, the best ORR was 32% [71].

Two phase II trials as part of the GAUGUIN study were subsequently undertaken. The first was in a heavily pre-treated, refractory, indolent NHL population comparing doses of 400 mg with 800 mg and increasing up to 1600 mg during the first cycle. Thirty one of forty patients completed the course, among which the nine discontinued due to insufficient response or pancreatitis. The observation period spanned 33.7 months. In the 1600/800 mg group, the ORR was 55% and PFS 11.9 months with steady state drug levels being achieved by cycle 3. The 400 mg group had consistently low serum concentrations, and their ORR was 17% and median PFS 6 months [72].

Combination studies were also undertaken where obinutuzumab was added to CHOP (G-CHOP) or FC (G-FC) in relapsed/refractory FL in the GAUDI study. Fifty six patients received G-CHOP (every 3 weeks for 6 to 8 cycles) or G-FC. The dose of obinutuzumab was randomly assigned to either 1,600/800-mg or 400/400-mg, followed by obinutuzumab maintenance if they responded initially. The 1,600/800-mg G-CHOP arm had a higher infection incidence than 400/400-mg arm (57% *vs.* 29%). Four patients in each arm required dose reductions and experienced delays in treatment cycles. The combinations had an acceptable safety profile, although G-FC was associated with more adverse events than G-CHOP. The end-of-induction ORR for G-CHOP was 93% (95% CI, 66.1% to

99.8%) in the low dose arm and 100% (95% CI, 76.8% to 100%) in the high dose arm. CR rate was 39% across the entire G-CHOP population. Within the G-FC cohort, the ORR was 100% (95% CI, 76.8% to 100%) in the low dose arm *vs.* 86% (95% CI 57.2 % to 98.2%) in the high dose arm. In total, 50% achieved CR. All the rituximab-resistant patients achieved at least PR despite a third of them having the unfavorable FcγRIIIA receptor variant [73]. These results proved the efficacy of obinutuzumab combinations even in patients with rituximab resistance.

The GAUGUIN study reported in 2014 evaluated obinutuzumab in the setting of relapsed/refractory indolent lymphoma. The infusional related reactions were more pronounced among CLL compared to patients in the NHL studies, which was thought to be secondary to increased cytokine release from malignant B-cells, but still deemed tolerable. The best response was PR, achieved in 62% in phase one and 30% in phase two, with a median PFS of 10.7 months [74].

In a multinational phase III trial conducted between 2010 to 2012, amongst untreated CLL who were symptomatic or Binet stage C, 781 patients were randomized to receive obinutuzumab-chlorambucil (G-Clb) or rituximab-chlorambucil (R-Clb), compared with chlorambucil (Clb) monotherapy. Patients were assessed at 3 months after the end of treatment with CT scans and bone marrow biopsies and followed up for 39 months. Minimal residual disease was analyzed centrally with allele-specific oligonucleotide PCR assays. The most common adverse events documented were again infusional with pyrexia and musculoskeletal pain, and cytopenias. The results predictably demonstrated that G-Clb group had increased ORR and median PFS compared to Clb (26.7 *vs.* 11.1 months; hazard ratio for progression or death 0.18). This benefit was seen across all analyzed subgroups except in patients with del (17p). G-Clb group had higher rates of CR compared to R-Clb (20.7% *vs.* 7%) and PFS (26.7 *vs.* 15.2 months). This large study concluded that obinutuzumab-chlorambucil provided an OS advantage over chlorambucil monotherapy as well as induced deeper remission compared to rituximab-chlorambucil, although not able to fully abrogate the del(17p) disadvantage [75]. With this trial result, obinutuzumab became the first drug with 'breakthrough therapy designation' to be approved by the FDA for use in combination with chlorambucil in untreated CLL patients.

Phase III trials looking at various combinations of obinutuzumab are currently ongoing, including, the GALLIUM study comparing G-chemotherapy against R-chemotherapy in indolent lymphoma (NCT01332968) and GADOLIN study using obinutuzumab and bendamustine or bendamustine alone in rituximab-resistant FL (NCT01059630). Results from other combination studies with kinase inhibitors like ibrutinib and idelalisib are eagerly awaited.

Aggressive NHL

Among the relapsed/refractory DLBCL or MCL population, a parallel study was conducted in 40 patients who were treated with the same dosing schedule of 400mg/400-mg (where they received 400mg infusions on day 1 and 8 of the first cycle and day 1 for the other seven cycles) in one cohort and 1,600/800-mg (where they received 1,600mg on day 1 and 8 of the first cycle and 800 mg on day 1 for the rest of the cycles) in another. The best ORR was 37% in 1,600/800-mg group *vs.* 24% in 400/400-mg group. Out of them, 32% (8 of 25) had DLBCL and 27% (4 of 15) had MCL. Among the DLBCL patients, 20% (5 of 25) were previously rituximab-resistant and 12% of this subset achieved CR. The median response duration was 9.8 months but there were 3 patients who had response duration of more than 2 years. In the 1,600/800-mg arm, serum concentrations of obinutuzumab increased up to cycle 2 where steady state was reached [76]. These results were encouraging as it demonstrated that obinutuzumab monotherapy had promising activity in heavily pretreated DLBCL and MCL.

Radiolabelled Anti-CD20 Monoclonal Antibody: Ibritumomab Tiuxetan

Lymphomas are inherently radiosensitive, but usually present at an advanced stage, precluding the delivery of conventional involved-field radiotherapy. In this regard, radiolabeled moAbs represent a novel way of targeting lymphoma cells and taking advantage of this radiosensitivity. Ibritumomab tiuxetan (IDEC-Y2B8; trade name Zevalin®) is the murine equivalent of rituximab and targets the same epitope on the CD20 molecule. It is covalently bonded to tiuxetan, which chelates the radioactive particle yttrium-90. ^{90}Y is more useful over other radioisotopes because it delivers high beta energy (maximum energy of 2.3 MeV with a half-life of 64 hours), and has a path length of 5 to 10mm - improving its ability to kill bulky, poorly vascularized tumors. It does not emit gamma rays and therefore can potentially be administered in the outpatient setting [77]. However, this drug should be used with caution in patients with significant lymphoma involvement of the bone marrow.

FL and Indolent NHL

Phase I/II studies have determined the maximum tolerated dose (MTD) to be 0.4 mCi/kg, achieving ORR and CR rates of 67% and 26%, respectively [78].

Responses have shown to be durable in indolent and large B-cell lymphomas, with a median TTP of 28.3 months, with some long-term responses lasting more than 5 years [79]. A comparison trial of IDEC-Y2B8 with rituximab was reported in 2002, where patients received either single dose IDEC-Y2B8 0.4 mCi/Kg (n=73) or 4 doses of rituximab 375mg/m^2 intravenously (n=70). The ORR was 80% and CR 30% in the IDEC-Y2B8 group *vs.* 56% and 16% in the rituximab group [80]. This prompted FDA to approve IDEC-Y2B8 for rituximab-naïve relapsed/refractory, low-grade follicular or transformed NHL [81]. The most common immediate toxicity is hematologic with cytopenias lasting 1 to 4 weeks. Myelodysplastic syndrome and acute myelogenous leukemia following treatment was observed in 1% of patients on follow up [82].

IDEC-Y2B8 monotherapy has demonstrated CR rates of 56% and median PFS of 26 months in patients with FL grade 1-3a when given as first-line therapy [83]. Based on previous studies, the optimal schedule of administration is one week after rituximab 250 mg/m^2, which is given to cytoreduce the tumors and circulating B-cells thereby aiding in delivery of the drug to the intended target. A trial using single agent therapy has also demonstrated activity in rituximab-resistant FL, in which 54 patients were given a single dose of IDEC-Y2B8. They achieved a 74% ORR (15% CR), which was especially encouraging since these patients previously received a median of four chemotherapy regimens prior to enrollment [84].

IDEC-Y2B8 was further assessed as a maintenance therapy after R-CHOP in a phase III trial of FL (*vs.* no maintenance therapy). Rituximab was given one week before and on the day of administration of IDEC-Y2B8 in the intervention arm. The median PFS with consolidation was prolonged and interestingly 77% of patients who achieved PR after induction converted to CR after consolidation, with a final CR of 87% [85]. These results were reproducible in another large phase III First-Line Indolent Trial (FIT) conducted in 414 patients in which researchers monitored minimal residual disease and reported that 90% of treated patients converted to bcl-2 PCR negative *vs.* only 36% in the non-IDEC-Y2B8 arm. At follow up, they achieved 8-year PFS of 41% with IDEC-Y2B8 *versus* 22% in the control population (hazard ratio 0.47; $P < .001$).

DLBCL and Aggressive NHL

Following the promising activity seen in indolent NHL, clinical trials assessed IDEC-Y2B8 in the setting of aggressive lymphoma/DLBCL patients who were transplant ineligible. Morschhauser *et al.* conducted a phase II trial of 104 patients who had first relapse or were primary refractory and over 60 years old. The therapeutic response to one dose of IDEC-Y2B8 produced ORR of 52% with CR/unconfirmed CR rates of 24% and 40% with response durations of up to 36 months. For MCL, IDEC-Y2B8 was associated with an ORR of 32% in heavily pre-treated, relapsed/refractory patients [86]. In limited stage aggressive NHL, the S0313 trial where patients were treated with 3 cycles of CHOP and 40 to 50 gray of involved field radiotherapy to the affected site followed by ibritumomab 3 to 6 weeks later showed PFS of 89% at 2 years, 82% at 5 years and 75% at 7 years [87].

Ibritumomab as Part of Transplant Conditioning Therapy

Ibritumomab has been extensively evaluated in combination with high dose chemotherapy regimens prior to myeloablative autologous stem cell transplants (ASCT) to obviate the need for total body irradiation (TBI), while aiming to lower the incidence of relapse. In combination with etoposide and cyclophosphamide followed by ASCT for DLBCL, preliminary results reported 2-year OS and PFS of 92% and 78%, respectively [88]. It was also proven effective and safe when administered to patients who were previously considered ineligible for TBI due to advanced age or prior radiotherapy. Forty one patients who had poor risk MCL (N = 13) or DLBCL (N = 20), or transformed lymphoma received IDEC-Y2B8then BEAM (carmustine, etoposide, cytarabine, melphalan) followed by ASCT. Median patient age was 60 years. The estimated 2-year OS and PFS were 88.9% and 69.8%, respectively. The median time to white cell engraftment was 11 days (range 9 – 26). Grade ≥ 3 pulmonary toxicity was seen in 10 patients [89]. This regimen also produced encouraging results in relapsed MCL patients with OS, event-free survival, and PFS at 4 years of 78%, 62%, and 71%, respectively [90]. A different trial that used myeloablative doses of IDEC-Y2B8 as high-dose therapy recently reported on their long-term follow up of 60 patients that were enrolled. At a median follow up of 5.9 years, the PFS and OS were 62.7% and 72.9%, respectively [91]. IDEC-Y2B8 has also been used as a part of conditioning therapy in reduced intensity allogeneic transplant regimens for advanced NHL. Date thus far indicate that the addition is well tolerated and does not compromise engraftment or increase transplant-related morbidity.

Overall, IDEC-Y2B8 has proven to be safe with relative efficacy when used as consolidation therapy and as a part of the transplant conditioning regimen.

Table 2. Summary of clinical trials with anti-CD20 moAbs.

Study	Drug	Condition	Dose	CR Rate	PFS	OS
Coiffer 1998	Rituximab	Low grade/ follicular lymphoma	375 or 500mg/m^2	9%	-	-
Colombat 2001	Rituximab	Follicular lymphoma	375mg/m^2	20%	74·9%	No significant difference
Salles 2011	Rituximab maintenance *vs.* observation	Follicular lymphoma	375mg/m^2	71.5% *vs.* 52%	75% *vs.* 58%	No significant difference
Czuczman 1999	RCHOP	Low-grade B cell relapse/refractory lymphoma	375mg/m^2	55%	-	95%
Vose 2001	RCHOP	Newly diagnosed aggressive NHL	375mg/m^2	IPI<2 67% IPI>2 56%	91%	88%
Feugier 2005	RCHOP *vs.* CHOP	Elderly DLBCL	375 mg/m^2	76% *vs.* 63%	-	47% *vs.* 29%
Pfreundschuh 2006	RCHOP *vs.* CHOP	Young untreated DLBCL	375 mg/m^2	86% *vs.* 68%	85% *vs.* 68%	93% *vs.* 84%
Hainsworth 2003	Rituximab	CLL	375 mg/m^2	9%	49%	-
Byrd 2005	fludarabine *vs.* rituximab/ fludarabine	CLL	375 mg/m^2	20% *vs.* 38%	45% *vs.* 67%	81% *vs.* 93%
Robak 2010	CR *vs.* FCR	Relapsed CLL	375 mg/m^2	13% *vs.* 24%	20.6 mths *vs.* 30.6 mths	No significant difference
Fischer 2011	bendamustine, rituximab	Relapsed CLL	375 and 500 mg/m^2	9%	15.2 mth	-
Coiffier 2008	ofatumumab	Relapse / refractory CLL, resistant to fludarabine, rituximab or alemtuzumab	Escalating dosage 100 mg to 2000 mg	0	15 wk	-
Wierda 2010	Ofatumumab	Refractory CLL	Dose 1 =300mg; doses 2 to 12=2,000m g).	1.7%	5.7months	13.7 months
Wierda 2011	O-FC	Untreated CLL	500mg or 1000mg	32% or 50%	-	-
Cortelezzi 2014	Ofatumumab and bendamustine	Relapse or refractory CLL	1000mg	17%	49%	83%
Coiffier 2013	Ofatumumab	Refractory DLCBCL	300mg then 1000mg	8%	9.5 mth	-
Czuczman 2012	O-CHOP	Untreated follicular lymphoma	500mg or 1000mg	24% or 38%	-	-

Table 2: contd…

Matasar 2013	O-ICE *vs* O-DHAP	Transplant eligible relapse DLBCL or transformed follicular lymphoma	1000mg	33% *vs* 42%	9.5 mth	16.7 mth
Byrd 2014	Ibrutinib *vs* ofatumumab	Relapse/refractory CLL	1000mg	2% *vs* 1%	9.4 mth *vs* 8.1 mth	90% *vs* 81%
Salles 2013	Obinutuzumab	Pre-treated/refractory indolent NHL	1600/400mg *vs* 1600/800mg	0 *vs* 9%	6mths *vs* 11.9 mths	-
Morschhauser 2013	Obinutuzumab	Relapse/refractory DLBCL and MCL	400/400mg *vs* 1600/800mg	12%	2.6 mth *vs* 2.7 mth	-
Radford 2013	G-CHOP	Relapse/refractory FL	400mg/400 mg or 1600/800mg	25%	-	-
	G-FC			30%		
Cartron 2014	Obinutuzumab	Relapse/refractory CLL	1000mg	-	10.7 mth	-
Goede 2014	G-Clb or R-Clb or Clb	Untreated CLL	1000mg	20.7% *vs* 7% *vs.* 0	26.7 mth *vs.* 15.2 mth *vs.* 11.1 mth	-
Witzig 1999	IDEC-Y2B8	Relapse/refractory indolent lymphoma	0.2-0.4mCi/kg	26%	-	-
Scholz 2013	IDEC-Y2B8	Untreated follicular lymphoma	250 mg/m^2 Rituximab, then 15 MBq/kg ^{90}Y	49%	57%	-
Emmanouilides 2006	IDEC-Y2B8	Rituximab resistant FL first relapse	250 mg/m^2 Rituximab, then 0.4 or 0.3 mC/kg	49%	-	-
		Second relapse		28%		
Morschhauser 2007	IDEC-Y2B8	Relapse/refractory DLBCL	0.4mCi/kg	12-39.5%	1.6-5.9 mth	4.6-22.4 mth
Morschhauser 2008	IDEC-Y2B8 *vs.* observation	Prev treated FL in first remission	14.8 MBq/kg	87%	36.5 *vs.* 13.3 mth	No significant difference
Nademanee 2005	IDEC-Y2B8, then ASCT	Poor risk/relapse NHL	Capped at 100 mCi	-	92%	78%
Krishnan 2008	IDEC Y2B8 then BEAM for ASCT	Lymphoma	0.4mCi/kg	66%	69.8%	88.9%
Kolstad 2014	IDEC Y2B8, then high dose BEAM ASCT	Relapse MCL	0.4mCi/kg	51%	71%	78%
Devizzi 2013	IDEC Y2B8, autologous SCT	Poor risk NHL, not candidate for BEAM protocol	Myeloablative dose	90%	62.7%	72.9%

CD30 Receptor

The CD30 receptor is a 120-kDa type I transmembrane protein that has extracellular domain sequence homology with members of the tumor necrosis factor (TNF) receptor superfamily [92]. It has an elongated structure which consists of six cysteine-rich pseudo-repeat motifs and a scaffold of disulphide bridges [93]. Its cytoplasmic tail contains a docking site for signaling molecules and several TNF receptor associated factor (TRAF)-binding motifs that activates a number of signaling pathways, including nuclear factor (NF)-κB.

Expression of CD30 is normally limited to activated B- and T- lymphocytes, as well as virally-infected lymphocytes. Its exact function is not clearly understood but induction of apoptosis has been seen upon activation in anaplastic large cell lymphoma (ALCL) cell lines. CD30 can undergo shedding into a soluble form (sCD30), which can be detected in the sera of patients with CD30+ tumors [94]. Hematologic malignancies that express CD30 include Hodgkin lymphoma (HL), ALCL, primary mediastinal B-cell lymphoma, primary effusion lymphoma associated with human herpes virus-8, adult T-cell leukemia/lymphoma, mycosis fungoides, multiple myeloma, and mast cell tumors [95-97]. The expression is particularly high in HL and ALCL [98-100].

Bare CD30-Antibody

An antagonistic CD30 murine antibody, C10 was licensed to Seattle Genetics where it was chimerized with the human gamma 1 heavy chain and the kappa light chain constant regions. The resulting chimeric antibody, SGN-30 resulted in *in vitro* growth arrest of HL cell lines in the G1 phase of the cell cycle and DNA fragmentation [101]. Furthermore, when combined with the chemotherapeutic agent's bleomycin, cytarabine, and etoposide, synergistic tumor cell killing was seen in HL tumor cell xenografts [102].

An open-label phase II study was conducted in patients with relapsed/refractory HL ($N = 38$) and ALCL ($N = 41$) whom had received a median of 3 prior regimens. SGN-30 was given in 6 weekly infusions but showed only modest clinical activity with objective responses seen in 5 ALCL patients and none in HL [103].

Anti-CD30 Drug Conjugate: Brentuximab Vedotin

Brentuximab vedotin (Adcetris®) is an anti-CD30 antibody (cAC10) conjugated with the anti-mitotic agent monomethyl auristatin E (MMAE) *via* a valine-citrulline peptide linker. Each antibody is conjugated with an average of 4 molecules of MMAE, which are released upon exposure to proteolytic enzymes *in vitro*. A similar mechanism is assumed to occur in cancer cells after internalization through CD30-receptor mediated endocytosis. MMAE causes tubulin polymerization and G2/M-phase growth arrest leading to apoptotic cell death [104, 105]. *In vitro* studies showed potent and selective activity against CD30+ tumor cell lines with a half maximum inhibitory concentration (IC_{50}) of less than 10 ng/mL. Subsequent *in vivo* testing in xenograft models of HL and ALCL tumors in SCID mice showed partial and complete responses respectively, at doses as low as 1 mg/kg [106].

In a phase I, multicenter, dose-escalation study of brentuximab vedotin 3 weekly given to 45 patients (42 with HL, 2 with ALCL, 1 with angioimmunoblastic T-cell lymphoma) with relapsed or refractory CD30-positive hematologic cancers, an objective response was observed in 17 patients. Seventy three percent of the cohort study had previously undergone autologous stem cell transplantation (ASCT). The MTD was found to be 1.8 mg/kg. Only one patient received 3.6 mg/kg and died from neutropenic sepsis. At dose level 2.7 mg/kg, there were 3 out of 12 patients with dose-limiting toxicities (one unrelated grade 3 acute renal failure, one grade 3 hyperglycemia, one unrelated grade 3 prostatitis). The most common adverse events were grade 1 or 2 - fatigue (36%), pyrexia (33%), diarrhoea (22%), nausea (22%), neutropenia (22%), peripheral neuropathy (22%). Serious adverse events deemed to be related to brentuximab vedotin given at ≤ 1.8 mg/kg were grade 3 hypercalcemia ($N = 1$), myocardial ischemia ($N = 1$), and anaphylactic reaction ($N = 1$).

U.S. Food and Drug Administration (FDA) Approval

The efficacy of brentuximab vedotin was subsequently evaluated in Hodgkin lymphoma and anaplastic large cell lymphoma. In HL, intravenous brentuximab vedotin 1.8 mg/kg every 3 weeks was studied in a multicenter trial (SG035-0003) in 102 patients who had relapsed at a median of 6.7 months after ASCT. Patients received a median of 9 cycles of treatment. The primary efficacy endpoint of objective response rate (ORR) was achieved in 75% (complete remissions (CR) of 34%) with a median duration of response (DoR) of 6.7 months. Patients achieving a CR had a median DoR of 20.5 months [107]. Follow up data of approximately 3 years showed median overall survival (OS) of 40.5 months and an estimated 36 month survival rate of 54%. Fourteen percent of patients have remained treatment-free (Nov 15, 2013; Blood 122(21)). A similar trial was carried out in 58 patients with CD30-positive ALCL (SG035-0004). The ORR as deemed by an Independent Review Facility was 86% with a median DoR of 12.6 months. The

CR rate was 57% (median DoR 13.2 months) [108]. Serious adverse events occurred in 31% of patients on both these trials (SG035-0003, SG035-0004) with the most common being peripheral neuropathy, urinary tract infection, and abdominal pain. Based on both these trials, on 19th August 2011, the FDA granted accelerated approval for brentuximab vedotin to be used intravenously at 1.8 mg/kg every 3 weeks (up to 16 cycles) for the treatment of:

- HL after failure of ASCT *or* after failure of at least 2 prior multi-agent chemotherapy regimens in transplant (ASCT) ineligible patients

- Systemic ALCL after failure of at least 1 prior multi-agent chemotherapy regimen

Post-approval, the FDA has issued a *Boxed Warning* highlighting the risk of development of progressive multifocal leukoencephalopathy as well as adding a new contraindication warning against the concurrent use of bleomycin, which can increase the risk of pulmonary toxicity.

There have also been recent reports of acute pancreatitis occurring as a rare but serious and potentially fatal complication related to BV use [109].

Combinations with Chemotherapy

In the L450cy HL xenograft model, a potential synergistic effect was shown between brentuximab vedotin and ABVD (adriamycin, bleomycin, vinblastine, dacarbazine) in tumour-bearing mice [110]. A phase I, dose-escalation study was conducted in fifty one newly diagnosed, treatment-naive stage IIA bulky disease or stage IIB - IV Hodgkin lymphoma using brentuximab vedotin plus ABVD or AVD (adriamycin, vinblastine, dacarbazine). The maximum tolerated dose of 1.2 mg/kg was not exceeded. The most common grade 3 or worse adverse events were neutropenia (78%), anemia (16%), dyspnea (8%), pulmonary embolism (6%), and fatigue (4%). Very importantly, grade 3 or worse pulmonary toxicity was seen exclusively in patients who received brentuximab vedotin with bleomycin (ABVD regimen) - 24% *versus* none in the AVD cohort. The CR rate achieved was 95%. In November 2012, a phase III, randomized trial of brentuximab vedotin plus AVD *versus* ABVD was initiated with the aim of increasing PFS and ultimately replace the current standard of care for frontline treatment of advanced stage HL (NCT01712490).

In patients with CD30+ peripheral T cell lymphomas, a recent phase I open label study combining BV 1.8 mg/kg with CHP (cyclophosphamide, doxorubicin, prednisolone) chemotherapy resulted in an objective response in all patients (CR rate 88%) and a 1-year PFS of 71%. Grade 3/4 adverse events were febrile neutropenia (31%), neutropenia (23%), anemia (15%) and pulmonary embolism (12%) [111]. A randomized phase III study is ongoing comparing BV plus CHP with the CHOP regimen in newly diagnosed patients (NCT01777152).

Consolidation Therapy After Autologous Transplant

The AETHERA trial was a multicenter, randomized, double-blind, placebo-controlled phase III trial conducted in unfavorable-risk relapsed or primary refractory classical HL in which patients were assigned to either 16 cycles of 1.8 mg/kg BV (*N* = 165) or placebo (*N* = 164) every 3 weeks after ASCT. The primary endpoint of PFS by independent review was significantly improved in patients receiving BV *versus* placebo at a median of 42.9 *versus* 24.1 months, respectively (HR 0.57). The estimated 2-year PFS was 63% with BV and 51% with placebo. Post-hoc analysis showed this benefit to be more in the presence of poor risk factors (primary refractory disease or relapse < 12 months from frontline therapy, PR or SD to most recent salvage therapy, extranodal disease, 'B' symptoms, and ≥ 2 previous salvage therapies). Both treatment arms continued for a median of 15 cycles. The most common adverse event with BV was peripheral neuropathy (67%) [112].

Safety After Allogeneic Transplantation

While not extensively studies, BV use has been reported in selected patients who relapse early after allogeneic transplantation. Gopal *et al.* reported their experience in 25 HL patients who suffered a relapse at a median of 12.5 months from their transplant. Patients received 1.2 mg/kg to 1.8 mg/kg based on which study protocol patients were enrolled on. The cohort studied was heavily pretreated for their HL with a median of 9 prior regimens. The ORR and CR rates were 50% and 38%, respectively. Median time to response was about 8 weeks after starting BV and the median PFS was 7.8 months. The most frequent adverse events reported were cough, fatigue, and pyrexia (52%), peripheral neuropathy (48%), dyspnoea (40%), neutropenia (24%), anemia (20%), thrombocytopenia (16%), and hyperglycaemia (12%). Notably, there was no increase in graft-*versus*-host disease attributable to BV [113].

CD52 Antigen

CD52 is a small glycopeptide composed of 12 amino acids containing an N-glycosylation site at its N-terminus and a glycosylphosphatidylinositol (GPI) anchor at its C-terminus [114]. It is tethered to the outer surface of the expressing cell membrane by the GPI-anchor. CD52 lacks an intracellular domain but is able to provide co-stimulatory signals when cross-linked [115]. The antigen is highly expressed on normal and malignant B and T lymphocytes. Radioisotope studies estimate about 5% of lymphocyte surfaces express the CD52 antigen. Lower level expression is found on monocytes, eosinophils, and cells of the male genital tract; haematological stem cells do not express of this molecule. The exact function of CD52 remains undefined but *in vitro* studies have shown a possible contribution towards induction of T-regulatory cells and inhibition of T-cell migration [116].

Anti-CD52 Antibody: Alemtuzumab (Campath®)

In the late 1970's and early 80's, the group at Cambridge developed a range of monoclonal antibodies to human cells, including a set that was able to destroy T-cells through complement activation [117]. The set of moAbs was named Campath-1 (CAMbridge PATHology) and through experiments with an IgM antibody (Campath-1M), it was shown that the target antigen was CD52. The original impetus for Campath-1 came from bone marrow transplantation and prevention of graft-*versus*-host disease. Replacing the IgM antibody with the rat IgG2b antibody (Campath-1G) showed an additional mechanism of action in ADCC through binding of human Fc receptors (FcR). A subsequent compound using the human IgG1 heavy chain was generated in lieu of the rat IgG2 in order to reduce immunogenicity [118]. This antibody, Campath-1H (now known as alemtuzumab) was the first antibody to undergo 'humanization'. Alemtuzumab binds to an epitope on the CD52 antigen consisting of the C-terminal peptide and a portion of the GPI-anchor [119] and causes lymphocyte lysis mainly through CDC and ADCC, although apoptosis (caspase-independent) induction of B-cell lymphoma cell lines has also been demonstrated [120].

Alemtuzumab for Relapsed/Refractory CLL and Accelerated FDA Approval

In 2001, single-agent alemtuzumab was granted accelerated approval by the U.S. FDA for use in chronic lymphocytic leukemia (CLL) patients who have received alkylating agents and failed fludarabine therapy. The approval was based on a prospective, phase II, international study of 93 patients with previously treated CLL [121]. These patients had a median age of 68 years, had received a median of 3 prior treatments (range 2 - 7) and 48 of them were considered fludarabine-refractory. Alemtuzumab was given at 30 mg three times per week up to maximum of 12 weeks and produced an ORR of 33% with a median duration of response of 8.7 month. Patients who responded to therapy had a better median OS of 32 months compared to 16 months in the entire cohort and 10 months in historical controls [122]. Grade 3/4 infections were seen in 26.9% of patients.

Apart from the ability to salvage patients with fludarabine-refractory CLL, this trial also highlighted two important features of alemtuzumab that were also seen in other earlier phase II studies.

The first was infusional toxicity of rigors, fevers, nausea, and rash, usually grade 1 or 2, seen as the most common adverse event across these early trials. This led to a series of phase II studies testing the efficacy of subcutaneous alemtuzumab. One of the largest of these was conducted by the German CLL Study Group in 109 fludarabine-refractory CLL patients which found similar efficacy compared to historical data of outcomes with the intravenous form (ORR 33%, PFS 7.7 months, and median OS of 19.1 months). Grade 3/4 infections were similar at 25% [123]. Based on these results, a registration study (NCT00328198) was initiated in 2006, results of which are now published on *clinicaltrials.gov*. This trial studied two starting doses of subcutaneous alemtuzumab 30 mg three times per week (up to 18 weeks) and an escalation dose starting at a lower dose and building up to the 30 mg dose in the first 1 to 2 weeks. In total, 86 patients with relapsed/ refractory B-cell CLL received SC alemtuzumab and achieved an ORR of 43% (CR 6%, PR 37%). Almost half of the cohort (N=38, 44%) did not complete the intended duration of treatment and was due to an adverse event in 21 of these patients.

The second important observation from the early phase II trials was the efficiency at eliminating CLL cells from the blood but to a much lesser extent, bone marrow, splenic and lymph node disease [124]. Although the proportion of patients with nodal disease who showed response to therapy in these trials differ, there is uniformity in the observation that larger lymph nodes respond poorly; in the pivotal phase II trial by Keating *et al.*, none of the patients with nodal disease of more than 5 cm had complete resolution [121].

Alemtuzumab in Upfront Treatment of CLL

In 2007, the FDA granted regular approval for single-agent alemtuzumab to be used for treatment of CLL based on a phase III randomized trial comparing it with chlorambucil. Two hundred ninety seven patients received as first-line treatment either alemtuzumab 30 mg three times per week (maximum 12 weeks) or chlorambucil 40 mg/m^2 every 28 days, up to 12 months. The primary endpoint of PFS was higher in the alemtuzumab group, 14.6 *versus* 11.7 months (hazard ratio

0.58, P =.0001) with a median time to next treatment of 23.3 *versus* 14.7 months for chlorambucil (P =.0001). There were also increased rates of ORR and CR of 83% (*versus* 55%, P <.0001) and 24% (*versus* 2%, P <.0001), respectively, in favor of alemtuzumab. There were a significant number of patients on the alemtuzumab arm that had positive cytomegalovirus (CMV)-PCR results *versus* none in the chlorambucil group. Twenty three patients who received alemtuzumab (N = 147) developed symptomatic CMV infection while 78 had asymptomatic PCR-positive results with 47 needing treatment interruption. No difference in OS was seen between the groups [125].

Alemtuzumab Combination with Chemotherapy

There have been small phase II studies done using alemtuzumab with chemotherapy in an attempt to further salvage relapsed/ refractory CLL patients. Elter *et al.* reported on their single-center experience of using 'FluCam' which is fludarabine 30 mg/m^2 (days 1 to 3) plus alemtuzumab 30 mg infusion (days 1 to 3) given on a 28-day cycle for up to 6 cycles to 36 patients with relapsed CLL who had received a median of 2 prior treatments (range 1 to 8). They observed an ORR of 83% (CR, N = 11; PR, N = 19) and noted responses in the 6 of 9 patients who were deemed fludarabine refractory at the time of enrolment [126].

The group from M.D. Anderson Cancer Center used the 'CFAR' regimen consisting of alemtuzumab (30 mg on day 1, 3, 5) and fludarabine, cyclophosphamide, rituximab (FCR) to treat 80 patients with relapsed/ refractory CLL in a single-center, phase II study. The ORR was 65% (CR in 29%) with an estimated PFS of 10.6 months and a median OS of 16.7 months. When compared to their historical data of similar patients treated with FCR alone, they noted a similar PFS but a worsened OS, mainly due to high rates of serious infections during (46%) and after (28%) therapy [127].

An Italian multicenter study that enrolled 50 patients with relapsed/ refractory CLL treated with combination bendamustine with subcutaneous alemtuzumab in a phase I/II study have recently been published. Twelve patients were enrolled onto the phase I dose-escalation study where the MTD corresponded with the highest dose level of bendamustine (70 mg/m^2 on days 1 to 2) and alemtuzumab (30 mg on days 1 to 3). Cycles were repeated every 28 days for up to 4 courses. The ORR achieved was 68% (CR in 24%) with a median time to re-treatment of 20.1 months in these patients. The median PFS and OS were 17.3 and 37 months, respectively [128].

Alemtuzumab in Deletion 17p CLL

Patients with del (17p) have a poorer response to chemotherapy than those without. This mutation is seen in about 5 to 7% of patients undergoing first-line treatment. In the German CLL8 trial that randomized patients between FCR (N = 408) and FC (fludarabine plus cyclophosphamide, N = 409) as upfront treatment, del 17p was detected in 51 patients. Outcomes of patients del 17p were poorer compared to the other patients with CR and PFS that were significantly lower at 2.3% and 11.3 months (*versus* 44% and 51.3 months in the FCR cohort) [129]. There is a high concordance rate of about 78 to 88% between the presence of del 17p and TP53 mutations, although the latter can be detected without the del 17p mutation [130] and is also associated with poorer outcomes [131].

In a prospective, multicenter, phase II trial conducted in the United Kingdom, 39 patients with TP53 deleted CLL (17 untreated and 22 previously treated) received the combination of intravenous alemtuzumab 30 mg/m^2 three times per week (for 16 weeks) and intravenous methylprednisolone 1 gram/m^2/day for 5 days every 28 days for 4 cycles. An ORR was seen in 82% of patients (CR 36%) with median PFS and OS of 12 and 23 months, respectively. Subgroup analysis showed that previously untreated patients derived the most benefit (CR 65%, median PFS 18.3 months, median OS 38.9 months). However, this regimen resulted in grade 3 to 4 hematologic toxicity and infection of 67% and 51%, respectively. CMV reactivation was common and 49% of patients required (val) ganciclovir treatment. Treatment related mortality was 5% and despite the relatively high CR rates, patients continued to relapse and die from disease. Therefore, as concluded by the authors of this trial, treatment with alemtuzumab in these patients may be a bridge to allogeneic transplantation. However, the newer kinase inhibitors have largely replaced alemtuzumab for this indication. In July 2014, single-agent ibrutinib (Ibruvica®) and idelalisib (Zydelig®) in combination with rituximab were approved for previously untreated CLL with del (17p). The FDA approval for ibrutinib was based on a randomized trial of ibrutinib *versus* ofatumumab in previously treated CLL patients (N = 391), of whom 127 had del (17p). In the latter, ibrutinib was associated with a 75% reduction in risk of disease progression or death [62]. The upfront data for idelalisib in this setting, upon which the European Medicines Agency approval is based, come from 9 patients who were treated as part of a larger study investigating the role of idelalisib plus rituximab in untreated CLL. Interim analysis from this trial reported a 100% ORR in 6 patients with del (17p) [132]. These kinase inhibitors require indefinite administration for a sustained response, in contrast to conventional chemotherapy and moAbs, where use is limited to a defined number of cycles per treatment regimen.

Alemtuzumab in T-Cell Prolymphocytic Leukemia (T-PLL)

Before the use of alemtuzumab in T-PLL, this aggressive T-cell disorder was associated with a shortened median survival of only 7.5 months despite therapy [133]. Phase II studies using alemtuzumab for T-PLL were initiated based on responses observed with the campath-1G compound. Dearden *et al.* reported the European and UK experience of using 30 mg three times per week in 39 patients with T-PLL, 37 of which had been previously treated. The ORR was 76% (CR in 60%) with a median disease-free interval of 7 months (range 4 – 45 months) [134]. These findings prompted studies using alemtuzumab as first-line therapy in this disease. The prospective UKCLL05 trial recruited untreated T-PLL patients across the UK with the aim of assessing ORR (primary objective), toxicity, PFS and OS using subcutaneous alemtuzumab 30 mg three times per week (up to 18 weeks). However, only 3 of 9 patients (33%) responded to subcutaneous therapy. There were also 2 deaths (of 9) from disease progression. The poor responses were postulated to be from the longer time needed to achieve peak drug levels after subcutaneous injection which may allow for uncontrolled replication of the tumor cells. Five of six patients who failed to respond with subcutaneous therapy were switched to intravenous dosing. Overall, 32 patients were treated with IV alemtuzumab and achieved an ORR of 91% (CR in 81%). The 12-month PFS was 67% and the 48-month OS 37%.

Based on currently available data, intravenous alemtuzumab is the treatment of choice for newly diagnosed as well as relapsed T-PLL. For patients who relapse after alemtuzumab, retreatment is an option especially in patients who have previously achieved a response of at least 6 months duration. Even when used as first-line therapy, relapse is almost invariable and median survival is approximately 20 months. Therefore, most specialized treatment centers recommend consolidation with allogeneic or autologous transplantation.

Infections with Alemtuzumab

Grade 3 and 4 infections have been a recurring issue in most of the phase II studies in CLL. Among these, CMV reactivation has been consistently observed, in addition to the other infections that have been reported, including aspergillosis, zygomycosis, listeria meningitis, and pneumocystis pneumonia [121]. The increased rate of infections is due to treatment-related neutropenia as well as induction of severe defects in cellular immunity. Peripheral blood NK, NK/T and

B lymphocytes have been shown to be significantly depleted after alemtuzumab with slow reconstitution to levels of less than 25% of baseline seen even more than 9 months after treatment [135].

Infectious complications remain a concern and have limited the use of alemtuzumab to specialized lymphoma treatment centers. However, alemtuzumab should be recommended for patients with del (17p) CLL who require treatment and have minimal nodal disease as well as T-PLL. In both these situations, alemtuzumab plays a major role in inducing a response before allogeneic stem cell transplantation.

CD22

CD22 is a type I transmembrane protein with a molecular weight of 140,000. The extracellular portion contains seven immunoglobulin (Ig) domains and the intracellular portion, six tyrosine residues. CD22 is exclusive to B-cells; expression is low on pre and immature B-cells, is maximal in mature B-cells, and is down regulated on plasma cells [136]. It functions mainly as a negative regulator of the B-cell receptor (BCR) and plays a key role in the setting of the threshold of the BCR response [137]. None of the following anti-CD22 compounds have been approved by the FDA thus far.

Unlabeled Anti-CD22 Monoclonal Antibody

Epratuzumab (LymphoCide®) is a humanized IgG1 monoclonal antibody that targets CD22. Binding of this drug results in rapid internalization of the antibody-receptor complex and induces colocalization of CD22 with the BCR, reduces spleen tyrosine kinase (Syk), phospholipase Cγ2, and intracellular calcium concentration which ultimately dampens B-cell activation.

Leonard *et al.* performed two phase I/II studies using single-agent epratuzumab in indolent B-cell and aggressive B-cell NHL, respectively. In the former, 55 patients with relapsed/refractory indolent lymphoma were treated with epratuzumab at doses ranging from 120 to 1000 mg/m2 weekly for 4 treatments. The mean serum half-life was found to be 23 days and doses of up to 1000 mg/m^2 were tolerated. However, a modest 18% of patients demonstrated an objective response and these were only patients with FL. The optimal dose established from this trial was 360 mg/m^2. Patients with other indolent lymphomas did not respond [138]. In the study of aggressive lymphoma ($N = 56$), only 5% of patients achieved an objective response. The modest response with single-agent epratuzumab led to clinical trials incorporating it into rituximab-based treatment with or without chemotherapy. One of the largest of these was a phase II multicenter study by the U.S. North Central Cancer Treatment Group (NCCTG) in patients with diffuse large B-cell lymphoma (DLBCL). Of 107 patients initially enrolled, 26 were excluded based on pathological findings. Among these included patients with discordant bone marrow histology (FL), high-grade FL, B-cell unclassifiable lymphoma, and disrupted c-myc locus as detected by fluorescence in-situ hybridization. The ORR was 96% (CR and CRu 74%) and the event-free survival (EFS) and OS at 3 years was 70% and 80%, respectively. Comparison of these numbers with a historical cohort of theirs suggested improved EFS with epratuzumab combination without added toxicity [139].

Anti-CD22 Antibody Drug Conjugates

Inotuzumab ozogamicin (CMC-544) is a humanized IgG4 anti-CD22 antibody covalently linked to N-acetyl gamma calicheamicin dimethyl hydrazide (calicheamicin) *via* an acid-labile 4-(4'-acetylphenoxy) butanoic acid linker. The drug is rapidly internalized upon CD22 binding releases calicheamicin following hydrolysis of its acid-labile linker under the acidic conditions of intracellular lysosomes. Calicheamicin diffuses into the nucleus and causes double-strand breaks leading to cell death [140, 141].

In phase I studies, the MTD was established at 1.8 mg/m^2. At this dose, no dose-limiting toxicities were identified. Grade 3 or higher adverse events were thrombocytopenia (50%), neutropenia (30%), and leukopenia (20%) [142]. A phase II study was done using a combination of rituximab 375 mg/m^2 (day 1) and inotuzumab 1.8 mg/m^2 (day 2) given every 28 days for up to 8 cycles in patients with relapsed FL (N = 38) and DLBCL (N = 40). The ORR was 87% and 80%, and median PFS 23.6 and 15.1 months in the FL and DLBCL groups, respectively. Patients who were rituximab-resistant (N = 25) achieved an ORR of only 20% and a median PFS of 2 months [143]. Given these promising results seen in DLBCL, a phase III trial (NCT01232556) was initiated in patients with relapsed/refractory CD22-positive aggressive B-NHL randomizing patients to receive either the combination of inotuzumab plus rituximab or investigator's choice of R-chemotherapy (bendamustine or gemcitabine). A report of the planned interim analysis of 338 enrolled patients resulted in termination of the study due to futility with 53% in the intervention arm (*versus* 57% in R-chemotherapy arm) discontinuing study treatment due to progressive disease [144].

Anti-CD22 Recombinant Immunotoxins

Apart from antibody-drug conjugation, protein toxins have also been used with moAbs to create chimeric immunotoxins. These usually contain only the Fv fragment of the antibody connected to a truncated toxin. The compound BL22 (CAT-8888) is a disulphide-stabilized Fv fragment of anti-CD22 fused to *Pseudomonas exotoxin* 38 (PE38) that was shown to induce ORR rates of 72 - 81% in relapsed/refractory hairy cell leukemia (HCL) and up to 50% ORR rates after a single dose [145, 146]. In a phase I trial that enrolled other CD22-positive NHLs ($N = 15$), BL22 showed much less activity compared to HCL, with none of the patients achieving even a PR [146]. This was attributed to the lower density of CD22 receptors on the surface of CLL (and ALL) cells and consequent lower affinity. An immunotoxin with a higher CD22 affinity has since been developed.

Moxetumomab pasudotox (CAT-8015) has an Fv fragment of anti-CD22 with a ~ 15-fold higher binding affinity to CD22 compared to BL22, which translated to a 50-fold increase in cytotoxicity towards HCL and CLL cell lines [147]. In a phase I dose-escalation study, doses ranging from 5 to 50 ug/kg every other day for 3 doses at 4-weekly intervals (up to 16 cycles) were given to 28 patients with relapsed/refractory HCL. Toxicities of grade 1 to 2 hypoalbuminemia, edema, hypotension, fatigue and nausea was observed. There were no DLTs and unlike BL22, no grade $\geq$ 3 hemolytic uremic syndrome was observed. Ten patients (38%) developed neutralizing antibodies towards moxetumomab. The ORR was 86% (CR in 46%) and at 26 months, the median disease-free survival had not yet been reached [148]. This compound is currently in phase I trials in patients with CLL, B-PLL, and other NHLs (NCT01030536), as well as in phase III trials in HCL (NCT01829711).

Other Monoclonal Antibodies

Anti-CD37

CD37 is a heavily glycosylated transmembrane protein of the tetraspanin family that is expressed on the surfaces of B-cells from the precursor B to mature peripheral B-cell stages [149]. It is also highly expressed in CLL, Burkitt lymphoma, MCL, and FL [150, 151]. CD37 is not expressed on normal progenitor B-cells or plasma cells. The function of CD37 is not fully understood but the molecule is highly expressed in endosomes and exosomes in B-lymphocytes, suggesting a possible involvement in intracellular trafficking and antigen presentation [152]. CD37 undergoes modest internalization and shedding in transformed B-cells [153].

A CD37-SMIP (small modular immunopharmaceutical) is a homodimeric protein with a binding region composed of a single-chain variable fragment derived from anti-CD37 antibody variable regions (V_L and V_H) and hinge and effector domains derived from engineered constant regions encoding human IgG_1 domains [154]. The size of the CD37-SMIP was specifically engineered to have a molecular weight too large to be filtered by the glomerulus, thereby avoiding rapid elimination and extending its half-life *in vivo* [154]. The compound demonstrated robust antitumor activity in human lymphoma xenograft molecules. Its humanized counterpart, otlertuzumab (TRU-016) which was built on the ADAPTIR (modular protein technology) platform also demonstrated similar pre-clinical activity and is currently in clinical trials targeting CD37-positive lymphoid malignancies. Otlertuzumab acts by two distinct mechanisms, the first, induction of direct caspase-independent apoptosis of malignant B-cells and the second, by ADCC. In a phase I, dose-escalation study, 57 patients with CLL received otlertuzumab at 0.03 to 20 mg/kg. The MTD was not reached. Pharmacokinetics was dose-dependent and the median terminal half-life ($t_{1/2}$) was 28 days. Twenty six patients subsequently enrolled onto an expansion phase. The most frequent grade ≥ 3 adverse events were thrombocytopenia, neutropenia, anemia, fatigue, and hypophosphatemia. The ORR was 23% and all responses were partial. PFS in for patients in PR was 289 days [155]. A multicenter phase I study in patients with other indolent lymphoma pathologies was also conducted. This trial treated 16 patients with relapsed/refractory FL, MCL, and Waldenström's macroglobulinemia with otlertuzumab 20 mg/kg once per week for up to 8 weeks followed by 4 monthly doses. The study cohort consisted of heavily pre-treated patients with 12 of them refractory to prior treatment and 5 were rituximab-refractory. The only grade ≥ 3 adverse event that occurred in more than one patient was neutropenia. The median $t_{1/2}$ documented was 9.5 days. Lymph node reduction of 50% or more was seen in 3 of 12 patients who had CT-scan assessments [156].

A phase Ib trial using otlertuzumab at 2 dose levels of 10 ($N = 6$) and 20 mg/kg ($N = 6$) given on days 1 and 15 together with rituximab 375 mg/m^2 (day 1) and bendamustine 90 mg/m^2 (days 1 and 2) for up to six 28-day cycles in heavily-pretreated relapsed/refractory NHL patients was subsequently conducted. The most frequent adverse events were neutropenia, nausea, fatigue, leukopenia, and insomnia. An impressive ORR of 83% (CR of 32%) was observed in these patients,

7 of whom were refractory to their last treatment before trial enrolment [157]. A phase II randomized study comparing otlertuzumab plus bendamustine ($N = 32$) *versus* bendamustine alone ($N = 33$) in relapsed CLL patients was done and interim analyses reported. Otlertuzumab was given at 20 mg/kg weekly for two 28-day cycles then every 14 days for four 28-day cycles. Bendamustine was given to patients on both arms at 70mg/m^2 on days 1 and 2 for up to six 28-day cycles. In 44 evaluable patients, the ORR and CR rates were 80% *versus* 42% and 20% *versus* 4% in the combination and bendamustine alone arms, respectively. The frequency of all adverse events was similar in both treatment groups [158].

Anti-CD79

CD79 is a covalent heterodimer consisting of CD79a and CD79b subunits and together with cell surface immunoglobulin (Ig), forms the BCR [159]. Both *a* and *b* subunits are transmembrane proteins consisting of a single extracellular Ig and an intracellular signaling domain. Cross-linking of the BCR antigen causes trafficking to a lysosomal-like compartment and triggers either, cell activation and division when supported by rescue signals from T-cells or apoptotic signaling. CD79 expression precedes Ig heavy-chain gene rearrangement and CD20 expression during B-cell development and disappears in the plasma-cell stage [160]. It is also expressed in most B-cell NHLs [161]. Anti-CD79 antibody drug conjugates have been developed that utilize the intracellular CD79 trafficking to deliver targeted cell death.

Antibody-drug conjugates against CD79 were shown to affect target-dependent killing of NHL cell lines *in vitro* [162]. Polatuzumab vedotin (DCDS4501A) is anti-CD79b moAb conjugated to MMAE that has moved in clinical testing. A phase I dose-escalation study using 0.1 to 2.4 mg/kg every 21 days until progression or unacceptable toxicity was carried out in patients with relapsed/refractory B-cell NHL. Interim results reported on 33 patients with different NHL histologies - FL ($N = 14$), DLBCL ($N = 11$), MCL ($N = 4$), MZL ($N = 2$), transformed FL ($N = 1$), and SLL ($N = 1$). All patients had received rituximab before and 9 had previously undergone high-dose therapy with autologous stem cell rescue. The MTD was not reached and the 2.4 mg/kg dose was selected as the recommended phase II dose. Grade ≥ 3 adverse events were neutropenia, leukopenia, and anemia. Serious adverse events occurred in 4 patients at the 2.4 mg/kg dose level. These were atrial fibrillation, neutropenia, pneumonia and cardiac failure. There was a dose proportional increase in antibody-conjugated MMAE and total antibody exposures. Five patients had a more than 50% reduction in target lesion burden at the first tumor assessment done after 3 to 4 treatment cycles [163]. Polatuzumab is currently in a phase II trial in combination with rituximab in patients with NHL (NCT01691898). Interim analysis of polatuzumab vedotin 2.4 mg/kg in combination with rituximab 375 mg/m^2 (every 21 days) in 57 patients with relapsed/refractory DLBCL and FL has recently been reported [164]. Grade ≥ 3 adverse events were neutropenia, diarrhea, hyperglycemia, and peripheral neuropathy. The ORR, CR, and PR rates in the DLBCL group ($N = 37$) was 51%, 14%, and 38%, respectively, while in FL group ($N = 20$), these rates were 60%, 30%, and 30%, respectively.

CONCLUSION

Monoclonal antibodies represent a novel way of delivering therapy to specific target antigens that are expressed on lymphoma cells while minimizing the collateral damage that is common with conventional chemotherapy. The paradigm of this approach is the targeting of CD20 in B-cell NHL. Rituximab has become one of the most widely prescribed agents, used alone in induction and maintenance therapy, in combination with chemotherapy or small molecules, in transplant conditioning regimens and even attempting to displace conventional cytotoxics as first-line treatment in indolent lymphoma. The potential of monoclonals have truly been realized with rituximab, a realization that came a mere decade from its inception as IDEC-C2B8. This has sparked intense research to find new potential cell surface targets and explore mechanisms to induce even more cytotoxicity. We hope that the current development is only the tip of the iceberg in a wave of new monoclonal antibodies in lymphoma.

ACKNOWLEDGEMENTS

Declared none.

CONFLICT OF INTEREST

The author confirms that he has no conflict of interest to declare for this publication.

REFERENCES

[1]   Koene HR, Kleijer M, Algra J, Roos D, von dem Borne AE, de Haas M. Fc gammaRIIIa-158V/F polymorphism influences the binding of IgG by natural killer cell Fc gammaRIIIa, independently of the Fc gammaRIIIa-48L/R/H phenotype. Blood 1997; 90(3):1109-14,

[2]   Ghesquieres H, Cartron G, Seymour JT, *et al.* Clinical outcome of patients with follicular lymphoma receiving chemoimmunotherapy in the PRIMA study is not affected by FCGR3A and FCGR2A polymorphisms. Blood 2012; 120(13):2650-7.

[3]   Persky DO, Dornan D, Goldman BH, *et al.* Fc gamma receptor 3a genotype predicts overall survival in follicular lymphoma patients treated on SWOG trials with combined monoclonal antibody plus chemotherapy but not chemotherapy alone. Haematologica 2012; 97(6):937-42.

[4]   Cartron G, Dacheux L, Salles G, *et al.* Therapeutic activity of humanized anti-CD20 monoclonal antibody and polymorphism in IgG Fc receptor FcgammaRIIIa gene. Blood 2002; 99(3):754-8.

[5]   Weng WK, Levy R. Two immunoglobulin G fragment C receptor polymorphisms independently predict response to rituximab in patients with follicular lymphoma. J Clin Oncol 2003; 21(21):3940-7.

[6]   Stashenko P, Nadler LM, Hardy R, Schlossman SF. Characterization of a human B lymphocyte-specific antigen. J Immunol 1980; 125(4):1678-85.

[7]   Beers SA, Chan CH, French RR, Cragg MS, Glennie MJ. CD20 as a target for therapeutic type I and II monoclonal antibodies. Seminars in hematology 2010; 47(2):107-14.

[8]   Cragg MS, Morgan SM, Chan HT, *et al.* Complement-mediated lysis by anti-CD20 mAb correlates with segregation into lipid rafts. Blood 2003; 101(3):1045-52.

[9]   Tedder TF, Engel P. CD20: a regulator of cell-cycle progression of B lymphocytes. Immunology today 1994; 15(9):450-4.

[10]   Rossmann ED, Lundin J, Lenkei R, Mellstedt H, Osterborg A. Variability in B-cell antigen expression: implications for the treatment of B-cell lymphomas and leukemias with monoclonal antibodies. Hematol J 2001; 2(5):300-6.

[11]   Liu AY, Robinson RR, Murray ED, Jr., Ledbetter JA, Hellstrom I, Hellstrom KE. Production of a mouse-human chimeric monoclonal antibody to CD20 with potent Fc-dependent biologic activity. J Immunol 1987; 139(10):3521-6.

[12]   Teeling JL, Mackus WJ, Wiegman LJ, *et al.* The biological activity of human CD20 monoclonal antibodies is linked to unique epitopes on CD20. J Immunol 2006; 177(1):362-71.

[13]   Buchsbaum DJ, Wahl RL, Normolle DP, Kaminski MS. Therapy with unlabeled and 131I-labeled pan-B-cell monoclonal antibodies in nude mice bearing Raji Burkitt's lymphoma xenografts. Cancer Res 1992; 52(23):6476-81.

[14]   Rudnicka D, Oszmiana A, Finch DK, *et al.* Rituximab causes a polarization of B cells that augments its therapeutic function in NK-cell-mediated antibody-dependent cellular cytotoxicity. Blood 2013; 121(23):4694-702.

[15]   Miller RA, Maloney DG, Warnke R, Levy R. Treatment of B-cell lymphoma with monoclonal anti-idiotype antibody. N Engl J Med 1982; 306(9):517-22.

[16]   Reff ME, Carner K, Chambers KS, *et al.* Depletion of B cells *in vivo* by a chimeric mouse human monoclonal antibody to CD20. Blood 1994; 83(2):435-45.

[17]   Maloney DG, Liles TM, Czerwinski DK, *et al.* Phase I clinical trial using escalating single-dose infusion of chimeric anti-CD20 monoclonal antibody (IDEC-C2B8) in patients with recurrent B-cell lymphoma. Blood 1994; 84(8):2457-66.

[18]   Berinstein NL, Grillo-Lopez AJ, White CA, *et al.* Association of serum Rituximab (IDEC-C2B8) concentration and anti-tumor response in the treatment of recurrent low-grade or follicular non-Hodgkin's lymphoma. Ann Oncol:1998; 9(9):995-1001.

[19]   McLaughlin P, Grillo-Lopez AJ, Link BK, *et al.* Rituximab chimeric anti-CD20 monoclonal antibody therapy for relapsed indolent lymphoma: half of patients respond to a four-dose treatment program. J Clin Oncol 1998; 16(8):2825-33.

[20]   Coiffier B, Haioun C, Ketterer N, *et al.* Rituximab (anti-CD20 monoclonal antibody) for the treatment of patients with relapsing or refractory aggressive lymphoma: a multicenter phase II study. Blood 1998; 92(6):1927-32.

[21]   Colombat P, Salles G, Brousse N, *et al.* Rituximab (anti-CD20 monoclonal antibody) as single first-line therapy for patients with follicular lymphoma with a low tumor burden: clinical and molecular evaluation. Blood 2001; 97(1):101-6.

[22]   Salles G, Seymour JF, Offner F, *et al.* Rituximab maintenance for 2 years in patients with high tumour burden follicular lymphoma responding to rituximab plus chemotherapy (PRIMA): a phase 3, randomised controlled trial. Lancet 2011; 377(9759):42-51.

[23]   Vidal L, Gafter-Gvili A, Salles G, *et al.* Rituximab maintenance for the treatment of patients with follicular lymphoma: an updated systematic review and meta-analysis of randomized trials. J Natl Cancer Inst 2011; 103(23):1799-806.

[24] Pettengell R, Schmitz N, Gisselbrecht C, *et al*. Rituximab purging and/or maintenance in patients undergoing autologous transplantation for relapsed follicular lymphoma: a prospective randomized trial from the lymphoma working party of the European group for blood and marrow transplantation. J Clin Oncol 2013; 31(13):1624-30.

[25] Kluin-Nelemans HC, Hoster E, Hermine O, *et al*. Treatment of older patients with mantle-cell lymphoma. N Engl J Med 2012; 367(6):520-31.

[26] Fisher RI, Gaynor ER, Dahlberg S, *et al*. Comparison of a standard regimen (CHOP) with three intensive chemotherapy regimens for advanced non-Hodgkin's lymphoma. N Engl J Med 1993; 328(14):1002-6.

[27] Coiffier B, Lepage E, Briere J, *et al*. CHOP chemotherapy plus rituximab compared with CHOP alone in elderly patients with diffuse large-B-cell lymphoma. N Engl J Med 2002; 346(4):235-42.

[28] Gribben JG, Freedman A, Woo SD, *et al*. All advanced stage non-Hodgkin's lymphomas with a polymerase chain reaction amplifiable breakpoint of bcl-2 have residual cells containing the bcl-2 rearrangement at evaluation and after treatment. Blood 1991; 78(12):3275-80.

[29] Freedman AS, Gribben JG, Neuberg D, *et al*. High-dose therapy and autologous bone marrow transplantation in patients with follicular lymphoma during first remission. Blood 1996; 88(7):2780-6.

[30] Rose AL, Smith BE, Maloney DG. Glucocorticoids and rituximab *in vitro*: synergistic direct antiproliferative and apoptotic effects. Blood 2002; 100(5):1765-73.

[31] Demidem A, Lam T, Alas S, Hariharan K, Hanna N, Bonavida B. Chimeric anti-CD20 (IDEC-C2B8) monoclonal antibody sensitizes a B cell lymphoma cell line to cell killing by cytotoxic drugs. Cancer Biother Radiopharm 1997; 12(3):177-86.

[32] Czuczman MS, Grillo-Lopez AJ, White CA, *et al*. Treatment of patients with low-grade B-cell lymphoma with the combination of chimeric anti-CD20 monoclonal antibody and CHOP chemotherapy. J Clin Oncol 1999; 17(1):268-76.

[33] Vose JM, Link BK, Grossbard ML, *et al*. Phase II study of rituximab in combination with chop chemotherapy in patients with previously untreated, aggressive non-Hodgkin's lymphoma. J Clin Oncol 2001; 19(2):389-97.

[34] Feugier P, Van Hoof A, Sebban C, *et al*. Long-term results of the R-CHOP study in the treatment of elderly patients with diffuse large B-cell lymphoma: a study by the Groupe d'Etude des Lymphomes de l'Adulte. J Clin Oncol 2005; 23(18):4117-26.

[35] Habermann TM, Weller EA, Morrison VA, *et al*. Rituximab-CHOP *versus* CHOP alone or with maintenance rituximab in older patients with diffuse large B-cell lymphoma. J Clin Oncol 2006; 24(19):3121-7.

[36] Pfreundschuh M, Trumper L, Osterborg A, *et al*. CHOP-like chemotherapy plus rituximab *versus* CHOP-like chemotherapy alone in young patients with good-prognosis diffuse large-B-cell lymphoma: a randomised controlled trial by the MabThera International Trial (MInT) Group. Lancet Oncol 2006; 7(5):379-91.

[37] Hainsworth JD, Litchy S, Barton JH, *et al*. Single-agent rituximab as first-line and maintenance treatment for patients with chronic lymphocytic leukemia or small lymphocytic lymphoma: a phase II trial of the Minnie Pearl Cancer Research Network. J Clin Oncol 2003; 21(9):1746-51.

[38] Byrd JC, Rai K, Peterson BL, *et al*. Addition of rituximab to fludarabine may prolong progression-free survival and overall survival in patients with previously untreated chronic lymphocytic leukemia: an updated retrospective comparative analysis of CALGB 9712 and CALGB 9011. Blood 2005; 105(1):49-53.

[39] Robak T, Dmoszynska A, Solal-Celigny P, *et al*. Rituximab plus fludarabine and cyclophosphamide prolongs progression-free survival compared with fludarabine and cyclophosphamide alone in previously treated chronic lymphocytic leukemia. J Clin Oncol 2010; 28(10):1756-65.

[40] Fischer K, Cramer P, Busch R, *et al*. Bendamustine combined with rituximab in patients with relapsed and/or refractory chronic lymphocytic leukemia: a multicenter phase II trial of the German Chronic Lymphocytic Leukemia Study Group. J Clin Oncol 2011; 29(26):3559-66.

[41] Teeling JL, French RR, Cragg MS, *et al*. Characterization of new human CD20 monoclonal antibodies with potent cytolytic activity against non-Hodgkin lymphomas. Blood 2004; 104(6):1793-800.

[42]   Pawluczkowycz AW, Beurskens FJ, Beum PV, *et al.* Binding of submaximal C1q promotes complement-dependent cytotoxicity (CDC) of B cells opsonized with anti-CD20 mAbs ofatumumab (OFA) or rituximab (RTX): considerably higher levels of CDC are induced by OFA than by RTX. J Immunol 2009; 183(1):749-58.

[43]   van Meerten T, van Rijn RS, Hol S, Hagenbeek A, Ebeling SB. Complement-induced cell death by rituximab depends on CD20 expression level and acts complementary to antibody-dependent cellular cytotoxicity. Clin Cancer Res 2006; 12(13):4027-35.

[44]   Almasri NM, Duque RE, Iturraspe J, Everett E, Braylan RC. Reduced expression of CD20 antigen as a characteristic marker for chronic lymphocytic leukemia. Am J Hematol 1992; 40(4):259-63.

[45]   Huhn D, von Schilling C, Wilhelm M, *et al.* Rituximab therapy of patients with B-cell chronic lymphocytic leukemia. Blood 2001; 98(5):1326-31.

[46]   Bowen DA, Call TG, Jenkins GD, *et al.* Methylprednisolone-rituximab is an effective salvage therapy for patients with relapsed chronic lymphocytic leukemia including those with unfavorable cytogenetic features. Leuk Lymphoma 2007; 48(12):2412-7.

[47]   Castro JE, Sandoval-Sus JD, Bole J, Rassenti L, Kipps TJ. Rituximab in combination with high-dose methylprednisolone for the treatment of fludarabine refractory high-risk chronic lymphocytic leukemia. Leukemia 2008; 22(11):2048-53.

[48]   Tam CS, O'Brien S, Lerner S, *et al.* The natural history of fludarabine-refractory chronic lymphocytic leukemia patients who fail alemtuzumab or have bulky lymphadenopathy. Leuk Lymphoma 2007; 48(10):1931-9.

[49]   Coiffier B, Lepretre S, Pedersen LM, *et al.* Safety and efficacy of ofatumumab, a fully human monoclonal anti-CD20 antibody, in patients with relapsed or refractory B-cell chronic lymphocytic leukemia: a phase 1-2 study. Blood 2008; 111(3):1094-100.

[50]   Wierda WG, Kipps TJ, Mayer J, *et al.* Ofatumumab as single-agent CD20 immunotherapy in fludarabine-refractory chronic lymphocytic leukemia. J Clin Oncol 2010; 28(10):1749-55.

[51]   Wierda WG, Padmanabhan S, Chan GW, *et al.* Ofatumumab is active in patients with fludarabine-refractory CLL irrespective of prior rituximab: results from the phase 2 international study. Blood 2011; 118(19):5126-9.

[52]   Wierda WG, Kipps TJ, Durig J, *et al.* Chemoimmunotherapy with O-FC in previously untreated patients with chronic lymphocytic leukemia. Blood 2011; 117(24):6450-8.

[53]   Cortelezzi A, Sciume M, Liberati AM, *et al.* Bendamustine in combination with ofatumumab in relapsed or refractory chronic lymphocytic leukemia: a GIMEMA Multicenter Phase II Trial. Leukemia 2014; 28(3):642-8.

[54]   Hagenbeek A, Gadeberg O, Johnson P, *et al.* First clinical use of ofatumumab, a novel fully human anti-CD20 monoclonal antibody in relapsed or refractory follicular lymphoma: results of a phase 1/2 trial. Blood 2008; 111(12):5486-95.

[55]   Czuczman MS, Fayad L, Delwail V, *et al.* Ofatumumab monotherapy in rituximab-refractory follicular lymphoma: results from a multicenter study. Blood 2012; 119(16):3698-704.

[56]   Coiffier B, Radford J, Bosly A, *et al.* A multicentre, phase II trial of ofatumumab monotherapy in relapsed/progressive diffuse large B-cell lymphoma. Br J Haematol 2013; 163(3):334-42.

[57]   Furtado M, Dyer MJ, Johnson R, Berrow M, Rule S. Ofatumumab monotherapy in relapsed/refractory mantle cell lymphoma--a phase II trial. Br J Haematol 2014; 165(4):575-8.

[58]   Czuczman MS, Hess G, Gadeberg OV, *et al.* Chemoimmunotherapy with ofatumumab in combination with CHOP in previously untreated follicular lymphoma. Br J Haematol 2012; 157(4):438-45.

[59]   Hiddemann W, Kneba M, Dreyling M, *et al.* Frontline therapy with rituximab added to the combination of cyclophosphamide, doxorubicin, vincristine, and prednisone (CHOP) significantly improves the outcome for patients with advanced-stage follicular lymphoma compared with therapy with CHOP alone: results of a prospective randomized study of the German Low-Grade Lymphoma Study Group. Blood 2005; 106(12):3725-32.

[60]   Matasar MJ, Czuczman MS, Rodriguez MA, *et al.* Ofatumumab in combination with ICE or DHAP chemotherapy in relapsed or refractory intermediate grade B-cell lymphoma. Blood 2013; 122(4):499-506.

[61]   van Imhoff GW MA, Matasar MJ, Radford J, Ardeshna KM, Kazimierz Kuliczkowski Ofatumumab *Versus* Rituximab Salvage Chemoimmunotherapy in Relapsed or Refractory Diffuse Large B-Cell

Lymphoma: The Orcharrd Study (OMB110928) American Society of Hematology Annual Meeting; San Francisco, CA 2012.

[62]   Byrd JC, Brown JR, O'Brien S, *et al*. Ibrutinib *versus* ofatumumab in previously treated chronic lymphoid leukemia. N Engl J Med 2014; 371(3):213-23.

[63]   Alduaij W, Ivanov A, Honeychurch J, *et al*. Novel type II anti-CD20 monoclonal antibody (GA101) evokes homotypic adhesion and actin-dependent, lysosome-mediated cell death in B-cell malignancies. Blood 2011; 117(17):4519-29.

[64]   Ivanov A, Beers SA, Walshe CA, *et al*. Monoclonal antibodies directed to CD20 and HLA-DR can elicit homotypic adhesion followed by lysosome-mediated cell death in human lymphoma and leukemia cells. J Clin Invest 2009; 119(8):2143-59.

[65]   Honeychurch J, Alduaij W, Azizyan M, *et al*. Antibody-induced nonapoptotic cell death in human lymphoma and leukemia cells is mediated through a novel reactive oxygen species-dependent pathway. Blood 2012; 119(15):3523-33.

[66]   Niederfellner G, Lammens A, Mundigl O, *et al*. Epitope characterization and crystal structure of GA101 provide insights into the molecular basis for type I/II distinction of CD20 antibodies. Blood 2011; 118(2):358-67.

[67]   Ferrara C, Stuart F, Sondermann P, Brunker P, Umana P. The carbohydrate at FcgammaRIIIa Asn-162. An element required for high affinity binding to non-fucosylated IgG glycoforms. J Biol Chem 2006; 281(8):5032-6.

[68]   Herter S, Herting F, Mundigl O, *et al*. Preclinical activity of the type II CD20 antibody GA101 (obinutuzumab) compared with rituximab and ofatumumab *in vitro* and in xenograft models. Mol Cancer Ther 2013; 12(10):2031-42.

[69]   Patz M, Isaeva P, Forcob N, *et al*. Comparison of the *in vitro* effects of the anti-CD20 antibodies rituximab and GA101 on chronic lymphocytic leukaemia cells. Br J Haematol 2011; 152(3):295-306.

[70]   Salles G, Morschhauser F, Lamy T, *et al*. Phase 1 study results of the type II glycoengineered humanized anti-CD20 monoclonal antibody obinutuzumab (GA101) in B-cell lymphoma patients. Blood 2012; 119(22):5126-32.

[71]   Sehn LH, Assouline SE, Stewart DA, *et al*. A phase 1 study of obinutuzumab induction followed by 2 years of maintenance in patients with relapsed CD20-positive B-cell malignancies. Blood 2012; 119(22):5118-25.

[72]   Salles GA, Morschhauser F, Solal-Celigny P, *et al*. Obinutuzumab (GA101) in patients with relapsed/refractory indolent non-Hodgkin lymphoma: results from the phase II GAUGUIN study. J Clin Oncol 2013; 31(23):2920-6.

[73]   Radford J, Davies A, Cartron G, *et al*. Obinutuzumab (GA101) plus CHOP or FC in relapsed/refractory follicular lymphoma: results of the GAUDI study (BO21000). Blood 2013; 122(7):1137-43.

[74]   Cartron G, de Guibert S, Dilhuydy MS, *et al*. Obinutuzumab (GA101) in relapsed/refractory chronic lymphocytic leukemia: final data from the phase 1/2 GAUGUIN study. Blood 2014; 124(14):2196-202.

[75]   Goede V, Fischer K, Busch R, *et al*. Obinutuzumab plus chlorambucil in patients with CLL and coexisting conditions. N Engl J Med 2014; 370(12):1101-10.

[76]   Morschhauser FA, Cartron G, Thieblemont C, *et al*. Obinutuzumab (GA101) monotherapy in relapsed/refractory diffuse large b-cell lymphoma or mantle-cell lymphoma: results from the phase II GAUGUIN study. J Clin Oncol 2013; 31(23):2912-9.

[77]   Knox SJ, Goris ML, Trisler K, *et al*. Yttrium-90-labeled anti-CD20 monoclonal antibody therapy of recurrent B-cell lymphoma. Clin Cancer Res : an official journal of the American Association for Cancer Research 1996; 2(3):457-70.

[78]   Witzig TE, White CA, Wiseman GA, *et al*. Phase I/II trial of IDEC-Y2B8 radioimmunotherapy for treatment of relapsed or refractory CD20(+) B-cell non-Hodgkin's lymphoma. J Clin Oncol 1999; 17(12):3793-803.

[79]   Gordon LI, Molina A, Witzig T, *et al*. Durable responses after ibritumomab tiuxetan radioimmunotherapy for CD20+ B-cell lymphoma: long-term follow-up of a phase 1/2 study. Blood 2004; 103(12):4429-31.

[80]   Witzig TE, Gordon LI, Cabanillas F, *et al.* Randomized controlled trial of yttrium-90-labeled ibritumomab tiuxetan radioimmunotherapy *versus* rituximab immunotherapy for patients with relapsed or refractory low-grade, follicular, or transformed B-cell non-Hodgkin's lymphoma. J Clin Oncol 2002; 20(10):2453-63.

[81]   Witzig TE, White CA, Gordon LI, *et al.* Safety of yttrium-90 ibritumomab tiuxetan radioimmunotherapy for relapsed low-grade, follicular, or transformed non-hodgkin's lymphoma. J Clin Oncol 2003; 21(7):1263-70.

[82]   Guidetti A, Carlo-Stella C, Ruella M, *et al.* Myeloablative doses of yttrium-90-ibritumomab tiuxetan and the risk of secondary myelodysplasia/acute myelogenous leukemia. Cancer 2011; 117(22):5074-84.

[83]   Scholz CW, Pinto A, Linkesch W, *et al.* (90)Yttrium-ibritumomab-tiuxetan as first-line treatment for follicular lymphoma: 30 months of follow-up data from an international multicenter phase II clinical trial. J Clin Oncol 2013; 31(3):308-13.

[84]   Emmanouilides C, Witzig TE, Gordon LI, *et al.* Treatment with yttrium 90 ibritumomab tiuxetan at early relapse is safe and effective in patients with previously treated B-cell non-Hodgkin's lymphoma. Leuk Lymphoma 2006; 47(4):629-36.

[85]   Morschhauser F, Radford J, Van Hoof A, *et al.* Phase III trial of consolidation therapy with yttrium-90-ibritumomab tiuxetan compared with no additional therapy after first remission in advanced follicular lymphoma. J Clin Oncol 2008; 26(32):5156-64.

[86]   Morschhauser F, Illidge T, Huglo D, *et al.* Efficacy and safety of yttrium-90 ibritumomab tiuxetan in patients with relapsed or refractory diffuse large B-cell lymphoma not appropriate for autologous stem-cell transplantation. Blood 2007; 110(1):54-8.

[87]   Goff L, Summers K, Iqbal S, *et al.* Quantitative PCR analysis for Bcl-2/IgH in a phase III study of Yttrium-90 Ibritumomab Tiuxetan as consolidation of first remission in patients with follicular lymphoma. J Clin Oncol 2009; 27(36):6094-100.

[88]   Nademanee A, Forman S, Molina A, *et al.* A phase 1/2 trial of high-dose yttrium-90-ibritumomab tiuxetan in combination with high-dose etoposide and cyclophosphamide followed by autologous stem cell transplantation in patients with poor-risk or relapsed non-Hodgkin lymphoma. Blood 2005; 106(8):2896-902.

[89]   Krishnan A, Nademanee A, Fung HC, *et al.* Phase II trial of a transplantation regimen of yttrium-90 ibritumomab tiuxetan and high-dose chemotherapy in patients with non-Hodgkin's lymphoma. J Clin Oncol 2008; 26(1):90-5.

[90]   Kolstad A, Laurell A, Jerkeman M, *et al.* Nordic MCL3 study: 90Y-ibritumomab-tiuxetan added to BEAM/C in non-CR patients before transplant in mantle cell lymphoma. Blood 2014; 123(19):2953-9.

[91]   Devizzi L, Guidetti A, Seregni E, *et al.* Long-Term Results of Autologous Hematopoietic Stem-Cell Transplantation After High-Dose 90Y-Ibritumomab Tiuxetan for Patients With Poor-Risk Non-Hodgkin Lymphoma Not Eligible for High-Dose BEAM. J Clin Oncol 2013; 31(23):2974-6.

[92]   Younes A, Kadin ME. Emerging applications of the tumor necrosis factor family of ligands and receptors in cancer therapy. J Clin Oncol 2003; 21(18):3526-34.

[93]   Locksley RM, Killeen N, Lenardo MJ. The TNF and TNF receptor superfamilies: integrating mammalian biology. Cell 2001; 104(4):487-501.

[94]   Gause A, Pohl C, Tschiersch A, *et al.* Clinical significance of soluble CD30 antigen in the sera of patients with untreated Hodgkin's disease. Blood 1991; 77(9):1983-8.

[95]   Swerdlow SH, International Agency for Research on Cancer., World Health Organization. WHO classification of tumours of haematopoietic and lymphoid tissues. 4th ed. Lyon, France: International Agency for Research on Cancer; 2008. 439 p. p.

[96]   Younes A, Carbone A. CD30/CD30 ligand and CD40/CD40 ligand in malignant lymphoid disorders. Int J Biol Markers 1999; 14(3):135-43.

[97]   Sotlar K, Cerny-Reiterer S, Petat-Dutter K, *et al.* Aberrant expression of CD30 in neoplastic mast cells in high-grade mastocytosis. Mod Pathol 2011; 24(4):585-95.

[98]   Schwab U, Stein H, Gerdes J, *et al.* Production of a monoclonal antibody specific for Hodgkin and Sternberg-Reed cells of Hodgkin's disease and a subset of normal lymphoid cells. Nature 1982; 299(5878):65-7.

[99]   Durkop H, Latza U, Hummel M, Eitelbach F, Seed B, Stein H. Molecular cloning and expression of a new member of the nerve growth factor receptor family that is characteristic for Hodgkin's disease. Cell 1992; 68(3):421-7.

[100]   Mir SS, Richter BW, Duckett CS. Differential effects of CD30 activation in anaplastic large cell lymphoma and Hodgkin disease cells. Blood 2000; 96(13):4307-12.

[101]   Wahl AF, Klussman K, Thompson JD, *et al.* The anti-CD30 monoclonal antibody SGN-30 promotes growth arrest and DNA fragmentation *in vitro* and affects antitumor activity in models of Hodgkin's disease. Cancer Res 2002; 62(13):3736-42.

[102]   Cerveny CG, Law CL, McCormick RS, *et al.* Signaling *via* the anti-CD30 mAb SGN-30 sensitizes Hodgkin's disease cells to conventional chemotherapeutics. Leukemia 2005; 19(9):1648-55.

[103]   Forero-Torres A, Leonard JP, Younes A, *et al.* A Phase II study of SGN-30 (anti-CD30 mAb) in Hodgkin lymphoma or systemic anaplastic large cell lymphoma. Br J Haematol 2009; 146(2):171-9.

[104]   Okeley NM, Miyamoto JB, Zhang X, *et al.* Intracellular activation of SGN-35, a potent anti-CD30 antibody-drug conjugate. Clin Cancer Res 2010; 16(3):888-97.

[105]   Doronina SO, Toki BE, Torgov MY, *et al.* Development of potent monoclonal antibody auristatin conjugates for cancer therapy. Nat Biotechnol 2003; 21(7):778-84.

[106]   Francisco JA, Cerveny CG, Meyer DL, *et al.* cAC10-vcMMAE, an anti-CD30-monomethyl auristatin E conjugate with potent and selective antitumor activity. Blood 2003; 102(4):1458-65.

[107]   Younes A, Gopal AK, Smith SE, *et al.* Results of a pivotal phase II study of brentuximab vedotin for patients with relapsed or refractory Hodgkin's lymphoma. J Clin Oncol 2012; 30(18):2183-9.

[108]   Pro B, Advani R, Brice P, *et al.* Brentuximab vedotin (SGN-35) in patients with relapsed or refractory systemic anaplastic large-cell lymphoma: results of a phase II study. J Clin Oncol 2012; 30(18):2190-6.

[109]   Mitul Gandhi AME, Timothy S. Fenske, Paul Hamlin, *et al*, editor Pancreatitis In Patients Treated With Brentuximab Vedotin: A Previously Unrecognized Serious Adverse Event American Society of Hematology Annual Meeting; 2013; New Orleans, LA.

[110]   McEarchern JA KD, McCormick R, Lewis TS, Anderson M, Zeng W, Sievers EL, Law C-L. Activity of SGN-35 in preclinical models of combination therapy and relapse prevention. 8th International Symposium on Hodgkin Lymphoma; Cologne, Germany: Haematologica; 2010.

[111]   Fanale MA, Horwitz SM, Forero-Torres A, *et al.* Brentuximab vedotin in the front-line treatment of patients with CD30+ peripheral T-cell lymphomas: results of a phase I study. J Clin Oncol 2014; 32(28):3137-43.

[112]   Moskowitz CH, Nademanee A, Masszi T, *et al.* Brentuximab vedotin as consolidation therapy after autologous stem-cell transplantation in patients with Hodgkin's lymphoma at risk of relapse or progression (AETHERA): a randomised, double-blind, placebo-controlled, phase 3 trial. Lancet 2015; 385(9980):1853-62.

[113]   Gopal AK, Ramchandren R, O'Connor OA, *et al.* Safety and efficacy of brentuximab vedotin for Hodgkin lymphoma recurring after allogeneic stem cell transplantation. Blood 2012; 120(3):560-8.

[114]   Domagala A, Kurpisz M. CD52 antigen--a review. Med Sci Monit 2001; 7(2):325-31.

[115]   Xia MQ, Tone M, Packman L, Hale G, Waldmann H. Characterization of the CAMPATH-1 (CDw52) antigen: biochemical analysis and cDNA cloning reveal an unusually small peptide backbone. Eur J Immunol 1991; 21(7):1677-84.

[116]   Watanabe T, Masuyama J, Sohma Y, *et al.* CD52 is a novel costimulatory molecule for induction of CD4+ regulatory T cells. Clin Immunol 2006; 120(3):247-59.

[117]   Hale G, Bright S, Chumbley G, *et al.* Removal of T cells from bone marrow for transplantation: a monoclonal antilymphocyte antibody that fixes human complement. Blood 1983; 62(4):873-82.

[118]   James LC, Hale G, Waldmann H, Bloomer AC. 1.9 A structure of the therapeutic antibody CAMPATH-1H fab in complex with a synthetic peptide antigen. J Mol Biol 1999; 289(2):293-301.

[119]   Xia MQ, Hale G, Waldmann H. Efficient complement-mediated lysis of cells containing the CAMPATH-1 (CDw52) antigen. Mol Immunol 1993; 30(12):1089-96.

[120]   Mone AP, Cheney C, Banks AL, *et al.* Alemtuzumab induces caspase-independent cell death in human chronic lymphocytic leukemia cells through a lipid raft-dependent mechanism. Leukemia 2006; 20(2):272-9.

[121]   Keating MJ, Flinn I, Jain V, *et al.* Therapeutic role of alemtuzumab (Campath-1H) in patients who have failed fludarabine: results of a large international study. Blood 2002; 99(10):3554-61.

[122] Keating MJ, O'Brien S, Kontoyiannis D, *et al.* Results of first salvage therapy for patients refractory to a fludarabine regimen in chronic lymphocytic leukemia. Leuk Lymphoma 2002; 43(9):1755-62.

[123] Stephan Stilgenbauer DW, Andreas Bühler, Thorsten Zenz, *et al.* Subcutaneous Alemtuzumab (MabCampath) in Fludarabine-Refractory CLL (CLL2H Trial of the GCLLSG). American Society of Hematology Annual Meeting Atlanta, Georgia: Blood; 2007.

[124] Osterborg A, Dyer MJ, Bunjes D, *et al.* Phase II multicenter study of human CD52 antibody in previously treated chronic lymphocytic leukemia. European Study Group of CAMPATH-1H Treatment in Chronic Lymphocytic Leukemia. J Clin Oncol 1997; 15(4):1567-74.

[125] Hillmen P, Skotnicki AB, Robak T, *et al.* Alemtuzumab compared with chlorambucil as first-line therapy for chronic lymphocytic leukemia. J Clin Oncol 2007; 25(35):5616-23.

[126] Elter T, Borchmann P, Schulz H, *et al.* Fludarabine in combination with alemtuzumab is effective and feasible in patients with relapsed or refractory B-cell chronic lymphocytic leukemia: results of a phase II trial. J Clin Oncol 2005; 23(28):7024-31.

[127] Badoux XC, Keating MJ, Wang X, *et al.* Cyclophosphamide, fludarabine, alemtuzumab, and rituximab as salvage therapy for heavily pretreated patients with chronic lymphocytic leukemia. Blood 2011; 118(8):2085-93.

[128] Montillo M, Tedeschi A, Gaidano G, *et al.* Bendamustine and subcutaneous alemtuzumab combination is an effective treatment in relapsed/refractory chronic lymphocytic leukemia patients. Haematologica 2014; 99(9):e159-61.

[129] Molica S. Progress in the treatment of chronic lymphocytic leukemia: results of the German CLL8 trial. Expert Rev Anticancer Ther 2011; 11(9):1333-40.

[130] Zenz T, Habe S, Denzel T, *et al.* Detailed analysis of p53 pathway defects in fludarabine-refractory chronic lymphocytic leukemia (CLL): dissecting the contribution of 17p deletion, TP53 mutation, p53-p21 dysfunction, and miR34a in a prospective clinical trial. Blood 2009; 114(13):2589-97.

[131] Zenz T, Eichhorst B, Busch R, *et al.* TP53 mutation and survival in chronic lymphocytic leukemia. J Clin Oncol 2010; 28(29):4473-9.

[132] Furman RR, Sharman JP, Coutre SE, *et al.* Idelalisib and rituximab in relapsed chronic lymphocytic leukemia. N Engl J Med 2014; 370(11):997-1007.

[133] Matutes E, Brito-Babapulle V, Swansbury J, *et al.* Clinical and laboratory features of 78 cases of T-prolymphocytic leukemia. Blood 1991; 78(12):3269-74.

[134] Dearden CE, Matutes E, Cazin B, *et al.* High remission rate in T-cell prolymphocytic leukemia with CAMPATH-1H. Blood 2001; 98(6):1721-6.

[135] Lundin J, Porwit-MacDonald A, Rossmann ED, *et al.* Cellular immune reconstitution after subcutaneous alemtuzumab (anti-CD52 monoclonal antibody, CAMPATH-1H) treatment as first-line therapy for B-cell chronic lymphocytic leukaemia. Leukemia 2004; 18(3):484-90.

[136] Nitschke L, Carsetti R, Ocker B, Kohler G, Lamers MC. CD22 is a negative regulator of B-cell receptor signalling. Curr Biol 1997; 7(2):133-43.

[137] Cyster JG, Goodnow CC. Tuning antigen receptor signaling by CD22: integrating cues from antigens and the microenvironment. Immunity 1997; 6(5):509-17.

[138] Leonard JP, Coleman M, Ketas JC, *et al.* Phase I/II trial of epratuzumab (humanized anti-CD22 antibody) in indolent non-Hodgkin's lymphoma. J Clin Oncol 2003; 21(16):3051-9.

[139] Micallef IN, Maurer MJ, Wiseman GA, *et al.* Epratuzumab with rituximab, cyclophosphamide, doxorubicin, vincristine, and prednisone chemotherapy in patients with previously untreated diffuse large B-cell lymphoma. Blood 2011; 118(15):4053-61.

[140] DiJoseph JF, Popplewell A, Tickle S, *et al.* Antibody-targeted chemotherapy of B-cell lymphoma using calicheamicin conjugated to murine or humanized antibody against CD22. Cancer Immunol Immunother 2005; 54(1):11-24.

[141] DiJoseph JF, Armellino DC, Boghaert ER, *et al.* Antibody-targeted chemotherapy with CMC-544: a CD22-targeted immunoconjugate of calicheamicin for the treatment of B-lymphoid malignancies. Blood 2004; 103(5):1807-14.

[142] Ogura M, Tobinai K, Hatake K, *et al.* Phase I study of inotuzumab ozogamicin (CMC-544) in Japanese patients with follicular lymphoma pretreated with rituximab-based therapy. Cancer Sci 2010; 101(8):1840-5.

[143] Nam H. Dang MRS, Fritz Offner, Gregor Verhoef, *et al.* Editor Anti-CD22 Immunoconjugate Inotuzumab Ozogamicin (CMC-544) + Rituximab: Clinical Activity Including Survival in Patients

with Recurrent/Refractory Follicular or 'Aggressive' Lymphoma American Society of Hematology Annual Meeting; 2009; New Orleans, LA.

[144]   Nam H. Dang MO, Sylvie Castaigne, Luis Fayad, *et al.* Editor Randomized, phase 3 trial of inotuzumab ozogamicin plus rituximab (R-InO) *versus* chemotherapy for relapsed/refractory aggressive B-cell non-Hodgkin lymphoma (B-NHL). American Society of Clinical Oncology Annual Meeting; 2014; Chicago, IL.

[145]   Kreitman RJ, Squires DR, Stetler-Stevenson M, *et al.* Phase I trial of recombinant immunotoxin RFB4(dsFv)-PE38 (BL22) in patients with B-cell malignancies. J Clin Oncol 2005; 23(27):6719-29.

[146]   Kreitman RJ, Wilson WH, Bergeron K, *et al.* Efficacy of the anti-CD22 recombinant immunotoxin BL22 in chemotherapy-resistant hairy-cell leukemia. N Engl J Med 2001; 345(4):241-7.

[147]   Salvatore G, Beers R, Margulies I, Kreitman RJ, Pastan I. Improved cytotoxic activity toward cell lines and fresh leukemia cells of a mutant anti-CD22 immunotoxin obtained by antibody phage display. Clin Cancer Res 2002; 8(4):995-1002.

[148]   Kreitman RJ, Tallman MS, Robak T, *et al.* Phase I trial of anti-CD22 recombinant immunotoxin moxetumomab pasudotox (CAT-8015 or HA22) in patients with hairy cell leukemia. J Clin Oncol 2012; 30(15):1822-8.

[149]   Barrena S, Almeida J, Yunta M, *et al.* Aberrant expression of tetraspanin molecules in B-cell chronic lymphoproliferative disorders and its correlation with normal B-cell maturation. Leukemia 2005; 19(8):1376-83.

[150]   Moore K, Cooper SA, Jones DB. Use of the monoclonal antibody WR17, identifying the CD37 gp40-45 Kd antigen complex, in the diagnosis of B-lymphoid malignancy. J Pathol 1987; 152(1):13-21.

[151]   Press OW, Howell-Clark J, Anderson S, Bernstein I. Retention of B-cell-specific monoclonal antibodies by human lymphoma cells. Blood 1994; 83(5):1390-7.

[152]   Schwartz-Albiez R, Dorken B, Hofmann W, Moldenhauer G. The B cell-associated CD37 antigen (gp40-52). Structure and subcellular expression of an extensively glycosylated glycoprotein. J Immunol 1988; 140(3):905-14.

[153]   Press OW, Farr AG, Borroz KI, Anderson SK, Martin PJ. Endocytosis and degradation of monoclonal antibodies targeting human B-cell malignancies. Cancer Res 1989; 49(17):4906-12.

[154]   Zhao X, Lapalombella R, Joshi T, *et al.* Targeting CD37-positive lymphoid malignancies with a novel engineered small modular immunopharmaceutical. Blood 2007; 110(7):2569-77.

[155]   Byrd JC, Pagel JM, Awan FT, *et al.* A phase 1 study evaluating the safety and tolerability of otlertuzumab, an anti-CD37 mono-specific ADAPTIR therapeutic protein in chronic lymphocytic leukemia. Blood 2014; 123(9):1302-8.

[156]   Pagel JM, Spurgeon SE, Byrd JC, *et al.* Otlertuzumab (TRU-016), an anti-CD37 monospecific ADAPTIR therapeutic protein, for relapsed or refractory NHL patients. Br J Haematol 2014.

[157]   Gopal AK, Tarantolo SR, Bellam N, *et al.* Phase 1b study of otlertuzumab (TRU-016), an anti-CD37 monospecific ADAPTIR therapeutic protein, in combination with rituximab and bendamustine in relapsed indolent lymphoma patients. Investigational new drugs 2014; 32(6):1213-25.

[158]   Tadeusz Robak AH, Janusz Kloczko, Javier Loscertales, *et al.* Editor Phase 2 Study Of Otlertuzumab (TRU-016), An Anti-CD37 ADAPTIRTM Protein, In Combination With Bendamustine *Vs* Bendamustine Alone In Patients With Relapsed Chronic Lymphocytic Leukemia (CLL) American Society of Hematology Annual Meeting; 2013; New Orleans, LA.

[159]   Koyama M, Ishihara K, Karasuyama H, Cordell JL, Iwamoto A, Nakamura T. CD79 alpha/CD79 beta heterodimers are expressed on pro-B cell surfaces without associated mu heavy chain. Int Immunol 1997; 9(11):1767-72.

[160]   Chu PG, Arber DA. CD79: a review. Applied immunohistochemistry & molecular morphology : AIMM / official publication of the Society for Applied Immunohistochemistry 2001; 9(2):97-106.

[161]   Cabezudo E, Carrara P, Morilla R, Matutes E. Quantitative analysis of CD79b, CD5 and CD19 in mature B-cell lymphoproliferative disorders. Haematologica 1999; 84(5):413-8.

[162]   Polson AG, Yu SF, Elkins K, *et al.* Antibody-drug conjugates targeted to CD79 for the treatment of non-Hodgkin lymphoma. Blood 2007; 110(2):616-23.

[163]   Maria Corinna Palanca-Wessels IWF, Laurie H. Sehn, Manish Patel, *et al.* editor A Phase I Study of the Anti-CD79b Antibody Drug Conjugate (ADC) DCDS4501A Targeting CD79b in Relapsed or Refractory B-Cell Non-Hodgkin's Lymphoma (NHL) American Society of Hematology Annual Meeting; 2012; Atlanta, GA.

[164]　Franck Morschhauser IF, Ranjana H. Advani, Laurie Helen Sehn, *et al.* Editor Preliminary results of a phase II randomized study (ROMULUS) of polatuzumab (PoV) or pinatuzumab vedotin (PiV) plus rituximab. American Society of Clinical Oncology Annual Meeting; 2014; Chicago, IL.

Subject Index

A

Accelerated phase (AP) 5, 30, 32, 33, 36
Accessory cells 97, 106, 111
Activated NK cells 109, 138
Activated T-cells (ATCs) 106, 107
Activate NK cells 146, 147
Activating KIRs 138, 141
Activating receptors 137, 138, 139, 155
Activation-Induced C-type Lectin (AICL) 138
Acute myeloid leukemia (AML) 4, 13, 36, 43, 86, 94, 95, 96, 141, 143, 145, 146, 147, 162
Adriamycin 162, 278
Adverse events (AEs) 85, 89, 90, 91, 92, 94, 96, 103, 115, 116, 120, 266, 269, 277, 278, 279, 281, 288, 289, 290
Agglomerates 181, 182, 183, 184, 210
Aggressive NHL 271, 272, 273
AICL on NK cells 139
Alemtuzumab 114, 258, 264, 265, 268, 274, 280, 281, 282, 283, 284, 285
Allo-reactive NK cells 143
AML cells 97, 143
Anaplastic large cell lymphoma (ALCL) 145, 146, 276, 277
Antibody-bound cells 257
Antibody dependent cellular cytotoxicity (ADCC) 97, 102, 138, 256, 257, 258, 260, 269, 280, 288
Antibody-dependent cellular phagocytosis (ADCP) 256, 257, 258, 268
Antibody-drug conjugates 256, 258, 289
Antibody formats 81, 94, 96, 97, 99, 123
Anti-CD20 moAbs 257, 259, 274
Antimicrobial nanomaterials 222
Anti-tumor activities 163
Anti-tumor cell activity 99
Anti-tumor responses 81, 82, 152
Apoptosis 3, 5, 10, 13, 17, 20, 24, 25, 27, 32, 37, 40, 44, 47, 48, 50, 101, 152, 161, 163, 209, 259, 276, 280
Apoptosis of CML cells 39, 50
Apoptotic cells 41, 152
Aqueous compartments 202, 203
Armed ATCs 106, 107
Autologous stem cell transplants (ASCT) 266, 267, 273, 275, 277, 278, 279
Autonomous control of NK cells 138, 139
Autophagic cell death 48, 153, 154, 162, 164
Autophagy inhibitors 49, 50, 164
Autophagy process 162

Auto-SCT 133, 134, 146, 147, 148, 152

B

Baseline toxicity 232
B-cell lines 106, 107, 108, 116
B-cell lymphomas 256, 262, 264
B cell malignancies 81
B-cell receptor (BCR) 4, 7, 8, 10, 15, 285, 289
BCR-ABL activates mTOR in CML cells 18
BCR-ABL activity 6, 11, 21, 35, 36, 42
BCR-ABL-expressing HSCs 38
BCR-ABL in CML patients 35
BCR-ABL inhibition 40, 49
BCR-ABL Kinase Domain 23
BCR-ABL mRNA levels 6, 7, 31
BCR-ABL oncoprotein 10, 11, 15, 16
BCR-ABL protein 3, 9, 10, 11
BCR-ABL stability in CML cells 25
BCR-ABL transcript levels 6, 7, 36
BCR-ABL-transduced cells 17
BCR-ABL-transduced human CD34+ cells 42
BCR-ABL-transformed haemopoietic cells 11
BCR-ABL tyrosine kinase activity 25, 26
BCR protein 8
Bendamustine 263, 266, 271, 274, 282, 286, 288, 289
BET cells 108
Biocompatibility 197, 205, 207, 208
Bispecific antibodies 81, 82, 101, 102, 105, 115, 116, 117, 123
Bispecific scFv 98, 99
Bispecific T-cell Engagers (BiTEs) 81, 82, 83, 84, 85, 94, 96, 98, 99, 117
BiTE-molecules 93, 94
Blastic phase (BP) 30
Blinatumomab 84, 85, 87, 88, 89, 90, 91, 92, 93, 104, 108, 110, 113, 117, 118, 119, 120
Blinatumomab infusions 88, 90, 108, 118, 119
Blinatumomab treatment 87, 88, 90, 91, 119
Block copolymer micelle products 201
Block copolymer micelles 201
Blood-derived NK cells 156
BM cells 4, 8
Bone marrow cells 37, 38
Bortezomib 44, 45, 154, 164
Bosutinib 19, 21, 22, 23, 29
Brentuximab vedotin 258, 276, 277, 278
Burkitt leukemia 121
Burkitt-lymphoma (BL) 118, 120, 121

www.ingramcontent.com/pod-product-compliance
Lightning Source LLC
Chambersburg PA
CBHW050811220326
41598CB00006B/177